T0134865

Physiology in Health and Disease

Published on behalf of the American Physiological Society by Springer

Physiology in Health and Disease

This book series is published on behalf of the American Physiological Society (APS) by Springer. Access to APS books published with Springer is free to APS members.

APS publishes three book series in partnership with Springer: *Physiology in Health and Disease* (formerly *Clinical Physiology*), *Methods in Physiology*, and *Perspectives in Physiology* (formerly *People and Ideas*), as well as general titles.

More information about this series at http://www.springer.com/series/11780

Jeppe Praetorius • Bonnie Blazer-Yost •
Helle Damkier

Editors

Role of the Choroid Plexus in Health and Disease

 Springer

american
physiological
society

Editors
Jeppe Praetorius
Department of Biomedicine, Faculty
of Health
Aarhus University
Aarhus, Denmark

Bonnie Blazer-Yost
Department of Biology
Indiana University – Purdue University
Indianapolis
Indianapolis, Indiana, USA

Helle Damkier
Department of Biomedicine, Faculty
of Health
Aarhus University
Aarhus, Denmark

ISSN 2625-252X ISSN 2625-2538 (electronic)
Physiology in Health and Disease
ISBN 978-1-0716-0538-7 ISBN 978-1-0716-0536-3 (eBook)
https://doi.org/10.1007/978-1-0716-0536-3

© The American Physiological Society 2020
This work is subject to copyright. All rights are reserved by the Publisher, whether the whole or part of the material is concerned, specifically the rights of translation, reprinting, reuse of illustrations, recitation, broadcasting, reproduction on microfilms or in any other physical way, and transmission or information storage and retrieval, electronic adaptation, computer software, or by similar or dissimilar methodology now known or hereafter developed.
The use of general descriptive names, registered names, trademarks, service marks, etc. in this publication does not imply, even in the absence of a specific statement, that such names are exempt from the relevant protective laws and regulations and therefore free for general use.
The publisher, the authors, and the editors are safe to assume that the advice and information in this book are believed to be true and accurate at the date of publication. Neither the publisher nor the authors or the editors give a warranty, expressed or implied, with respect to the material contained herein or for any errors or omissions that may have been made. The publisher remains neutral with regard to jurisdictional claims in published maps and institutional affiliations.

This Springer imprint is published by the registered company Springer Science+Business Media, LLC part of Springer Nature.
The registered company address is: 1 New York Plaza, New York, NY 10004, U.S.A.

Contents

Chapter 1
Structure of the Mammalian Choroid Plexus

Helle Damkier and Jeppe Praetorius

Abstract The human choroid plexus (CP) is a highly vascularized epithelial structure, weighing approximately 1 g and residing inside the brain ventricles. The CP secretes the majority of the daily production of 500 ml cerebrospinal fluid (CSF) with a transport rate, which is unsurpassed by other human epithelia. The CP is a key structure for (1) providing the CSF for buoyancy decreasing the effective weight of the brain from 1.5 kg to only 50 g, (2) delivering local mediators and hormones to the brain parenchyma via the CSF, (3) maintaining a suitable ionic microenvironment, and (4) forming a barrier against toxins, drugs, microorganisms, and immune cells. Choroid plexus dysfunctions are described in a wide range of clinical conditions such as aging, Alzheimer's disease, brain edema, stroke, neoplasms, and several types of hydrocephalus. Knowledge on the structure and ultrastructure of the choroid plexus is essential for generating hypotheses on the mechanisms involved in the normal function of the CP and in diseases and conditions with deranged CSF secretion, in inflammation, or in drug delivery.

1.1 Development of the Choroid Plexus

The choroid plexus (CP) is a highly vascularized epithelial structure that protrudes into the brain ventricles. The development of the CP is initiated after neural tube closure in the fourth week of gestation in humans. The CP in the fourth brain ventricle is the first to appear from an invagination of the dorsal roof plate in the midline of the neural tube (Fig. 1.1). The lateral ventricles thereafter extend from the area choroidea and form the anterior and posterior domains of the lateral (telencephalic) CP. The last CP to emerge is the third ventricle CP in the diencephalon. Although the CP in the third ventricle is the last to develop, this structure is the first to complete its differentiation (Lun et al. 2015). Also, the CP in the third ventricle develops as an individual structure but eventually it bifurcates and merges into one continuous structure with the lateral ventricle CP.

H. Damkier (✉) · J. Praetorius
Department of Biomedicine, Faculty of Health, Aarhus University, Aarhus, Denmark
e-mail: hd@biomed.au.dk

© The American Physiological Society 2020
J. Praetorius et al. (eds.), *Role of the Choroid Plexus in Health and Disease,*
Physiology in Health and Disease, https://doi.org/10.1007/978-1-0716-0536-3_1

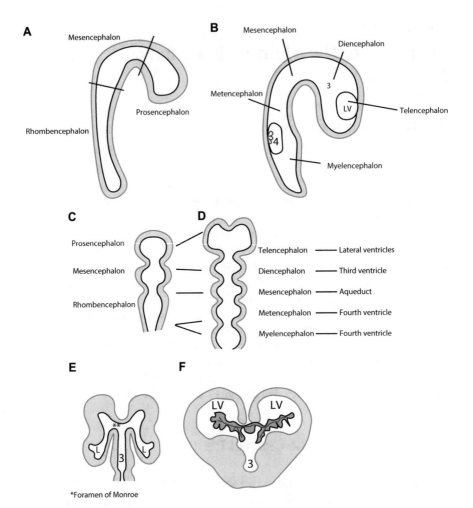

Fig. 1.1 Development of the choroid plexus. Initially following closure of the neural tube the brain is divided into the rhombencephalon (hindbrain), the mesencephalon (midbrain) and the prosencephalon (forebrain) (**a**: sagittal section and **c**: frontal section). At 5 weeks the prosencephalon consists of the telencephalon and the diencephalon; the rhombencephalon develops into the metencephalon and the myelencephalon (**b** and **d**). The fourth ventricle CP is the first to appear in the hindbrain (**b**). The lateral ventricle CP arises next in the telencephalon and finally the third ventricle CP in the diencephalon. The mesencephalon contains the aquaduct which connects the third and fourth ventricle. As the telencephalon expands the lateral ventricle elongates but remains in contact with the third ventricle through the foramen of Monroe (**e**). The third ventricle CP initially develops as a separate structure but eventually comes into contact with the lateral ventricle CP through the foramen of Monroe (**f**)

The epithelial cells of the choroid plexus are derived from neuroepithelial cells that arise from the neural ectoderm. The signal for induction of differentiation of the CP is facilitated by bone morphogenetic protein (BMP) signaling (Hebert et al. 2002). In the hindbrain growth factors such as Wnt1 (Bally-Cuif et al. 1995) and Gdf7 (Currle et al. 2005) as well as the transcription factor Lmx1a (Chizhikov et al. 2010) contribute to development of CP. The CP cells emerge as non-mitotic cells in the human roof plate epithelium. The cells first appear as pseudostratified and densely packed cells that later form an elongated single layer of cuboidal cells expanding not only by conformational changes in the cells but also by addition of new cells by the rhombic lip through activation of the non-mitotic cells via NOTCH1 signaling (Hunter and Dymecki 2007). The CP contains very few mitotic cells. In humans, mitotic cells are only observed until the seventh week of development (Korzhevskii 2000), thereafter they are only found in the root of the CP where they reside as dormant cells that proliferate following a response caused by e.g. injury (Barkho and Monuki 2015; Chouaf-Lakhdar et al. 2003).

The developing fourth ventricle CP epithelial cells produce sonic hedgehog signaling that mediates the development of the CP vascular outgrowth through pericyte-activation (Nielsen and Dymecki 2010) as well as proliferation of CP progenitor cells (Huang et al. 2009) thereby contributing to the continued growth of the fourth ventricle CP. Factors produced by the embryonic CP not only have an effect on continued CP development, but also appear to have a role in brain-development by stimulating the proliferation of distinct progenitor cells. Examples of such factors are the previously mentioned SHH (Huang et al. 2010) as well as retinoic acid (Lehtinen et al. 2011; Parada et al. 2008), insulin-like growth factors (Lehtinen et al. 2011; Salehi et al. 2009), fibroblast growth factors (Martin et al. 2006; Raballo et al. 2000), and bone morphogenic proteins (Lehtinen et al. 2011).

In the beginning, the CPs take up most of the space in the ventricles and play an important role in brain expansion. Early studies indicate that following closure of the neural tube the choroid plexus participates in expansion of the brain by creation of a fluid pressure increase (Desmond and Jacobson 1977). This fluid pressure increase is believed to be caused by the osmotic action of chondroitin sulphate proteoglycans in the embryonic CSF that leads to water retention in the brain cavities (Alonso et al. 1999).

During embryonic development, the CP epithelium undergoes substantial structural changes. Kappers and later, Netsky, described four stages of lateral ventricle CP development [reviewed in Dziegielewska et al. (2001) and Ghersi-Egea et al. (2018)]. In stage I (human gestational week 7–9) the cells appear tall and pseudostratified with central nuclei, in stage II (9–16 weeks) the cells are low columnar cells with nuclei closer to the luminal membrane and abundant glycogen. In stage III (17–28 weeks) the cells become cuboidal with central nuclei and less glycogen and start to develop numerous villi and the mesenchyme starts to accumulate collagen fibers. Finally, in stage IV (from week 29 and until gestation) the cells remain cuboidal, have fully developed microvilli, and central to basal nuclei. However, they contain less or no glycogen at this stage. The amount of collagen continues to increase in the stroma. Similar to the lateral ventricle CP, the third and fourth ventricle CP also contain glycogen but for briefer periods. The role of glycogen is presumably to compensate for the before mentioned poor vascularization in the brain and CP during development. The surface area of the plasma membrane also increases during development. Keep and Jones described an 82% increase in luminal surface area in rats from 16 days of gestation to 30 days after birth (Keep and Jones 1990). This large growth in surface area of the luminal membrane corresponds to the specialization of the CPE (choroid plexus epithelium) to become an actively secreting epithelium.

The mature choroid plexus epithelial cells are connected by tight junctions (see below). These cell junctions and the epithelial cells per se create a barrier from blood to CSF: the blood-cerebrospinal fluid barrier. The barrier matures very early during development, as the tight junctions appear and as the epithelial cells differentiate (Mollgard and Saunders 1986). The secretion of CSF is undertaken by a range of transporters expressed in the membranes of the CP where they mediate the net movement of Na^+, HCO_3^- and Cl^- from blood to CSF creating a gradient for water movement. As will be described in detail below and in the following chapters, much is known about these transport proteins involved in the secretion of CSF in adults, and much less is known about their role during embryonic CSF secretion. In rats, the water channel aquaporin 1 is present almost immediately after the appearance of the CP (Johansson et al. 2005). The Na^+/K^+ ATPase which in the adult is the driving force for CSF secretion is expressed at the same time as AQP1 but at a low level in early stages (Johansson et al. 2008). The level of expression of this transporter increases during development. Similarly, in the adult, the secretion of CSF is dependent on the activity of carbonic anhydrases. The carbonic anhydrase II is expressed at low levels in early CP development, and with increasing abundance during maturation (Johansson et al. 2008). The Na^+, K^+, 2Cl- transporter, NKCC1, is involved in CSF secretion by the mature CP and is strongly expressed early during development in mice (Li et al. 2002).

The brain is poorly vascularized at the time of CP development, which implies that the CP has an important nutritive function in this period. For instance, the fetal CSF contains the same protein levels as plasma, which indicates either efficient vectorial transport or filtering of plasma/interstitial fluid. The amount of protein decreases as the fetus grows. The proteins of the fetal CSF are identical to proteins in blood plasma, which suggests transfer across the blood-CSF barrier by carrier proteins, since the BCSFB (blood cerebrospinal fluid barrier) has already been formed. Thus, the majority of proteins in the developing sheep CSF consists of albumin, fetuin and α-fetoprotein (Dziegielewska et al. 1980a). Albumin initially amounts to about 25% of total protein but later increases to 60%. Albumin is primarily believed to be transported across the CP, though there is evidence suggesting local synthesis. In the adult sheep, albumin amounts to 40% of total protein in CSF. Tracer-analysis in sheep revealed that proteins are transferred from blood to CSF at different degrees correlating to their levels in CSF (Dziegielewska et al. 1980b). The presence of a high number of carrier proteins in the CSF in the developing brain can possibly be explained by the fact that the brain is poorly vascularized during development.

Elucidating the role of the CP during embryonic development and the continued role of the secreted factors by the CP in the adult is a field of great interest in current research. Given the important role of the embryonic CSF on brain development there could be great potential for using a strategy of treatment aimed at rejuvenating the CP to treat, for instance, neurodegenerative disorders such as Alzheimer's disease.

1.2 Gross Anatomy of the Human Choroid Plexus and Ventricle System

1.2.1 The Choroid Plexus and Circulation of CSF

The choroid plexi reside in the four brain ventricles as either sheet-like structures in the lateral ventricles or more villi-like structures in the third and fourth ventricle (Damkier et al. 2013). The choroid plexus (CP) is comprised of a monolayer of tight epithelial cells (see below) residing on a vascularized stroma that also contains dendritic cells and macrophages. The structures are linked on each side of the brain by connective tissue containing the blood vessels and nerves that supply and inner-vate the tissue, respectively. The choroid plexus is also referred to as the blood cerebrospinal fluid barrier (BCSFB) and serves as the first barrier between the blood and the cerebrospinal fluid. Unlike the blood-brain barrier (BBB), the BCSFB has leaky capillaries. Thus, the barrier function between the blood side and the cerebrospinal fluid in the brain ventricles is comprised of the epithelial cell layer. As will be described later, a major function of the choroid plexus is the secretion of cerebrospinal fluid (CSF). The mammalian choroid plexus secretes approximately 500 ml cerebrospinal fluid per day (Damkier et al. 2013; Cserr 1971). CSF secreted by the CP in the lateral ventricles flows through the foramen of Monroe to the third ventricle where it is mixed with the CSF produced by the CP located in this ventricle. From the third ventricle, CSF flows to the fourth ventricle through the aqueduct of Sylvius and is mixed with the CSF secreted by the CP located here (Fig. 1.2a, b). From the fourth ventricle, CSF finally flows through the foramen of Magendie or Luschka to the cisterna magna where it enters the subarachnoid space covering the exterior surfaces of the brain and spinal cord. The pathways for return of CSF to the venous system is still much debated. Classically, the CSF is reabsorbed directly through arachnoid granulations or villi into the venous sinuses in the dura of the brain (Damkier et al. 2013). Alternative pathways for CSF to return to the systemic circulation are via meningeal lymph vessels lining the dura (Louveau et al. 2015), along cranial nerves (e.g. the olfactory nerve) or along the spinal nerves (Cserr and Knopf 1992). The CSF has also been suggested to flow along the perivascular space surrounding the arteries and veins that enter the brain from the surface of the cortex. The perivascular space is also referred to as the Virchow-Robin space (Virchow 1851; Robin 1859). Solutes and water are, thus, believed to enter the brain tissue at the arterial side by convectional flow brought on by the pulsatile movement of the blood

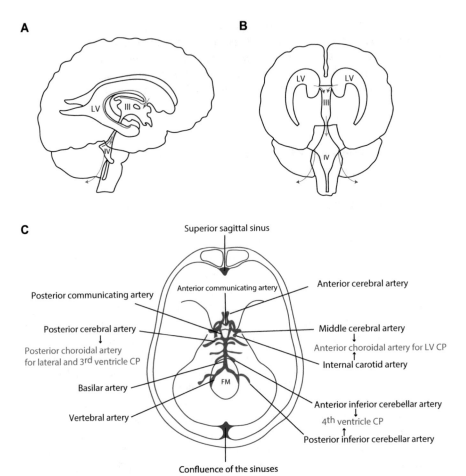

Fig. 1.2 Circulation of cerebrospinal fluid (CSF) in the brain ventricles and blood supply to the choroid plexus. (**a**) and (**b**) The CSF is secreted in the 4 brain ventricles that all contain choroid plexi (red line). The CSF from the lateral ventricles (LV) flows through the foramen of Monroe (blue arrow) to the third ventricle (III). From here it flow through the cerebral aqueduct (Aqueduct of Sylvius, green arrow) into the fourth ventricle (IV). The CSF leaves the fourth ventricle through the foramen of Magendie (pink arrow) or foramen of Luschka (purple arrow) to the cisterna magna where it flows into the subarachnoid space. From the subarachnoid space the CSF drains to the systemic circulation by different pathways. The majority is drained by the superior sagital sinus in the found between the dural layers in the top of the skull (**c**). (**c**) The blood supply to the CP arises from the arteries in the base of the skull (the circle of Willis). The anterior choroidal artery that supplies the lateral ventricle CP arises from either the internal carotid artery or the middle cerebral artery. The remaining CPs are supplied by blood primarily from the posterior circulation. The posterior choroidal arteries that supply both the lateral and third ventricle CP arise from the posterior cerebral artery. Finally, the fourth ventricle CP is supplied by blood vessels from the anterior and posterior inferior cerebellar arteries. These arteries branch from the basilar and vertebral artery, respectively. The venous drainage from the CP is very variable but generally drains towards the confluence of the sinuses in the posterior part of the skull. This confluence is the site of drainage of venous blood from the superior sagital and straight sinus that drain the majority of the brain as well as CSF. *FM* foramen magnum

and return to the venous side thereby creating what is known as the "glymphatic" pathway (Iliff et al. 2013) or a means of clearing the brain of waste products.

1.2.2 Arterial Blood Supply of the Choroid Plexus

The arterial blood supply to the choroid plexus arises from either the anterior or posterior circulation. The anterior choroidal artery originates from either the internal carotid or the middle cerebral artery. The posterior choroidal artery branches from the posterior cerebral artery, which is part of the posterior circulation through the basilar artery and eventually the vertebral artery (Fig. 1.2c). The anterior choroidal artery supplies only the lateral ventricle CP whereas the posterior choroidal artery supplies both the lateral and the third ventricle CP. The fourth ventricle CP receives blood from the posterior circulation but through the anterior and posterior inferior cerebellar arteries which branch from the basilar and vertebral artery, respectively (Netter 2014). Some studies indicate that the capillaries form a mesh like network around the arteries and veins (Weiger et al. 1986). The capillary network is proposed to release vasoactive substances that regulate the blood flow in the choroid plexus by acting on receptors on the smooth muscle cells surrounding the blood vessels (Chodobski and Szmydynger-Chodobska 2001).

1.2.3 Venous Drainage of the Choroid Plexus

At the venous side, the mesh-like capillaries collect into venules in the rat as spiral and vine-like collecting veins, as described by Sun and Hashimoto (Sun and Hashimoto 1991). The venous drainage of the choroid plexus is—much like the veins of the systemic circulation in general—very variable and connected via numerous collateral vessels. In most cases, the blood drains towards the confluence of the sinuses in the posterior base of the skull either via the internal cerebral vein, the great vein of Galen and the straight sinus, or directly to the confluence by a number of collateral pathways. The blood can also bypass the confluence and drain via the superior petrosal sinus to the transverse sinus. In all cases, the blood from the brain as well as most of the CSF taken up through the superior sagittal sinus and exits the skull via the internal jugular veins (Netter 2014).

1.2.4 Lymphatic Vessels in the CNS

The presence of lymph vessels in the CNS and meninges has been greatly debated. In the eighteenth century, Paolo Mascagni described the presence of lymph vessels in connection to the dura but although this was shown again in later studies, the CNS was considered devoid of lymph [reviewed in Da Mesquita et al. (2018)]. Nearly 200 years after the first description of lymph vessels in the dura, a study in Nature in 2015 (Louveau et al. 2015), convincingly showed the presence of lymphatic vessels lining the dural sinuses. This rather distinct and remote localization perhaps is the reason for the concept that the CNS was previously seen as without lymphatic drainage. In the healthy choroid plexus, to our knowledge, no lymph vessels have been described, but tertiary lymphoid structures have been described in the choroid plexus in the presence of disease e.g. systemic lupus erythematosus (Stock et al. 2019). Tertiary lymphoid structures are ectopic lymphoid structures that appear during chronic inflammation and cancer in non-lymphoid structures.

1.2.5 *Innervation of the Choroid Plexus*

Benedict first demonstrated nerve fibers in the choroid plexus (Benedikt 1874). Later several other authors verified and expanded the observations (Stöhr 1922; Clark 1928; Junet 1926; Bakay 1941; Hworostuchin 1911; Schapiro 1931). In general, the nerves follow the path of the vessels. They originate from the periarterial sympathetic plexuses, with parasympathetic contribution from the vagus nerve (Benedikt 1874; Clark 1928) and suggestedly with contribution from the glossopharyngeal nerve (Stöhr 1922). The majority of fibers in the choroid plexus are non-myelinated vasomotor fibers, although myelinated sensory fibers exist (Voetmann 1949). Pericapillary nerve cells in the choroid plexus villi branch into several rather long processes and reach the base of the epithelial cells (Andia 1935). Three types of sensory structures have been described in the choroid plexus (Stöhr 1922): End-bulbs, Meissner's corpuscles-like structures, and a network of fine nerve fibers. It is believed that the structures function as pressure receptors (Stöhr 1922). Finally, it has been suggested that sensory processes from a subepithelial nervous plexus of very fine fibers project to the space between individual epithelial cells of the choroid plexus (Junet 1926).

1.3 **Histology and Ultrastructure of the Choroid Plexus**

In a doctoral thesis from 1949, Edel Voetmann (1949) cites Jan E. Purkinje (1836) as the first—in 1836—to observe and describe the choroid plexus epithelium, and refers to Faivre, Luschka, and Haeckl for contributing a more detailed description in the 1850s (Faivre 1854; Luschka 1855; Haeckl 1859). The authors describe two morphologically separate forms of the choroid plexus: as tongue-like sheet (lateral and third ventricles) and as complex tree-like villi (fourth ventricle), respectively. In both forms, a thin layer of connective tissue with capillaries and nerves is covered by a continuous monolayer of cuboidal epithelial cells (Fig. 1.3a, b) (Dempsey and

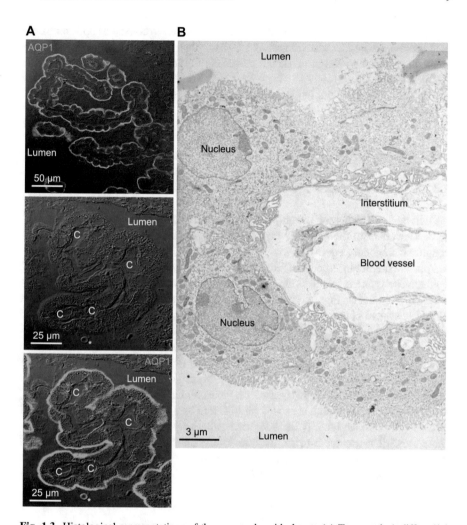

Fig. 1.3 Histological representations of the mouse choroid plexus. (**a**) Top panel: A differential interference contrast micrograph of the choroid plexus at lower magnification. The image is overlaid immuno-fluorescence signal for markers of the luminal membrane (the water channel AQP1, green) and the basolateral membrane (The Na$^+$:HCO$_3^-$ cotransporter NBCn1, red). Mid panel: A differential interference contrast image at high magnification. Bottom panel: the same micrograph overlaid the immunoreactivity for the water channel AQP1 (green). Note that AQP1 immunoreactivity marks both the luminal brush border and—to a lesser degree—the basolateral domains of the epithelial cells, as well as the capillary "C" endothelial cells. The asterix denotes a red blood cell (positive for AQP1). (**b**) Overview transmission electron micrograph of the epithelial monolayer of a choroid plexus villus, indicating the brush border membrane facing the ventricle lumen, the nuclei, basal labyrinths, basement membranes, the interstitium and a central capillary (blood vessel). See previous publications for methods (Praetorius et al. 2004a)

Wislocki 1955). The human choroid plexus is estimated to consist of approximately 100 million epithelial cells (Dohrmann 1970). Ultrastructural studies published in 1955 and 1956 revealed more details of the choroid plexus epithelium cytology as detailed below (Dempsey and Wislocki 1955; Van Breemen and Clemente 1955; Maxwell and Pease 1956; Millen and Rogers 1956; Pease 1956; Luse 1956).

1.3.1 Cell Surfaces of the Choroid Plexus Epithelial Cells

1.3.1.1 General Plasma Membrane Features

As will be detailed below, the choroid plexus epithelial cell has a complex morphology. Viewed from the exterior, the CSF facing surfaces of the epithelial cells are roughly hexagonal (Voetmann 1949). The roughly cuboidal cells present with luminal plasma membrane specializations, they share cell-cell contacts with neighboring epithelial cells on the lateral sides, they are attached to a basement membrane by basal cell-matrix contacts (Figs. 1.3b and 1.4a, schematic representation in Fig. 1.4b). An elaborate system of membrane protrusions extends from the cells to interdigitate with similar protrusions from the adjacent cells, the basal labyrinth. The total luminal surface area was originally estimated by light microscope morphometry to 600 cm^2 in humans (Voetmann 1949) probably underestimating the microvillus surface area expansion, and by electron microscopy to 75 cm^2 or 213 cm^2 in rat (Keep and Jones 1990; Keep et al. 1986). Based on measurements cell capacitance, however, the total plasma membrane area was estimated to as much as 5400 cm^2 (Kotera and Brown 1994), which notably includes the basolateral surface area. By scaling up the morphometric estimates for rats, the human luminal surface area may amount to as much as 2–5 m^2 (Spector et al. 2015) or approximately 1/3 of the surface of the human blood-brain-barrier (Nag and Begley 2005).

1.3.1.2 Luminal Specialization

The epithelial cells of the choroid plexus contain a luminal brush border connected with a dense terminal web underneath (Kalwaryiski 1924; Cobb 1932). Initially, repeated observations of luminal polypose bleb-like extrusions and apparent release into the lumen led to the theory of apocrine secretion (Dempsey and Wislocki 1955; Wislocki and Ladman 1958). Maxwell described the luminal surface as a polypoid border or brush border resembling structures (Maxwell and Pease 1956), while Millen and Luse both observed brush border and actual microvilli (Millen and Rogers 1956; Luse 1956). As pointed out by Tennyson and Pappas, the formation of polypoid extrusions is most likely to be a result of cell injury and fixation artifact (Tennyson and Pappas 1961). Thereby, the purpose of the observed structure is surface area extension to promote non-vesicular transport of solutes rather than apocrine secretion.

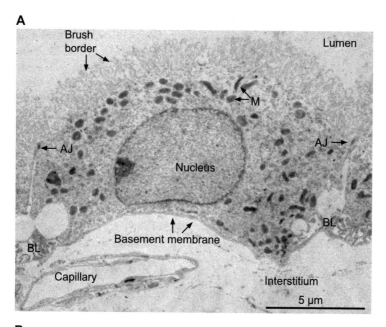

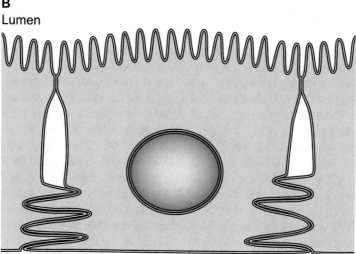

Fig. 1.4 The ultrastructure of a choroid plexus epithelial cell. (**a**) Transmission electron micrograph (TEM) of a choroid plexus epithelial cell, indicating the brush border membrane facing the ventricle lumen, adherence junctions (AJ), the nucleus, Mitochondria (M), the basal labyrinths (BL), basement membrane, the interstitium, and a capillary. See previous publications for methods (Praetorius et al. 2004a). (**b**) Schematic diagram of the cell as applied in later figures

Tufts of cilia project from the luminal surface of the mammalian choroid plexus epithelium (Studnicka 1900). The cilia were confirmed by early electron microscopy in several species (Maxwell and Pease 1956; Millen and Rogers 1956; Luse 1956). The number of cilia in individual cells varies among species from 3–4 in rabbit, 4–8 in rat, to 11–16 in monkey choroid plexus (Millen and Rogers 1956; Wislocki and Ladman 1958). The cilia display a typical ultrastructure for cilia with nine pairs of peripheral double microtubules perhaps with two central single microtubules, as elsewhere in beating cilia from various mammalian epithelia, i.e. a 9 + 2 structure (Maxwell and Pease 1956; Fawcett and Porter 1954). Although the structural indications for ciliary movement of CSF at the luminal surface of the choroid plexus seems well documented, primary cilia with a 9 + 0 structure lacking the two central microtubules may be more prevalent in the adult choroid plexus than motile cilia (Narita et al. 2012). The 9 + 2 structure disappears shortly after birth in mice and they do not seem to beat systematically for directional flow. Instead the prevalent non-motile cilia may serve an important sensory function that modulate transcytosis in the choroid plexus (Narita et al. 2010). The sensory function of primary cilia was first described in renal cells (Praetorius and Spring 2001) and has since proven a feature of a wide array of cellular sensory functions.

1.3.1.3 Lateral Intercellular Spaces

The space between the lateral cell membranes of neighboring choroid plexus epithelial cells is normally quite narrow 11–17 nm (Maxwell and Pease 1956). Tight junctions were described in early ultrastructural work, where a continuous belt of 4-stranded tight junctions (Fig. 1.5) separate the luminal space from the basolateral space as demonstrated by restriction to peroxidase or lanthanum infusion (Becker et al. 1967; Brightman and Reese 1969). The tight junctions contain both occludins and claudins, while the associated cytosolic ZO-1 found along the entire lateral surface (Vorbrodt and Dobrogowska 2003). Claudin-1, -2, -3, and -11 are all demonstrated in the tight junctions of the choroid plexus epithelium (Lippoldt et al. 2000; Wolburg et al. 2001; Steinemann et al. 2016). Notably, claudin-2 displays a unique transport permeation of monovalent cations as well as H_2O (Rosenthal et al. 2010, 2017) and probably contribute to the intermediate transepithelial resistance of the choroid plexus. In epithelia, zonula adherens are systematically expressed beneath the tight junctions, but are not expressed constantly in the CPE (Peters et al. 1976). Both α-catenin and β-catenin are distributed along the lateral cell surface, probably along with non-epithelial cadherin forms (see below) (Vorbrodt and Dobrogowska 2003). Desmosomes are demonstrated both close to the luminal surface beneath the tight junction and in the basal labyrinth (see below) (Segal and Burgess 1974). Desmosomes provide mechanical strength in and between cells by linking transmembrane adhesive proteins via plaque proteins to the cellular intermediate filaments. In the choroid plexus epithelium the intermediate filaments are keratins 8, 18, and possibly 19 (Kasper et al. 1986; Miettinen et al. 1986).

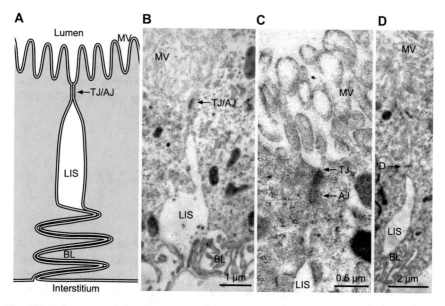

Fig. 1.5 Ultrastructure of the lateral intercellular space of choroid plexus epithelial cells. (**a**) Schematic diagram of the lateral intercellular space (LIS) with marked lumen, interstitium, microvilli (MV), basal labyrinth (BL), and a zone with tight junctions and adherence junctions (TJ/AJ). (**b**) TEM of a similar region between neighboring epithelial cells. (**c**) Higher magnification of the cell junctions beneath the luminal microvilli. (**d**) The rare observation of a desmosome

1.3.1.4   The Basement Membrane and a Basal Labyrinth

A laminated thin basement membrane or basal lamina anchors the base of the choroid plexus epithelial cells to the underlying connective tissue (Askanazy 1914; Franceschini 1929). Electron micrographs by van Breemen were the first to describe a basement membrane at the ultrastructural level (Van Breemen and Clemente 1955), while Maxwell was the first to describe a basal labyrinth between the basal parts of adjacent cells that increases the basolateral membrane surface area massively (Maxwell and Pease 1956). Both findings were soon confirmed by Millen and Rogers (1956), Luse (1956), and Pease (1956). In the basal labyrinth, branched protrusions extend and interdigitate with protrusions from the joining cells and represents a large increase in the basolateral membrane area. The space between the membranes of adjacent protrusions of the labyrinth is fairly constant and narrow. Pease estimated the distance between plasma membranes of adhering cells in basal labyrinth to 11–17 nm (Pease 1956). These are important features for modeling water and solute transport across the epithelium (Diamond and Bossert 1968). ·Focal adhesions and hemi-desmosomes are sporadically described in the literature, although the involved adhesive proteins integrin $\alpha 3$, $\alpha 6$, $\beta 1$, $\beta 4$ are demonstrated to be expressed in the choroid plexus (Paulus et al. 1993). The anchoring of basolateral membrane proteins, such as integrins to the cytoskeleton seems to occur through the proteins utrophin A, utrophin, and syntrophins $\alpha 1$ and β2 (Gorecki et al. 1997; Haenggi and Fritschy 2006; Weir et al. 2002). This could also be the case for ion transport proteins in the basolateral membrane, as the usual transporter anchoring protein ankyrin is expressed only in the luminal domain of the cytoplasm (see below). The basement membrane seems to include typical lamina densa proteins, such as laminins and type IV collagen consisting of α3, α4, and α5 molecules, which suggests permeability properties similar to the basement membrane of renal glomeruli (Urabe et al. 2002).

1.3.2   Organelles of the Choroid Plexus Epithelial Cells

1.3.2.1   The Nucleus

The nucleus of choroid plexus epithelial cells is most often spherical or slightly oval and positioned near the cell center (Fig. 1.4a) (Pettit and Girard 1901). Lobated nuclei have been described in young mice but not in older mice (Dohrmann and Herdson 1969). The nucleus is enclosed by a normal nuclear envelope with a double membrane system with nuclear pores (Millen and Rogers 1956). Both heterochromatin and euchromatin is clearly visible in the nucleus (Van Breemen and Clemente 1955), which contains 1–3 nucleoli (Millen and Rogers 1956) indicating ongoing synthesis of mRNA and rRNA, respectively.

1.3.2.2   The Endoplasmatic Reticulum and Ribosomes

Polyribosomes and the cisterns belonging to the rough endoplasmic reticulum are distributed throughout the cytoplasm of the choroid plexus epithelial cells (Van Breemen and Clemente 1955) (Fig. 1.6) indicating a quite high level continuous protein synthesis. The cisterns of the rough endoplasmic reticulum appear both separate from and to be continuous with the nuclear envelope (Millen and Rogers 1956).

1.3.2.3   The Golgi Apparatus

The Golgi apparatus of choroid plexus epithelial cells resembles elliptical vesicles with a smooth membrane and is located close to the nucleus, but seems to lack luminal-basal orientation (Maxwell and Pease 1956). The lack of the normal epithelial supranuclear Golgi apparatus indicates an atypical pattern of directing the trans-Golgi vesicles to their intracellular or plasma membrane destinations.

Fig. 1.6 TEM ultrastructure of the epoxy embedded choroid plexus epithelium. (**a**) Overview image of the entire cell with indications for the ventricle lumen, ciliar basal bodies (BB), microvilli (MV), basal labyrinth (BL), and the interstitium. (**b**) Higher magnification of details of the same cell. The cell nucleus (N), MV, BB and mitochondria (M) are clearly visible. (**c**) Further magnification allowing identification of rough endoplasmc reticululum (RER), free ribosomes (R), multivesicular body (MVB), mitochondrion (M), clathrin coated pit (CCP), and an undefined vesicle (V). The increasing magnification of the same cell is indicated by red boxes

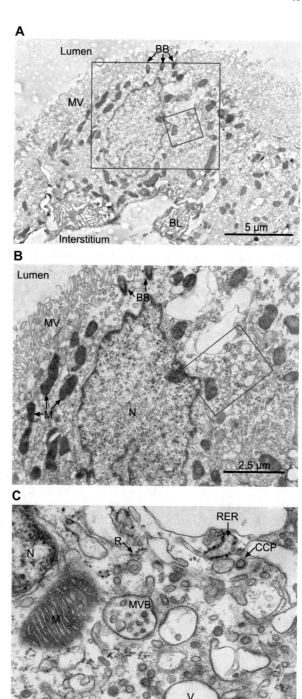

1.3.2.4 The Endolysosomal System

Pinocytotic vesicles are primarily situated just beneath the luminal, basal and lateral membranes. Epithelial cell lysosomes were demonstrated by acid phosphatase enzyme histochemistry on the rat choroid plexus (Cancilla et al. 1966) but are not found in higher abundance. Multivesicular bodies and cytoplasmic areas of degradation—probably proteasomes—have been noted in the choroid plexus epithelium by electron microscopy analysis (Fig. 1.6). The epithelial cells possess the known protein degradation machineries in quantities that are sufficient for cellular turnover rather than breaking down large quantities of exogenous material.

1.3.2.5 Mitochondria

Numerous mitochondria are found scattered throughout the cytoplasm of the choroid plexus epithelial cells. The mitochondria of cells were first visualized by light microscopy using selective staining (Sundwall 1917). They have a typical ultrastructure with an outer membrane and inner system of cristae as assessed by electron microscopy (Van Breemen and Clemente 1955) (Fig. 1.6). The mitochondria of the epithelial cells seem to be randomly distributed in the cytoplasm (Maxwell and Pease 1956), where most are small and rounded mitochondria and only few appear elongated (Millen and Rogers 1956; Luse 1956). The choroid plexus epithelium seems to have less mitochondrial area compared to e.g. the renal proximal tubule indicating a lower cellular capacity for ATP production than expected for sustaining the high transport rate.

1.3.3 Cell Polarity of the Choroid Plexus Epithelial Cells

1.3.3.1 Cell Polarity

Epithelial cell polarity is established and maintained by an array of specific polarity proteins or *apical and basolateral determinants*. The apical-basolateral polarity of epithelial cells is initiated by their cell-cell and cell-matrix interactions and is formed as the polarity proteins as well as two plasma membrane lipid compositions induce and maintain a specific cellular orientation (Rodriguez-Boulan and Macara 2014; Yeaman et al. 1999; Di Paolo and De Camilli 2006; Shewan et al. 2011).

1.3.3.2 Crumbs, PAR, and Scribble Complexes

In mammalian cells, the apical plasma membrane is determined by the arrangement of the Crumbs complex (Crumbs, Pals1, and Patj) and the Par protein complexes (Par-3, Par-6, PKCζ, and Cdc42) at this site (Assemat et al. 2008). We recently localized the components of both the Crumbs and Par complexes at the luminal membrane domain of the CPECs (Christensen et al. 2018a). The basolateral membrane domain is developed by the Scribble complex (Scribble, Lgl-1/Lgl-2, and Dlg1) and Par-1 (Goehring 2014; Horikoshi et al. 2009; Hurov et al. 2004; Makarova et al. 2003). In the choroid plexus epithelium, we mapped components of both the Scribble complex and Par-1 corresponding to the basolateral membrane domain. Thus, this epithelium seems to adhere to the general paradigm for proteinous apical-basal determinants.

1.3.3.3   Phosphatidylinositol-Phosphates

The spatially separated accumulation of PIP_2 in the luminal membrane and PIP_3 in the basolateral membrane is pivotal in establishing cell polarity (Gassama-Diagne et al. 2006; Kierbel et al. 2007). The exclusion of PIP_2 from the basolateral membrane is governed by PI3K (Shewan et al. 2011; Gassama-Diagne et al. 2006), whereas the exclusion of PIP_3 from the luminal membrane is mediated by the phosphatase Pten (Wu et al. 2007). Phosphatidylinositol-(3,4,5)-trisphosphate (PIP_3) was recently immunolocalized to the basolateral membrane domain of choroid plexus epithelial cells, while phosphatidylinositol-(4,5)-bisphosphate (PIP_2) staining was most prominent in the luminal membrane domain along with the PIP_3 phosphatase, Pten (Christensen et al. 2018a). As for the polarity proteins, the choroid plexus epithelium also displays a normal distribution of basic polarity defining phospholipids. This underscores the previous observations of a normally polarized cell type, as judged from above-mentioned ultrastructural analysis with juxta-luminal tight junctions, a basal basement membrane and luminal specialization such as microvilli and cilia.

1.3.3.4   SNARE Proteins

Vesicles from the Golgi apparatus targeting a specific plasma membrane domain undergo membrane-selective tethering and insertion into the plasma membrane. The processes are controlled by Rab proteins, the exocyst complex, and SNARE proteins (Grindstaff et al. 1998; Stoops and Caplan 2014; Weisz and Rodriguez-Boulan 2009; Wu and Guo 2015). Syntaxin-3 in the luminal plasma membrane and syntaxin-4 in the basolateral plasma membrane represents two such SNARE proteins in polarized epithelia (Low et al. 1996; Sharma et al. 2006; ter Beest et al. 2005). Interestingly, both Syntaxin-3 and Syntaxin-4 were localized to the luminal membrane of choroid plexus epithelial cells (Christensen et al. 2018a). The basolateral accumulation of certain membrane proteins requires basolateral recycling (Folsch et al. 1999; Ohno et al. 1999). Here, internalized proteins are re-exocytosed to the basolateral membrane in a process depending on the epithelium-specific clathrin adaptor AP-1B. Choroid plexus epithelial cells, are among the few exceptions of epithelia, which lack expression of AP-1B (Christensen et al. 2018a). Both the luminal membrane expression of syntaxin-4 and the lack of AP-1B expression may have consequences for the asymmetrical delivery and retention of plasma membrane receptors and transport proteins in this epithelium.

1.3.3.5 Cadherins

The expression of cadherins influences the specific plasma membrane accumulation of basic polarity proteins and other membrane proteins. The expression of cadherins in the choroid plexus has been demonstrated before, however, the specific forms of cadherin expressed was uncertain until recently (Christensen et al. 2013; Kaji et al. 2012; Lagunowich et al. 1992; Marrs et al. 1993). Based on mass spectrometry analysis and immunolocalization, we reported the expression of P-cadherin and -with less confidence- N-cadherin in the lateral membrane and in the basal labyrinth of the choroid plexus epithelium (Christensen et al. 2018a). Notably, the typical epithelial cadherin, E-cadherin, is absent from the choroid plexus.

In conclusion, the choroid plexus epithelial cells are normally polarized. However, the molecular architecture of the choroid plexus epithelium is exceptional in three ways, which could affect plasma membrane protein distribution: (1) the basolateral membrane SNARE protein, syntaxin-4, is expressed in the luminal plasma membrane with syntaxin-3, (2) the non-epithelial cadherins, P-and N-cadherin, are expressed rather than E-cadherin, and finally (3) the epithelial cells lack the basolateral recycling clathrin adaptor protein AP-1B.

1.3.3.6 Spectrin and Ankyrin

The preferential luminal domain expression of both ankyrin and spectrin in the choroid plexus epithelial cells was demonstrated in two studies in the 1993–94 (Marrs et al. 1993; Alper et al. 1994). Later several spectrin forms and the specific ankyrin isoform were detected in these cells (Christensen et al. 2013, 2018a). The spectrin cytoskeleton consists of αII-spectrin in combination with βI-, βII-, and βIII-spectrin in the luminal membrane domain and at the basolateral infoldings. Ankyrin-3 was the only ankyrin form expressed in the choroid plexus, where its localization was restricted to the unusual position in the luminal membrane domain (Christensen et al. 2018a). Interestingly, ankyrin-3 protruded into the core of the microvilli, but none of the spectrin forms were localized inside the microvilli and, instead, took a subluminal position. The lack of subcellular colocalization of ankyrin and spectrin forms was unexpected as certain luminal membrane proteins of the choroid plexus are thought to link to spectrins via ankyrin-3.

1.3.4   Plasma Membrane Solute and Water Transport Proteins

Multiple studies have shown that CSF is not produced as an ultrafiltrate. The ionic composition of the CSF as well as the selection of ion transport inhibitors affecting the secretion rate are solid arguments for active and secondary active transport by the choroid plexus (Fig. 1.7). A large number of transport proteins and receptors are expressed in the choroid plexus epithelium for mediating and controlling the many functions of this tissue (Fig. 1.8). Probably as a consequence of the aforementioned deviations from normal epithelial cell biology, more basolateral-type transport proteins are expressed at the luminal membrane (Damkier et al. 2013). Thus, the epithelium operates with an alternative orchestration of the transcellular and paracellular transport than other secretory (and absorptive) epithelia and has been described as a backwards fluid-transporting epithelium (Diamond and Bossert 1968).

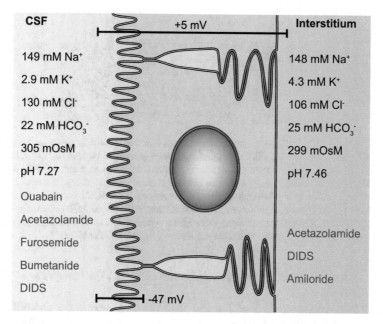

Fig. 1.7 Sites of action for inhibitors of CSF secretion by the choroid plexus epithelium and the ionic composition of the interstitial fluid and the nascent cerebrospinal fluid. The inhibitors ouabain, acetazolamide, furosemide, bumetanide, and DIDS are reported to inhibit CSF secretion from the ventricle facing/luminal membrane, while acetazolamide, DIDS, and amiloride, are inhibitors acting from the blood facing/basolateral membrane. The transepithelial potential difference amounts to 5 mV lumen positive and the membrane potential in mouse CPECs is -47 mV

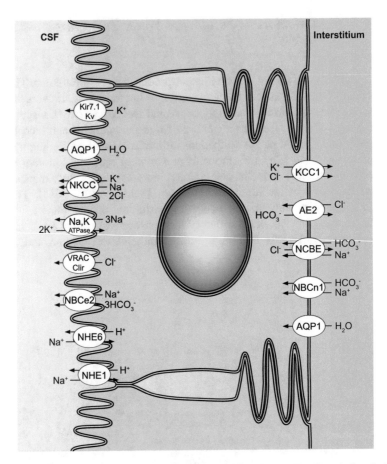

Fig. 1.8 Cellular localization of transport mechanisms believed to be directly or indirectly implicated in CSF secretion by the CPE. The luminal membrane is occupied by the Na^+-K^+-ATPase, the
water channel AQP1, K^+ channels, Cl^- conductances, and the cotransporters NKCC1. Acid/base
transporters NHE1 and NBCe2 are also localized to the luminal membrane. Fewer types of transport
proteins seem present in the basolateral membrane domain. The KCC1 cotransporter has been
suggested at the basolateral membrane along with modest numbers of AQP1 channels. Multiple
acid/base transporters are expressed in the basolateral membrane: the anion exchanger AE2, the
cotransporter NBCn1 (sometimes a luminal membrane protein), and the Ncbe

1.3.4.1 The Na^+,K^+-ATPase and the Monovalent Ion Cotransporters

Ouabain is a potent inhibitor of ion and fluid transport across the choroid plexus
when injected into the CSF (Welch 1963; Davson and Segal 1970; Wright 1972),
indicating both strong dependence of CSF secretion on this pump and a luminal
membrane localization. The Na^+,K^+-ATPase was directly immunolocalized to the
luminal membrane of choroid plexus epithelial cells (Masuzawa et al. 1984; Siegel
et al. 1984).

The Na^+, K^+, $2Cl^-$ cotransporter NKCC1 is also a luminal membrane protein in the choroid plexus epithelium (Plotkin et al. 1997; Praetorius and Nielsen 2006). Bumetanide applied to the ventricle side of the epithelium inhibits CSF secretion (Bairamian et al. 1991; Javaheri and Wagner 1993; Keep et al. 1994). Some controversy remains to whether NKCC1 directly transports ions into the CSF or is necessary for cell volume and Cl^- regulation to allow sustained secretion (Steffensen et al. 2018; Gregoriades et al. 2019). NKCC1 usually transport ions into cells, driven by the inward Na^+ and Cl^- gradients. In the choroid plexus, such inward transport was indeed suggested (Bairamian et al. 1991), and a role in regulatory volume increase found in isolated CPECs from rats (Gregoriades et al. 2019; Wu et al. 1998). Not all studies are supportive of a role for NKCC1 in regulatory volume increase (Hughes et al. 2010). It can be argued that the gradients existing across the luminal membrane of CPECs will favor the opposite direction of transport (Keep et al. 1994; Steffensen et al. 2018), but no reliable data are available on the exact ionic grandients operating in the tissue in vivo (Delpire and Gagnon 2018).

The K^+,Cl^- cotransporters (KCCs) transport the ions outward driven by the K^+ gradient. Luminal K^+,Cl^- cotransport was described in the amphibian CPE (Zeuthen 1991, 1994; Zeuthen and Wright 1981), but their relevance in mammalian systems remains to be demonstrated. At the molecular level, KCC1 mRNA and protein has been detected in the mammalian choroid plexus (Kanaka et al. 2001; Damkier et al. 2018) and probably resides in the luminal membrane (Steffensen et al. 2018). KCC4 is expressed in the CPE and localized to the luminal membrane (Li et al. 2002; Karadsheh et al. 2004), where both KCC1 and KCC4 could contribute to K^+ recycling. The previous detection of KCC3a in the basolateral membrane (Pearson et al. 2001) has not been verified by others.

1.3.4.2 Acid-Base Transporters

Implication of Na^+/H^+ exchangers in CSF secretion was indicated by an inhibitory effect of the inhibitor amiloride applied from both the blood and the CSF side of the (Segal and Burgess 1974; Davson and Segal 1970; Murphy and Johanson 1989a, b). The mRNA encoding the Na^+/H^+ exchanger NHE1 was detected by RT-PCR analysis (Kalaria et al. 1998), and the prevailing notion was that was that the resulting protein was expressed in the basolateral membrane. However, both NHE1, and more robustly NHE6, seem to reside in the luminal membrane (Damkier et al. 2009; Christensen et al. 2017), and are, thus, unlikely to contribute to the secretion of CSF.

CSF secretion requires continuous uptake of Cl^- across the basolateral membrane of the choroid plexus epithelium. The only known Cl^- import mechanism at this position is the anion exchanger AE2 (Alper et al. 1994; Praetorius and Nielsen 2006; Lindsey et al. 1990). This finding corresponds well with the functional evidence that the anion transport blocker DIDS inhibited Cl^- transport into the CSF when applied from the basolateral side (Deng and Johanson 1989; Frankel and Kazemi 1983). The function of a Cl^-/HCO_3^- exchange mechanism was further supported by the

HCO_3^- dependence of cell volume regulation in mouse choroid plexus (Hughes et al. 2010).

No less than three Na^+ dependent HCO_3^- transporters are expressed in the choroid plexus epithelium. First, the Na^+-dependent Cl^-/HCO_3^- exchanger Ncbe was detected and localized to the luminal membrane (Praetorius et al. 2004b). Ncbe activity accounts for the vast majority of the DIDS sensitive Na^+-HCO_3^- import in isolated choroid plexus epithelial cells (Jacobs et al. 2008). The brain ventricle volume in Ncbe knockout mice was greatly reduced in the same study, indicating a diminished CSF secretion compared to wild type littermates and thereby a central role in vectorial transport in the choroid plexus. Another Na^+ dependent HCO_3^- transporter, the NBCn1 is also expressed in the choroid plexus epithelium (Bouzinova et al. 2005), but seem to play less of a role in the cellular Na^+ dependent HCO_3^- uptake than Ncbe.

The presence of CO_2/HCO_3^- is required for CSF secretion (Haselbach et al. 2001). Thus, it is not surprising that application of the inhibitor DIDS to the luminal side of the epithelium suggested the existence of luminal membrane HCO_3^- transporters as well as their role in CSF pH regulation (Nattie and Adams 1988). Transport of HCO_3^- across the luminal membrane was shown to be electrogenic (Saito and Wright 1984) and depend on Na^+ (Johanson et al. 1992) indicating outward luminal membrane $Na^+:HCO_3^-$ cotransport. Previous molecular and functional evidence pointed to the involvement of the electrogenic $Na^+:HCO_3^-$ cotransporter 2, NBCe2 in this process (Bouzinova et al. 2005; Millar and Brown 2008; Banizs et al. 2007). A recent study provided strong support for such involvement in vitro in an NCBe2 gene knockout model (Christensen et al. 2018b). The same study provided evidence for a direct CSF pH regulatory role of NBCe2 in vivo.

Carbonic anhydrase inhibition reduces CSF secretion within the range of 50–100% (Welch 1963; Davson and Segal 1970; Ames et al. 1965) by affecting the HCO_3^- levels needed for secretion of HCO_3^-. Investigators ascribed the effect to cytosolic carbonic anhydrase activity. However, enzyme histochemistry analysis demonstrated this activity to be confined to the brush border membrane end the basolateral surface particularly at the basal labyrinth (Masuzawa et al. 1981). Indeed, the expression of two membrane bound carbonic anhydrases, CAXII and CAIX, have been demonstrated in the CPE (Kallio et al. 2006) in addition to the cytosolic CAII. More information on the acid-base transporters and carbonic anhydrase activity in the choroid plexus and roles in CSF production and pH regulation is found in the dedicated Chap. 6.

1.3.4.3   Ion Channels

Only the luminal membrane of the choroid plexus epithelial cells harbors K^+ conductances (Zeuthen and Wright 1981). These conductances probably both set the membrane potential and provide luminal membrane K^+ recycling. An inward-rectifying conductance, Kir, as well as an outward-rectifying conductance Kv have been described (Kotera and Brown 1994). The respective molecular identities for these conductances are Kir7.1 channels (Döring et al. 1998), and Kv1.1 as well as Kv1.3 channels (Speake et al. 2004). KCNQ1 and KCNE2 channels also seems to contribute to the K conductance (Roepke et al. 2011). All the K^+ channels were immunolocalized to the luminal surface of rodent choroid plexus (Speake et al. 2004; Roepke et al. 2011; Nakamura et al. 1999).

Extrusion of Cl^- across the luminal membrane in the CPE is electrogenic. Both a PKA activated inward-rectifying conductance and a volume sensitive conductance are detected (Kajita and Brown 1997; Kibble et al. 1996, 1997). In addition to the Cl^- conductance, Clir channels display a high HCO_3^- conductance in the mammalian CPE (Kibble et al. 1996), which may contribute to CSF secretion in two ways. By contrast, the volume-regulated anion conductances seem not to significantly take part in the baseline Cl^- conductance (Kibble et al. 1996, 1997; Millar et al. 2007). The molecular identities of the Cl^- and HCO_3^- channels are not fully established. The CFTR and Clc-2 have been excluded as mediators of the Clir activity (Kibble et al. 1996, 1997; Speake et al. 2002). Electrogenic Cl^- pathways identified in the mouse choroid plexus include several proteins belonging to the voltage-dependent anion channels (Vdac), intracelluar Cl^- channels (Clic), H^+/Cl^- exchangers (Clcn), and subunits from a volume regulated Lrrc channel by mass spectrometry (Damkier et al. 2018).

The non-selective mechanosensitive cation channel TRPV4 (transient receptor potential, vanilloid 4) is expressed in the choroid plexus (Liedtke et al. 2000) has been immunolocalized to the luminal membrane of choroid plexus epithelial cells (Narita et al. 2015). TRPV4 agonists induce calcium influxes and probably play roles in the regulation of epithelial permeability to ions and proteins (Narita et al. 2015; Preston et al. 2018). Several proteins belonging to the ionotropic purineric P2X receptor family are expressed in the choroid plexus (Johansson et al. 2007; Xiang and Burnstock 2005), where they may be involved in ATP sensing or like the TRPV4 channels mechano-sensation in the control of CSF secretion A detailed review of TRPV4 is found in Chap. 7. The expression of selective Na^+ channels (ENaCs) is a matter of continued debate. Although some researchers find evidence of ENaC subunit expression and function (Van Huysse et al. 2012), Brown and Miller found no evidence of the hallmark amiloride sensitive whole cell conductance (Millar et al. 2007).

1.3.4.4 Xeno- and Endobiotic Efflux Systems

As reviewed recently, the choroid plexus epithelium expresses a host of molecular mechanisms to protect the CSF against xeno- and endobiotics arising from the family of ATP-binding cassette transporters and families of specific solute carriers (Ghersi-Egea et al. 2018). In brief, the transporters ABCC1 and ABCC4 are basolateral energy-dependent extruders of lipid soluble compounds, whereas the luminal transporters *Slc15a2*, *Slc22A8*, *Slc29a4*, *Slco1a5*, and *Slco1a5* facilitate the epithelial uptake of potentially harmful substances and drugs from the CSF. The multidrug extruder *SLC47A1* and/or the organic cation transporter SLC29A4 mediate uptake of endogenous monoamines. A dedicated chapter on these transporters is given in the dedicated Chap. 8.

1.3.4.5 Transepithelial Water Transport

The molecular entities that constitute the routes for H_2O across the choroid plexus epithelium are still not entirely unraveled. Two key observations have helped hypothesis generation on this topic. Firstly, the transepithelial difference in osmolarity has been estimated to approximately 5 mOsM with the CSF slightly hyperosmolar (Davson and Segal 1970), yielding a gradient for H_2O of similar magnitude as (but smaller than) across the renal proximal tubules. Secondly, H_2O transport across the choroid plexus epithelium is closely tied to the secretion of Na^+ and the Na^+,K^+-ATPase activity (Welch 1963; Ames et al. 1965; Pollay and Curl 1967). A transcellular pathway for H_2O is indicated by a high abundance expression of aquaporin 1, AQP1, in the luminal plasma membrane and, although at lower abundance, in the basolateral membrane (Praetorius and Nielsen 2006; Nielsen et al. 1993). This may reflect that the difference in osmolarity between the interstitial fluid and the intracellular compartment is sufficiently steep to allow a smaller basolateral H_2O permeability. Conversely, the difference in osmolarity across the luminal membrane may be minute if the AQP1 abundance is proportional to the H_2O permeability. Functional evidence supporting a central role of AQP1 in CSF secretion came from studies of AQP1 knockout mice displaying an approximately 80% reduction in transepithelial H_2O permeability (Oshio et al. 2005). The resulting reduction in the rate of CSF secretion was approximately 35% (Oshio et al. 2005). This may appear as a lower than expected reduction, but one has to take into account that the choroid plexus secretes only 70–80% of the CSF and that similar numbers were reported for renal proximal tubules, where compensatory changes partly made up for the knockout of AQP1 (Schnermann et al. 1998; Vallon et al. 2000). It seems unlikely that the choroid plexus epithelia express other aquaporins in significant abundances although this has been reported sporadically. The possibility of solute-coupled H_2O movement through cotransporters (Zeuthen 1991; Zeuthen et al. 2016; Charron et al. 2006) is still debated and is discussed separately in the dedicated Chap. 4.

The combination of leaky TJs and the complex morphology of the basal labyrinth are probably the explanation for an electrical resistance in the intermediate range across the choroid plexus epithelium (Wright 1972). Claudin-2 is the only pore forming claudin expressed in the choroid plexus (Wolburg et al. 2001; Kratzer et al. 2012; Krug et al. 2012). As mentioned above, tight junctions of the claudin-2 is cation selective and also permeates H_2O (Rosenthal et al. 2010, 2017), and may well contribute for the AQP1-independent epithelial H_2O permeability and secretion. Nevertheless, the relative importance of the transepithelial H_2O pathways of the choroid plexus CSF secretion remains to be fully established.

1.3.5   Connective Tissue and Microvasculature of the Choroid Plexus

A layer of connective tissue with fibroblasts and collagen fibrils is found between the choroid plexus epithelium and the capillaries (Voetmann 1949). In rodents and young human beings this layer is quite thin, while it becomes considerably thicker in older individuals. Macrophages, fibroblasts and several leukocytes have been seen by Wislocki and Ladman (Wislocki and Ladman 1958).

The arteries feeding the choroid plexus and the draining veins contain the typical components of an intima, media and adventitia. An exception is venous vessels observed in the glomus, where large diameter vessels seem to connect with small diameter vessels through very narrow anastomoses. The walls of the venous cavities consist only of endothelium and connective tissue (Luschka 1855; Schmid 1929; Becker 1939; Meek 1907; Findlay 1899). Arterioles are most abundant close to the base of the choroid plexus, where the arteries enter the tissue. They contain typical squamous endothelial cells surrounded by smooth muscle cells and an amorphous matrix (Fawcett 1959). The capillaries in the choroid plexus are wide in comparison with capillaries elsewhere (10–20 μm) (Voetmann 1949; Maxwell and Pease 1956), and thus described as "sinusoid capillaries" (Andia 1935). They present with a fenestrated endothelium and a thin basement membrane (Maxwell and Pease 1956). The pores measure 30–50 nm in diameter resembling the structure of capillaries in other transporting epithelia (Maxwell and Pease 1956; Pease 1956; Palay and Karlin 1959).

References

Alonso MI et al (1999) Involvement of sulfated proteoglycans in embryonic brain expansion at earliest stages of development in rat embryos. Cells Tissues Organs 165(1):1–9

Alper SL et al (1994) The fodrin-ankyrin cytoskeleton of choroid plexus preferentially colocalizes with apical Na+K(+)-ATPase rather than with basolateral anion exchanger AE2. J Clin Invest 93 (4):1430–1438

Ames A III, Higashi K, Nesbett FB (1965) Effects of Pco2 acetazolamide and ouabain on volume and composition of choroid-plexus fluid. J Physiol 181(3):516–524

Andia ED (1935) Plexos coroideos de los ventriculos laterales. Virtus, Buenos Aires

Askanazy M (1914) Zur Physiologie und Pathologie der Plexus chorioidei. Zentralbl Allg Pathol 25:390–391

Assemat E et al (2008) Polarity complex proteins. Biochim Biophys Acta 1778(3):614–630

Bairamian D et al (1991) Potassium cotransport with sodium and chloride in the choroid plexus. J Neurochem 56(5):1623–1629

Bakay Lv (1941) Die Innervation der Pia Mater, der Plexus chorioideus und der Hirngefässen, mit Rücksicht auf den Einfluss des sympatischen Nervensystems auf die Liquorsekretion. Arch Psychiatr Nervenkr 113:412–427

Bally-Cuif L, Cholley B, Wassef M (1995) Involvement of Wnt-1 in the formation of the mes/metencephalic boundary. Mech Dev 53(1):23–34

Banizs B et al (2007) Altered pH(i) regulation and Na(+)/HCO3(-) transporter activity in choroid plexus of cilia-defective Tg737(orpk) mutant mouse. Am J Physiol Cell Physiol 292(4):C1409–C1416

Barkho BZ, Monuki ES (2015) Proliferation of cultured mouse choroid plexus epithelial cells. PLoS One 10(3):e0121738

Becker G (1939) Beiträge zur Orthologie und Pathologie der Plexus chorioidei und des Ependyms. Beiträge zur Pathologische Anatomie und zur Allgemeine Pathologie 103:457–478

Becker NH, Novikoff AB, Zimmerman HM (1967) Fine structure observations of the uptake of intravenously injected peroxidase by the rat choroid plexus. J Histochem Cytochem 15 (3):160–165

Benedikt M (1874) Über die innervation des plexus chorioideus inferior. Archiv Pathol Anat Physiol 59:395–400

Bouzinova EV et al (2005) Na+-dependent HCO_3^- uptake into the rat choroid plexus epithelium is partially DIDS sensitive. Am J Physiol Cell Physiol 289(6):C1448–C1456

Brightman MW, Reese TS (1969) Junctions between intimately apposed cell membranes in the vertebrate brain. J Cell Biol 40(3):648–677

Cancilla PA, Zimmerman HM, Becker NH (1966) A histochemical and fine structure study of the developing rat choroid plexus. Acta Neuropathol 6(2):188–200

Charron FM, Blanchard MG, Lapointe JY (2006) Intracellular hypertonicity is responsible for water flux associated with Na+/glucose cotransport. Biophys J 90(10):3546–3554

Chizhikov VV et al (2010) Lmx1a regulates fates and location of cells originating from the cerebellar rhombic lip and telencephalic cortical hem. Proc Natl Acad Sci U S A 107 (23):10725–10730

Chodobski A, Szmydynger-Chodobska J (2001) Choroid plexus: target for polypeptides and site of their synthesis. Microsc Res Tech 52(1):65–82

Chouaf-Lakhdar L et al (2003) Proliferative activity and nestin expression in periventricular cells of the adult rat brain. Neuroreport 14(4):633–636

Christensen IB et al (2013) Polarization of membrane associated proteins in the choroid plexus epithelium from normal and slc4a10 knockout mice. Front Physiol 4:344

Christensen HL et al (2017) The V-ATPase is expressed in the choroid plexus and mediates cAMP-induced intracellular pH alterations. Physiol Rep 5(1):e13072

Christensen IB et al (2018a) Choroid plexus epithelial cells express the adhesion protein P-cadherin at cell-cell contacts and syntaxin-4 in the luminal membrane domain. Am J Physiol Cell Physiol 314(5):C519–C533

Christensen HL et al (2018b) The choroid plexus sodium-bicarbonate cotransporter NBCe2 regulates mouse cerebrospinal fluid pH. J Physiol 596(19):4709–4728

Clark SL (1928) Nerve endings in the choroid plexuses of the fourth ventricle. J Comp Neurol 47:1–21

Cobb S (1932) The cerebrospinal blood vessels. In: Penfield W (ed) Cytology and cellular pathology of the nervous system. Hoeber, New York, pp 575–577

Cserr HF (1971) Physiology of the choroid plexus. Physiol Rev 51(2):273–311

Cserr HF, Knopf PM (1992) Cervical lymphatics, the blood-brain barrier and the immunoreactivity of the brain: a new view. Immunol Today 13(12):507–512

Currle DS et al (2005) Direct and indirect roles of CNS dorsal midline cells in choroid plexus epithelia formation. Development 132(15):3549–3559

Da Mesquita S, Fu Z, Kipnis J (2018) The meningeal lymphatic system: a new player in neurophysiology. Neuron 100(2):375–388

Damkier HH et al (2009) Nhe1 is a luminal Na+/H+ exchanger in mouse choroid plexus and is targeted to the basolateral membrane in Ncbe/Nbcn2-null mice. Am J Physiol Cell Physiol 296 (6):C1291–C1300

Damkier HH, Brown PD, Praetorius J (2013) Cerebrospinal fluid secretion by the choroid plexus. Physiol Rev 93(4):1847–1892

Damkier HH et al (2018) The murine choroid plexus epithelium expresses the 2Cl(-)/H(+) exchanger ClC-7 and Na(+)/H(+) exchanger NHE6 in the luminal membrane domain. Am J Physiol Cell Physiol 314(4):C439–C448

Davson H, Segal MB (1970) The effects of some inhibitors and accelerators of sodium transport on the turnover of 22Na in the cerebrospinal fluid and the brain. J Physiol 209(1):131–153

Delpire E, Gagnon KB (2018) Na(+) -K(+) -2Cl(-) cotransporter (NKCC) physiological function in nonpolarized cells and transporting epithelia. Compr Physiol 8(2):871–901

Dempsey EW, Wislocki GB (1955) An electron microscopic study of the blood-brain barrier in the rat, employing silver nitrate as a vital stain. J Biophys Biochem Cytol 1(3):245–256

Deng QS, Johanson CE (1989) Stilbenes inhibit exchange of chloride between blood, choroid plexus and the cerebrospinal fluid. Brain Res 510:183–187

Desmond ME, Jacobson AG (1977) Embryonic brain enlargement requires cerebrospinal fluid pressure. Dev Biol 57(1):188–198

Di Paolo G, De Camilli P (2006) Phosphoinositides in cell regulation and membrane dynamics. Nature 443(7112):651–657

Diamond JM, Bossert WH (1968) Functional consequences of ultrastructural geometry in "backwards" fluid-transporting epithelia. J Cell Biol 37(3):694–702

Dohrmann GJ (1970) The choroid plexus: a historical review. Brain Res 18(2):197–218

Dohrmann GJ, Herdson PB (1969) Lobated nuclei in epithelial cells of the choroid plexus of young mice. J Ultrastruct Res 29(3):218–223

Döring F et al (1998) The epithelial inward rectifier channel Kir 7.1 displays unusual K^+ permeation properties. J Neurosci 18:8625–8636

Dziegielewska KM et al (1980a) Proteins in cerebrospinal fluid and plasma of fetal sheep during development. J Physiol 300:441–455

Dziegielewska KM et al (1980b) Blood-cerebrospinal fluid transfer of plasma proteins during fetal development in the sheep. J Physiol 300:457–465

Dziegielewska KM et al (2001) Development of the choroid plexus. Microsc Res Tech 52(1):5–20

Faivre J (1854) Recherches sur la structure du coronarimn et des plexus choroides chez l'homme et les animaux. C R Acad Sci 39:424–427

Fawcett DW (1959) The fine structure of capillaries, arterioles and small arteries. In: Reynolds SRM, Zweifach BW (eds) The microcirculation. Univeristy of Illinois Press, Urbana, pp 1–27

Fawcett DW, Porter KR (1954) A study of the fine structure of ciliated epithelia. J Morphol 94:221–282

Findlay W (1899) The choroid plexus of the lateral ventricles of the brain, their histology, normal and pathological. Brain 22:161–202

Folsch H et al (1999) A novel clathrin adaptor complex mediates basolateral targeting in polarized epithelial cells. Cell 99(2):189–198

Franceschini P (1929) Presence of connective tissue elements in the central nervous system; peculiarities in structure of the pia-arachnoid and choroid plexi; so-called 'hematoencephalic barrier'. Sperimentale 83:419–445

Frankel H, Kazemi H (1983) Regulation of CSF composition--blocking chloride-bicarbonate exchange. J Appl Physiol 55(1 Pt 1):177–182

Gassama-Diagne A et al (2006) Phosphatidylinositol-3,4,5-trisphosphate regulates the formation of the basolateral plasma membrane in epithelial cells. Nat Cell Biol 8(9):963–970

Ghersi-Egea JF et al (2018) Molecular anatomy and functions of the choroidal blood-cerebrospinal fluid barrier in health and disease. Acta Neuropathol 135(3):337–361

Goehring NW (2014) PAR polarity: from complexity to design principles. Exp Cell Res 328 (2):258–266

Gorecki DC et al (1997) Differential expression of syntrophins and analysis of alternatively spliced dystrophin transcripts in the mouse brain. Eur J Neurosci 9(5):965–976

Gregoriades JMC et al (2019) Genetic and pharmacologic inactivation of apical NKCC1 in choroid plexus epithelial cells reveals the physiological function of the cotransporter. Am J Physiol Cell Physiol 316(4):C525–C544

Grindstaff KK et al (1998) Sec6/8 complex is recruited to cell-cell contacts and specifies transport vesicle delivery to the basal-lateral membrane in epithelial cells. Cell 93(5):731–740

Haeckl E (1859) Beiträge zur nurmalen und pathologischen Anatomie des Plexus chorioid. Virchows Archiv A Pathol Anat 16:253–289

Haenggi T, Fritschy JM (2006) Role of dystrophin and utrophin for assembly and function of the dystrophin glycoprotein complex in non-muscle tissue. Cell Mol Life Sci 63(14):1614–1631

Haselbach M et al (2001) Porcine Choroid plexus epithelial cells in culture: regulation of barrier properties and transport processes. Microsc Res Tech 52(1):137–152

Hebert JM, Mishina Y, McConnell SK (2002) BMP signaling is required locally to pattern the dorsal telencephalic midline. Neuron 35(6):1029–1041

Horikoshi Y et al (2009) Interaction between PAR-3 and the aPKC-PAR-6 complex is indispensable for apical domain development of epithelial cells. J Cell Sci 122(Pt 10):1595–1606

Huang X et al (2009) Sonic hedgehog signaling regulates a novel epithelial progenitor domain of the hindbrain choroid plexus. Development 136(15):2535–2543

Huang X et al (2010) Transventricular delivery of sonic hedgehog is essential to cerebellar ventricular zone development. Proc Natl Acad Sci U S A 107(18):8422–8427

Hughes AL, Pakhomova A, Brown PD (2010) Regulatory volume increase in epithelial cells isolated from the mouse fourth ventricle choroid plexus involves Na^+-H^+ exchange but not Na^+-K^+-$2Cl^-$ cotransport. Brain Res 1323:1–10

Hunter NL, Dymecki SM (2007) Molecularly and temporally separable lineages form the hindbrain roof plate and contribute differentially to the choroid plexus. Development 134(19):3449–3460

Hurov JB, Watkins JL, Piwnica-Worms H (2004) Atypical PKC phosphorylates PAR-1 kinases to regulate localization and activity. Curr Biol 14(8):736–741

Hworostuchin W (1911) Zur Frage über den Bau des Plexus chorioideus. Arch Mikrosk Anat 77:232–244

Iliff JJ et al (2013) Brain-wide pathway for waste clearance captured by contrast-enhanced MRI. J Clin Invest 123(3):1299–1309

Jacobs S et al (2008) Mice with targeted Slc4a10 gene disruption have small brain ventricles and show reduced neuronal excitability. Proc Natl Acad Sci U S A 105(1):311–316

Javaheri S, Wagner KR (1993) Bumetanide decreases canine cerebrospinal fluid production. In vivo evidence for NaCl cotransport in the central nervous system. J Clin Invest 92(5):2257–2261

Johanson CE, Parandoosh Z, Dyas ML (1992) Maturational differences in acetazolamide-altered pH and HCO3 of choroid plexus, cerebrospinal fluid, and brain. Am J Phys 262(5 Pt 2):R909–R914

Johansson PA et al (2005) Aquaporin-1 in the choroid plexuses of developing mammalian brain. Cell Tissue Res 322(3):353–364

Johansson PA et al (2007) Expression and localization of P2 nucleotide receptor subtypes during development of the lateral ventricular choroid plexus of the rat. Eur J Neurosci 25 (11):3319–3331

Johansson P, Dziegielewska K, Saunders N (2008) Low levels of Na, K-ATPase and carbonic anhydrase II during choroid plexus development suggest limited involvement in early CSF secretion. Neurosci Lett 442(1):77–80

Junet W (1926) Terminaisons nerveuses intraépithéliales dans les plexus choroides de la souris. C R Soc Biol 95:1397–1398

Kaji C et al (2012) The expression of podoplanin and classic cadherins in the mouse brain. J Anat 220(5):435–446

Kajita H, Brown PD (1997) Inhibition of the inward-rectifying Cl- channel in rat choroid plexus by a decrease in extracellular pH. J Physiol 498(Pt 3):703–707

Kalaria RN et al (1998) Identification and expression of the Na^+/H^+ exchanger in mammalian cerebrovascular and choroidal tissues: characterisation by amiloride-sensitive [^{3}H]MIA binding and RT-PCR analysis. Brain Res Mol Brain Res 58:178–187

Kallio H et al (2006) Expression of carbonic anhydrases IX and XII during mouse embryonic development. BMC Dev Biol 6:22

Kalwaryiski EB (1924) Sur la membrane basale et la bordure en brosse de cellules épithéliales des plexus choroides. C R Soc Biol 90:903–904

Kanaka C et al (2001) The differential expression patterns of messenger RNAs encoding K-Cl cotransporters (KCC1,2) and Na-K-2Cl cotransporter (NKCC1) in the rat nervous system. Neuroscience 104(4):933–946

Karadsheh MF et al (2004) Localization of the KCC4 potassium-chloride cotransporter in the nervous system. Neuroscience 123(2):381–391

Kasper M, Karsten U, Stosiek P (1986) Detection of cytokeratin(s) in epithelium of human plexus choroideus by monoclonal antibodies. Acta Histochem 78(1):101–103

Keep RF, Jones HC (1990) A morphometric study on the development of the lateral ventricle choroid plexus, choroid plexus capillaries and ventricular ependyma in the rat. Brain Res Dev Brain Res 56(1):47–53

Keep RF, Jones HC, Cawkwell RD (1986) A morphometric analysis of the development of the fourth ventricle choroid plexus in the rat. Brain Res 392(1–2):77–85

Keep RF, Xiang J, Betz AL (1994) Potassium cotransport at the rat choroid plexus. Am J Phys 267 (6 Pt 1):C1616–C1622

Kibble JD, Tresize AO, Brown PD (1996) Properties of the cAMP-activated Cl⁻ conductance in choroid plexus epithelial cells isolated from the rat. J Physiol 496:69–80

Kibble JD et al (1997) Whole-cell Cl⁻ conductances in mouse choroid plexus epithelial cells do not require CFTR expression. Am J Phys 272:C1899–C1907

Kierbel A et al (2007) Pseudomonas aeruginosa exploits a PIP3-dependent pathway to transform apical into basolateral membrane. J Cell Biol 177(1):21–27

Korzhevskii DE (2000) Proliferative zones in the epithelium of the choroid plexuses of the human embryo brain. Neurosci Behav Physiol 30(5):509–512

Kotera T, Brown PD (1994) Two types of potassium current in rat choroid plexus epithelial cells. Pflugers Arch 237:317–324

Kratzer I et al (2012) Complexity and developmental changes in the expression pattern of claudins at the blood-CSF barrier. Histochem Cell Biol 138(6):861–879

Krug SM et al (2012) Charge-selective claudin channels. Ann N Y Acad Sci 1257:20–28

Lagunowich LA et al (1992) Immunohistochemical and biochemical analysis of N-cadherin expression during CNS development. J Neurosci Res 32(2):202–208

Lehtinen MK et al (2011) The cerebrospinal fluid provides a proliferative niche for neural progenitor cells. Neuron 69(5):893–905

Li H et al (2002) Patterns of cation-chloride cotransporter expression during embryonic rodent CNS development. Eur J Neurosci 16(12):2358–2370

Liedtke W et al (2000) Vanilloid receptor-related osmotically activated channel (VR-OAC), a candidate vertebrate osmoreceptor. Cell 103(3):525–535

Lindsey AE et al (1990) Functional expression and subcellular localization of an anion exchanger cloned from choroid plexus. Proc Natl Acad Sci U S A 87(14):5278–5282

Lippoldt A et al (2000) Organization of choroid plexus epithelial and endothelial cell tight junctions and regulation of claudin-1, -2 and -5 expression by protein kinase C. Neuroreport 11 (7):1427–1431

Louveau A et al (2015) Structural and functional features of central nervous system lymphatic vessels. Nature 523(7560):337–341

Low SH et al (1996) Differential localization of syntaxin isoforms in polarized Madin-Darby canine kidney cells. Mol Biol Cell 7(12):2007–2018

Lun MP, Monuki ES, Lehtinen MK (2015) Development and functions of the choroid plexus-cerebrospinal fluid system. Nat Rev Neurosci 16(8):445–457

Luschka H (1855) Die Adergeflechte des menschlichen Gehirns. Georg Reimer, Berlin

Luse SA (1956) Electron microscopic observations of the central nervous system. J Biophys Biochem Cytol 2(5):531–542

Makarova O et al (2003) Mammalian Crumbs3 is a small transmembrane protein linked to protein associated with Lin-7 (Pals1). Gene 302(1–2):21–29

Marrs JA et al (1993) Distinguishing roles of the membrane-cytoskeleton and cadherin mediated cell-cell adhesion in generating different Na+,K(+)-ATPase distributions in polarized epithelia. J Cell Biol 123(1):149–164

Martin C et al (2006) FGF2 plays a key role in embryonic cerebrospinal fluid trophic properties over chick embryo neuroepithelial stem cells. Dev Biol 297(2):402–416

Masuzawa T et al (1981) Ultrastructural localization of carbonic anhydrase activity in the rat choroid plexus epithelial cell. Histochemistry 73(2):201–209

Masuzawa T et al (1984) Immunohistochemical localization of Na^+, K^+-ATPase in the choroid plexus. Brain Res 302(2):357–362

Maxwell DS, Pease DC (1956) The electron microscopy of the choroid plexus. J Biophys Biochem Cytol 2(4):467–474

Meek WJ (1907) A study of the choroid plexus. J Comp Neurol Psychol 17:286–306

Miettinen M, Clark R, Virtanen I (1986) Intermediate filament proteins in choroid plexus and ependyma and their tumors. Am J Pathol 123(2):231–240

Millar ID, Brown PD (2008) NBCe2 exhibits a 3 HCO_3^- :1 Na^+ stoichiometry in mouse choroid plexus epithelial cells. Biochem Biophys Res Commun 373:550–554

Millar ID, Bruce JI, Brown PD (2007) Ion channel diversity, channel expression and function in the choroid plexuses. Cerebrospinal Fluid Res 4:8

Millen JW, Rogers GE (1956) An electron microscopic study of the chorioid plexus in the rabbit. J Biophys Biochem Cytol 2(4):407–416

Mollgard K, Saunders NR (1986) The development of the human blood-brain and blood-CSF barriers. Neuropathol Appl Neurobiol 12(4):337–358

Murphy VA, Johanson CE (1989a) Alteration of sodium transport by the choroid plexus with amiloride. Biochim Biophys Acta 979(2):187–192

Murphy VA, Johanson CE (1989b) Acidosis, acetazolamide, and amiloride: effects on 22Na transfer across the blood-brain and blood-CSF barriers. J Neurochem 52(4):1058–1063

Nag S, Begley DJ (2005) Blood–brain barrier, exchange of metabolites and gases. In: Kalimo H (ed) Pathology and genetics: cerebrovascular diseases. ISN Neuropathology Press, Basel, pp 22–29

Nakamura N et al (1999) Inwardly rectifying K^+ channel Kir7.1 is highly expressed in thyroid follicular cells, intestinal epithelial cells and choroid plexus epithelial cells: implication for a functional coupling with Na^+,K^+-ATPase. Biochem J 342:329–336

Narita K et al (2010) Multiple primary cilia modulate the fluid transcytosis in choroid plexus epithelium. Traffic 11(2):287–301

Narita K et al (2012) Proteomic analysis of multiple primary cilia reveals a novel mode of ciliary development in mammals. Biol Open 1(8):815–825

Narita K et al (2015) TRPV4 regulates the integrity of the blood-cerebrospinal fluid barrier and modulates transepithelial protein transport. FASEB J 29(6):2247–2259

Nattie EE, Adams JM (1988) DIDS decreases CSF HCO3- and increases breathing in response to CO2 in awake rabbits. J Appl Physiol (1985) 64(1):397–403

Netter FH (2014) Atlas of human anatomy, 6th edn. Saunders/Elsevier, Philadelphia, PA

Nielsen CM, Dymecki SM (2010) Sonic hedgehog is required for vascular outgrowth in the hindbrain choroid plexus. Dev Biol 340(2):430–437

Nielsen S et al (1993) Distribution of the aquaporin CHIP in secretory and resorptive epithelia and capillary endothelia. Proc Natl Acad Sci U S A 90(15):7275–7279

Ohno H et al (1999) Mu1B, a novel adaptor medium chain expressed in polarized epithelial cells. FEBS Lett 449(2–3):215–220

Oshio K et al (2005) Reduced cerebrospinal fluid production and intracranial pressure in mice lacking choroid plexus water channel Aquaporin-1. FASEB J 19(1):76–78

Palay SL, Karlin LJ (1959) An electron microscopic study of the intestinal villus. I. The fasting animal. J Biophys Biochem Cytol 5(3):363–372

Parada C, Gato A, Bueno D (2008) All-trans retinol and retinol-binding protein from embryonic cerebrospinal fluid exhibit dynamic behaviour during early central nervous system development. Neuroreport 19(9):945–950

Paulus W et al (1993) Characterization of integrin receptors in normal and neoplastic human brain. Am J Pathol 143(1):154–163

Pearson MM et al (2001) Localization of the K^+-Cl^- cotransporter, KCC3, in the central and peripheral nervous systems: expression in the choroid plexus, large neurons and white matter tracts. Neuroscience 103(2):481–491

Pease DC (1956) Infolded basal plasma membranes found in epithelia noted for their water transport. J Biophys Biochem Cytol 2(4 Suppl):203–208

Peters A, Palay SL, Webster HD (1976) The fine structure of the nervous system. Saunders, Philadelphia

Pettit A, Girard J (1901) Processus sécrétoires dans les cellules de revêtement des plexus choroïdes des ventricules latéraux, consécutifs à l'administration de la muscarine et de l'éther. Comptes Rendus de la Société de Biologie 53:825–828

Plotkin MD et al (1997) Expression of the Na^+-K^+-$2Cl^-$ cotransporter BSC2 in the nervous system. Am J Phys 272(1 Pt 1):C173–C183

Pollay M, Curl F (1967) Secretion of cerebrospinal fluid by the ventricular ependyma of the rabbit. Am J Phys 213(4):1031–1038

Praetorius J, Nielsen S (2006) Distribution of sodium transporters and aquaporin-1 in the human choroid plexus. Am J Physiol Cell Physiol 291(1):C59–c67

Praetorius HA, Spring KR (2001) Bending the MDCK cell primary cilium increases intracellular calcium. J Membr Biol 184(1):71–79

Praetorius J, Nejsum LN, Nielsen S (2004a) A SCL4A10 gene product maps selectively to the basolateral plasma membrane of choroid plexus epithelial cells. Am J Physiol Cell Physiol 286 (3):C601–C610

Praetorius J, Nejsum LN, Nielsen S (2004b) A SLC4A10 gene product maps selectively to the basolateral membrane of choroid plexus epithelial cells. Am J Phys 286:C601–C610

Preston D et al (2018) Activation of TRPV4 stimulates transepithelial ion flux in a porcine choroid plexus cell line. Am J Physiol Cell Physiol 315(3):C357–C366

Purkinje JE (1836) Ueber Flimmerbewegungen im Gehirn. Archiv für Anatomie, Physiologie und Wissenschaftliche Medicin, pp 289–291

Raballo R et al (2000) Basic fibroblast growth factor (Fgf2) is necessary for cell proliferation and neurogenesis in the developing cerebral cortex. J Neurosci 20(13):5012–5023

Robin C (1859) Recherches sur quelques particularite's de la structure des capillaires de l'encephale. J Physiol Homme Anim 2:537–548

Rodriguez-Boulan E, Macara IG (2014) Organization and execution of the epithelial polarity programme. Nat Rev Mol Cell Biol 15(4):225–242

Roepke TK et al (2011) KCNE2 forms potassium channels with KCNA3 and KCNQ1 in the choroid plexus epithelium. FASEB J 25:4264–4273

Rosenthal R et al (2010) Claudin-2, a component of the tight junction, forms a paracellular water channel. J Cell Sci 123(Pt 11):1913–1921

Rosenthal R et al (2017) Claudin-2-mediated cation and water transport share a common pore. Acta Physiol (Oxf) 219(2):521–536

Saito Y, Wright E (1984) Regulation of bicarbonate transport across the brush border membrane of the bull-frog choroid plexus. J Physiol 350:327–342

Salehi Z et al (2009) Insulin-like growth factor-1 and insulin-like growth factor binding proteins in cerebrospinal fluid during the development of mouse embryos. J Clin Neurosci 16(7):950–953

Schapiro B (1931) Über die Innervation des Plexus chorioideus. Zeitschrift für die Gesamte Neurologie und Psychiatrie 136:539–547

Schmid H (1929) Anatomischer Bau und Entwicklung der Plexus chorioidei in der Wirbeltierreihe und beim Menschen. Zeitschrift fur Mikroskopisch-Anatomische Forschung 16:413–498

Schnermann J et al (1998) Defective proximal tubular fluid reabsorption in transgenic aquaporin-1 null mice. Proc Natl Acad Sci U S A 95(16):9660–9664

Segal MB, Burgess AM (1974) A combined physiological and morphological study of the secretory process in the rabbit choroid plexus. J Cell Sci 14(2):339–350

Sharma N et al (2006) Apical targeting of syntaxin 3 is essential for epithelial cell polarity. J Cell Biol 173(6):937–948

Shewan A, Eastburn DJ, Mostov K (2011) Phosphoinositides in cell architecture. Cold Spring Harb Perspect Biol 3(8):a004796

Siegel GJ et al (1984) Purification of mouse brain ($Na^+ + K^+$)-ATPase catalytic unit, characterization of antiserum, and immunocytochemical localization in cerebellum, choroid plexus, and kidney. J Histochem Cytochem 32(12):1309–1318

Speake T et al (2002) Inward-rectifying anion channels are expressed in the epithelial cells of choroid plexus isolated from ClC-2 'knock-out' mice. J Physiol 539:385–390

Speake T, Kibble JD, Brown PD (2004) Kv1.1 and Kv1.3 channels contribute to the delayed-rectifying K^+ conductance in rat choroid plexus epithelial cells. Am J Phys 286:C611–C620

Spector R et al (2015) A balanced view of choroid plexus structure and function: focus on adult humans. Exp Neurol 267:78–86

Steffensen AB et al (2018) Cotransporter-mediated water transport underlying cerebrospinal fluid formation. Nat Commun 9(1):2167

Steinemann A et al (2016) Claudin-1, -2 and -3 are selectively expressed in the epithelia of the choroid plexus of the mouse from early development and into adulthood while Claudin-5 is restricted to endothelial cells. Front Neuroanat 10:16

Stock AD et al (2019) Tertiary lymphoid structures in the choroid plexus in neuropsychiatric lupus. JCI Insight 4(11):e124203

Stöhr P (1922) Über die Innervation der Pia mater und des Plexus chorioideus des Menschen. Z Anat Entwicklungsgesch 63:562–607

Stoops EH, Caplan MJ (2014) Trafficking to the apical and basolateral membranes in polarized epithelial cells. J Am Soc Nephrol 25(7):1375–1386

Studnicka FK (1900) Untersuchungen über den Bau des Ependyms der nervösen Zentralorgane. Anatomische Hefte 15:303–430

Sun SQ, Hashimoto PH (1991) Venous microvasculature of the pineal body and choroid plexus in the rat. J Electron Microsc 40(1):29–33

Sundwall I (1917) The chorioid plexus with special reference to interstitial granular cells. Anat Rec 12:221–254

Tennyson VM, Pappas GD (1961) Electron microscope studies of the developing telencephalic choroid plexus in normal and hydrocephalic rabbits. In: Fields WS, Desmond MM (eds) Disorders of the developing nervous system. Thomas, Springfield, pp 267–318

ter Beest MB et al (2005) The role of syntaxins in the specificity of vesicle targeting in polarized epithelial cells. Mol Biol Cell 16(12):5784–5792

Urabe N et al (2002) Basement membrane type IV collagen molecules in the choroid plexus, pia mater and capillaries in the mouse brain. Arch Histol Cytol 65(2):133–143

Vallon V, Verkman AS, Schnermann J (2000) Luminal hypotonicity in proximal tubules of aquaporin-1-knockout mice. Am J Physiol Renal Physiol 278(6):F1030–F1033

Van Breemen VL, Clemente CD (1955) Silver deposition in the central nervous system and the hematoencephalic barrier studied with the electron microscope. J Biophys Biochem Cytol 1(2):161–166

Van Huysse JW et al (2012) Salt-induced hypertension in a mouse model of Liddle syndrome is mediated by epithelial sodium channels in the brain. Hypertension 60(3):691–696

Virchow R (1851) Ueber die Erweiterung kleinerer Gefaesse. Archiv Pathol Anat Physiol Klin Med 3:427–462

Voetmann E (1949) On the structure and surface area of the human choroid plexuses – a quantitative anatomivcal study. Acta Anat 8:20–32

Vorbrodt AW, Dobrogowska DH (2003) Molecular anatomy of intercellular junctions in brain endothelial and epithelial barriers: electron microscopist's view. Brain Res Brain Res Rev 42(3):221–242

Weiger T et al (1986) The angioarchitecture of the choroid plexus of the lateral ventricle of the rabbit. A scanning electron microscopic study of vascular corrosion casts. Brain Res 378(2):285–296

Weir AP et al (2002) A- and B-utrophin have different expression patterns and are differentially up-regulated in mdx muscle. J Biol Chem 277(47):45285–45290

Weisz OA, Rodriguez-Boulan E (2009) Apical trafficking in epithelial cells: signals, clusters and motors. J Cell Sci 122(Pt 23):4253–4266

Welch K (1963) Secretion of cerebrospinal fluid by choroid plexus of the rabbit. Am J Phys 205:617–624

Wislocki GB, Ladman AJ (1958) The fine structure of the mammalian choroid plexus. In: Wolsteinholme GEW, O'Connor CM (eds) The cerebrospinal fluid. Little, Brown, & Company, Boston, pp 55–79

Wolburg H et al (2001) Claudin-1, claudin-2 and claudin-11 are present in tight junctions of choroid plexus epithelium of the mouse. Neurosci Lett 307(2):77–80

Wright EM (1972) Mechanisms of ion transport across the choroid plexus. J Physiol 226 (2):545–571

Wu B, Guo W (2015) The exocyst at a glance. J Cell Sci 128(16):2957–2964

Wu Q et al (1998) Functional demonstration of Na^+-K^+-$2Cl^-$ cotransporter activity in isolated, polarized choroid plexus cells. Am J Phys 275(6 Pt 1):C1565–C1572

Wu H et al (2007) PDZ domains of Par-3 as potential phosphoinositide signaling integrators. Mol Cell 28(5):886–898

Xiang Z, Burnstock G (2005) Expression of P2X receptors in rat choroid plexus. Neuroreport 16 (9):903–907

Yeaman C, Grindstaff KK, Nelson WJ (1999) New perspectives on mechanisms involved in generating epithelial cell polarity. Physiol Rev 79(1):73–98

Zeuthen T (1991) Secondary active transport of water across ventricular cell membrane of choroid plexus epithelium of *Necturus maculosus*. J Physiol 444:153–173

Zeuthen T (1994) Cotransport of K^+, Cl^- and H_2O by membrane proteins from choroid plexus epithelium of *Necturus maculosus*. J Physiol 478(Pt 2):203–219

Zeuthen T, Wright EM (1981) Epithelial potassium transport: tracer and electrophysiological studies in choroid plexus. J Membr Biol 60:105–128

Zeuthen T et al (2016) Structural and functional significance of water permeation through cotransporters. Proc Natl Acad Sci U S A 113(44):E6887–E6894

Chapter 2
Blending Established and New Perspectives on Choroid Plexus-CSF Dynamics

Conrad E. Johanson and Richard F. Keep

Abstract Modern cerebrospinal fluid (CSF) physiologic investigation accelerated after World War II with the advent of radio-isotopes to quantify choroid plexus (CP) ion transport and CSF flow dynamics. Hugh Davson, Malcolm Segal, Michael Bradbury, Michael Pollay, Keasley Welch, Helen Cserr, John Pappenheimer, and colleagues, developed laboratory preparations to assess tracer uptake/release by CP and associated flow dynamics within the ventricular-brain system. This canonical research established several basic physiologic concepts: differences between CP and blood-brain barrier, CP as primary site of CSF formation, ependymal permeability, intimate CSF and brain interstitial fluid association, CSF sink action for excretion, and a quasi-lymphatic function of CSF flow/drainage to cervical lymph. Their seminal findings constitute a reliable foundation on which to build contemporary models of CP transport/CSF dynamics, including functional interaction with the newly-described glymphatic system. The 1980s–1990s provided research findings on CSF regulation by neurotransmitters and neuropeptides; also, many new pharmacologic agents were tested for controlling CSF formation. Over the past 2–3 decades, the advent/refinement of diverse immuno-histochemical, neuroendocrine and molecular techniques has delineated the expression and function of basolateral and apical transporters at the blood-CSF interface. Recently, CP gene knock-out models and transcriptomic approaches have engendered specific analysis of CP transport and metabolism. Increasing attention is being paid to the CP role in diseases such as Alzheimer's, Parkinson's, stroke, intracranial hypertension and hydrocephalus. The physiologic impact of CP-CSF fluid generation, pressure and homeostasis on the putative glymphatic system is a promising topic. There is an increasing need to blend transport/fluid phenomena at the BCSFB with the BBB and ependymal CSF-brain interface. Integrated models that incorporate all CNS transport interfaces will provide a key impetus for advances in translational neuromedicine.

C. E. Johanson (✉)
Department of Neurosurgery, Alpert Medical School at Brown University, Providence RI USA
e-mail: Conrad_Johanson@Brown.edu

R. F. Keep
Department of Neurosurgery, University of Michigan, Ann Arbor MI USA
e-mail: rkeep@med.umich.edu

© The American Physiological Society 2020
J. Praetorius et al. (eds.), *Role of the Choroid Plexus in Health and Disease*,
Physiology in Health and Disease, https://doi.org/10.1007/978-1-0716-0536-3_2

2.1 The Place of the Choroid Plexus in CNS Fluid Flow and Solute Transport

A global view of fluid dynamics in the central nervous system (CNS) includes a key role for the solute transporters and fluid-manufacturing machinery in choroid plexus (CP) (Johanson et al. 2008). As a prominent part of the blood-cerebrospinal fluid barrier (BCSFB), the CP has a spectrum of ion and organic solute transporters (Janssen et al. 2013) distinguishing it from the blood-brain barrier (BBB). At both the neuroanatomical and ultrastructural levels, the constituent cells of the BCSFB and BBB differ greatly. Figure 2.1 schematizes the configuration of the major CNS transport interfaces, including the permeable ependymal layer that partitions the ventricular CSF from cerebral interstitial fluid (CSF-brain interface). Most substances, even macromolecules, that enter CSF can access the brain parenchyma across the ependyma. In addition, CSF may rapidly enter brain parenchyma from the subarachnoid space (SAS) via the perivascular space of penetrating arterioles—part of the proposed glymphatic system (Iliff et al. 2012). Thus, SAS CSF solute flux across the 'external' surface of the cortex, by the pial/perivascular route, may be analogous to the relatively unimpeded ventricular CSF solute movement transependymally across the 'internal' surface of the brain.

While the endothelium of cerebral microvessels (the BBB) has a prodigious role in supplying neurons with glucose, amino acids, and free fatty acids, the choroid plexus epithelium (CPE) furnishes hormones, growth factors, neurotrophins, and neuroprotective proteins to the CSF-brain nexus. The CP-CSF route also uniquely supplies neuronal networks and glia with micronutrients such as vitamin C (Fig. 2.2) and folate, both critical for brain development and maintenance functions such as DNA methylation (Spector and Johanson 2014). Together, the CP and brain microvessels provide complementary regulatory factors and substrates to meet the needs of cerebral metabolism. Both barrier systems are essential to extracellular homeostasis and brain wellbeing.

The most familiar role of CP is to elaborate and secrete CSF. CSF formation is linked to vascular perfusion of the plexus. Choroidal blood flow, ~4 ml/min/g, outstrips the brain perfusion rate and coupled with a high rate of epithelial metabolism, this enables a fluid production that greatly exceeds fluid formation at the BBB. CSF formation rate is ~0.4 ml/min/g CP in adult mammals, i.e., considerably greater than in perinatal life (Johanson and Woodbury 1974) when the CP transport systems are incompletely-developed. Consistent with slower fluid turnover at the infant BCSFB (Spector and Johanson 2014), the lower transport capacities of Na^+-K^+-ATPase (Parmelee and Johanson 1989) and carbonic anhydrase in the developing rat CPE steadily increase over several postnatal weeks to adult levels.

Fluid generation by CP epithelial cells in adulthood is pivotal for: (1) maintaining the body of the ventricular system (preventing collapse), (2) helping to regulate intracranial pressure (ICP), and (3) mediating volume transmission of CSF (bulk flow) that carries solutes/cells throughout the ventriculo-subarachnoid system (Agnati et al. 1995). CSF distribution enables diverse roles to be carried out by the endocrine, reparative and immune systems within the CNS.

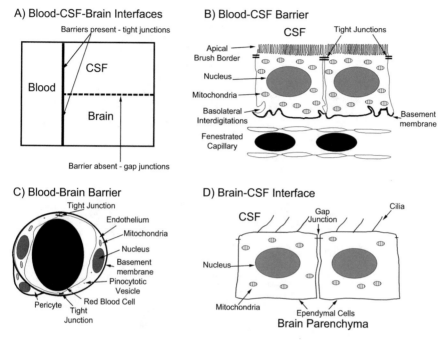

Fig. 2.1 Structures of CNS transport interfaces in adult mammals: (**a**) Blood is separated from ventricular CSF by restrictive junctions at the CP blood-CSF barrier. An even tighter barrier (by ~10 times) demarcates the cerebral capillary blood from brain tissue proper (bold solid line). A third prominent transport interface (with gap junctions) is the permeable ependyma, that separates ventricular CSF from the internal surface of brain tissue (broken line); correspondingly, overlying the cortical (external) surface is the permeable pia-glial (thin meninges) membrane. (**b**) The BCSFB is a single layer of epithelium welded together by *zonulae occludentes* (tight junctions) at the apex of adjacent cells. By contrast the choroidal capillaries are leaky, allowing large molecules to penetrate into the interstitium right up to the basolateral membrane of CP epithelium. Extensive epithelial surface area is present at the interdigitating basolateral side (facing plasma), and even more so at the lush microvilli on the CSF-facing apical membrane. Abundant mitochondria (Fig. 2.3) energize the dynamic, polarized transport of ions (Fig. 2.4) and organic substrates (Fig. 2.6). (**c**) The BBB resides in tight junctions between endothelial cells. The capillary endothelium contains many mitochondria, to power bi-directional solute transport across the luminal and abluminal (brain-facing) membranes. Pinocytotic vesicles also mediate trans-endothelial movement of molecules into brain. Pericytes and astrocyte foot processes cover much of the microvessel wall, providing additional regulation of permeability and active transport. (**d**) The ependymal cells of the CSF-brain interface, containing gap junctions, are highly permeable allowing even large proteins and cytokines to distribute readily between the ventricles and brain ISF. Extensive, bidirectional distribution of materials across the ependyma enables the CSF to serve as both a nutritional source and an excretory sink for the brain. Diagram from Smith et al., Adv Drug Deliv Rev 56:1765–91, 2004, with permission

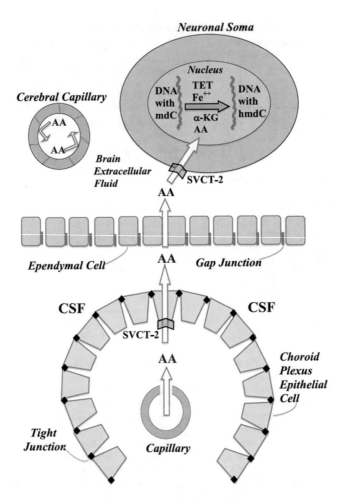

Fig. 2.2 Distribution of ascorbic acid (AA) by CP-CSF into brain for co-factoring neuronal DNA reactions: AA in plasma is reflected by the inside surface of the impervious endothelium and restricting tight junction (*upper left*); thus, AA does not penetrate the BBB. Rather, AA reaches neurons circuitously, by way of the CP, CSF and ependyma. Accordingly, AA moves (unfilled arrow) from leaky vessels in CP to epithelial cells (*bottom panel*), for transport into CSF by the Na-dependent vitamin C transporter 2 (SVCT-2). From CSF, AA diffuses through the permeable ependyma into ISF; and also likely enters the cortical perivascular spaces by bulk flow (Fig. 2.7). Within neuronal nucleoplasm (*upper right*), AA is a co-factor for the 10–11 translocation enzymes (TET), with Fe^{++} and α-ketoglutarate (KG), to oxidize methyldeoxycytidine (mdC) molecules in DNA to hydroxymethyldeoxycytidine (hmdC). SVCT-2 transporters concentrate AA in CSF and neurons, maintaining neuronal concentrations at millimolar levels. CSF also likely distributes AA to spinal cord, that also lacks SVCT2 at the blood-spinal cord barrier. From Spector and Johanson (2014)

2.2 CP Epithelial Properties: Comparison to Other Fluid-Transporting Epithelia

Transport epithelia possess many structural and functional common denominators. Choroid epithelial cells bear many features and execute various roles performed by the ciliary body and renal tubular cells. The ciliary body secretes aqueous humor, thereby helping to control intraocular pressure. Although the ciliary is a double-layered epithelium compared to the single layer at the CP, they both engage in fluid pressure regulation and homeostatic secretion of hormones, growth factors and micronutrient vitamins. Steroids and fluid-regulating hormones such as cortisol and atrial natriuretic peptide exert similar pressure effects on the aqueous and CSF. Therefore, due to analogous structure and functions in the eye, the modeling of CP-CSF transport/secretion has benefited from research findings on the ciliary-aqueous nexus.

Interestingly, the CP has been described as the 'kidney' of the brain (Spector and Johanson 1989). Renal function, similar to CP, purifies extracellular fluid (plasma and CSF, respectively). CPE ultrastructure mimics the proximal tubule. Like CP, the proximal tubule transfers a relatively large volume of fluid; and expresses several acid-base transporters that engage in extracellular fluid pH regulation. Common denominators of basic Na^+ pump activity and carbonic anhydrase generation of HCO_3^- are similar in CP and proximal tubule; however, the respective *physiologic functions* are reversed (CP secretion vs. proximal reabsorption) due to different Na^+-K^+-ATPase, NKCC1 and AQP1 polarizations (apical in CP vs. basolateral in proximal tubule) (Damkier et al. 2013). Still, renal transport data, as modulated by hormones and drugs, contribute to understanding BCSFB homeostatic phenomena. In the distal tubule, the effects of steroid and sex hormones on Na^+ transport mimic those at the CP. Research on modulated Na^+ transport in the collecting duct, as it homeostatically affects arterial pressure, furnishes insight on hormonal regulation of CP secretion and ICP.

The ultrastructure and transporters of the CPE are intimately related to CSF secretion. Figure 2.3 presents the ultrastructure of an epithelial cell from lateral ventricle CP. There is a lush apical (CSF-facing) brush border. The abundant microvillus membrane with numerous transporters/channels allows tremendous solute and H_2O movement between the CPE and CSF. The BCSFB surface area is larger than previously appreciated, with estimates approaching the order-of-magnitude of the BBB transport area (Keep and Jones 1990). Another factor promoting solute-H_2O transport is CP blood flow, i.e., ~5 times the mean perfusion rate of brain. Three factors, then, render the BCSFB a very dynamic interface: high blood flow, great epithelial surface of apical/basolateral membranes, and abundantly-expressed transporters/channels.

CP, ciliary and renal epithelial cells share a substantial capacity for transferring fluid. The CPE has a high metabolic rate to support profuse protein synthesis (especially transthyretin), considerable CSF production (~500 ml/d in adult humans) and regulation of CSF composition. This is reflected by an above-average density of

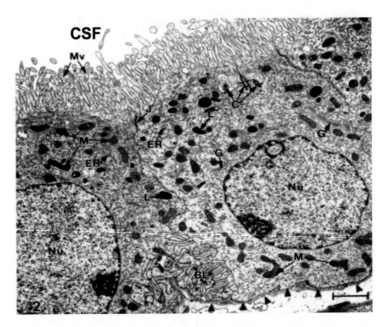

Fig. 2.3 Ultrastructure of choroidal epithelium: CP is from a lateral ventricle of an untreated adult Sprague-Dawley rat. The apical membrane (CSF-facing) is lush with microvilli (Mv). J refers to the tight junction welding of two epithelial cells at their apical poles. C = centriole. Mitochondria (M) are abundant. Nucleus (Nu) is oval with a nucleolus. G and ER, Golgi apparatus and endoplasmic reticulum. Arrowheads point to basal lamina at the plasma face of the epithelial cell; the basal lamina separates the epithelial cell above from ISF below. The basal labyrinth (BL) configuration intertwines the basolateral membranes of adjacent cells. CP ultrastructure resembles proximal tubule, consistent with both epithelia extensively turning over fluid. Tissue fixative for electron microscopy was OsO4. Scale bar is 2 μm. Adapted from Johanson et al. (2008)

mitochondria, endoplasmic reticulum and golgi apparati. Ultrastructurally and biochemically, the CPE is geared to manufacture substantial ATP to energize protein/peptide synthesis and CSF formation.

2.3 CP Epithelium as the Primary Generator of Ventricular CSF

2.3.1 Overview of CSF Formation and Composition

A large battery of techniques has characterized ionic fluxes into and out of CP (Johanson 1988). In vivo, in situ, in vitro and cultured CP approaches consistently demonstrate a strong net inward flux of Na^+, Cl^- and HCO_3^- across the plasma membrane into the epithelial cells. CPE monolayers in transwells enable vectorial analyses, i.e., basolateral vs. apical transport phenomena (Gath et al. 1997). Figure 2.4

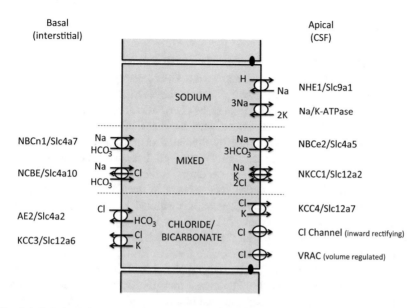

Fig. 2.4 Polarized distribution of transporters and channels involved in Na^+, Cl^- and HCO_3^- movement at the CP epithelium: Note that many of the Na^+ transporters also transport Cl^- and/or HCO_3^-, particularly at the basolateral membrane. For the transporters, the protein and solute carrier (Slc) names are given. *VRAC* volume regulated anion channel. The inwardly rectifying Cl^- channel (s) have not been molecularly identified. See Praetorius and Damkier (Praetorius and Damkier 2017) for detailed, descriptive evidence for different transporters/channels

delineates numerous CPE transporters, identified by multiple physiologic methods. Excepting a few differences in rat vs. human xenobiotic transporter expression (Uchida et al. 2015), the human CPE transport information (localization of transporters/channels, enzyme activities, and receptors), along with CSF dynamics, usually mirror animal model data (Janssen et al. 2013; Praetorius and Nielsen 2006).

CSF composition reflects the transported ions and H_2O across CPE. A cardinal, well-documented feature of CP is secretion into the ventricles of a fluid rich in Na^+, Cl^- and HCO_3^- (Spector et al. 2015a). Typical adult mammalian CSF contains ~145, 115 and 25 mmol/L, respectively, for Na^+, Cl^- and HCO_3^-. These are the predominant ions in CSF and, consequently, the discussion below focuses largely on Na^+, Cl^- and HCO_3^- transport across the BCSFB. CSF is ~99% H_2O. Therefore, it is incumbent to understand the driving forces behind the H_2O as well as ion fluxes across the CPE.

Mammalian CP transport data inform on key factors in CSF dynamics, useful for strategies to reduce choroidal fluid turnover in disorders such as hydrocephalus and intracranial hypertension. Although there is considerable progress in identifying CPE transporters/channels that mediate CSF secretion, informational gaps remain. Regulatory manipulation of CSF will be improved by attaining deeper molecular insight of forces that drive ion-transport-associated fluid formation by the CP.

2.3.2 Na⁺ Transport

The lead player in CSF formation is Na^+. CSF formation rate is more-or-less proportional to net Na^+ transport across the CPE into the ventricles (Johanson et al. 2008). Bidirectional flux of Na^+ occurs at the BCSFB, but the net Na^+ flux is predominantly into CSF. Following passive distribution across the CP capillaries, Na^+ ions diffuse through the interstitium to the CPE basolateral membrane (Johanson et al. 2011a). Na^+ uptake by epithelial cells from CP interstitial fluid (plasma ultrafiltrate) has been thoroughly quantified, but awaits mechanistic clarification as to the exact basolateral transporters and/or channels involved. Rapid amiloride-sensitive uptake of radio-$^{22}Na^+$ occurs in the CP in vivo and in vitro (Murphy and Johanson 1989a, b), perhaps mediated by the sodium-hydrogen antiporter, NHE1 (Damkier et al. 2013), or by the epithelial sodium channel, ENaC, expressed in human CP (Janssen et al. 2013; Bergen et al. 2015). Obtaining additional immunohistochemical staining to localize ENaC would help to evaluate this working model. Moreover, the basolateral Na^+-linked Cl^--HCO_3^- exchanger is another candidate for Na^+ uptake. Functional transport data for Na^+ are needed to complement the mRNA data and basolateral immunolocalization of Na^+-linked Cl^--HCO_3^- exchange. Na^+ exits the CPE into CSF at the apical membrane via three mechanisms: the Na^+ pump (Na^+-K^+-ATPase), Na-K-Cl cotransport, and electrogenic $NaHCO_3$ cotransport (Christensen et al. 2013).

2.3.3 Cl⁻ Transport

Cl^- is actively accumulated by CPE cells above its electrochemical equilibrium. Basolateral Cl/HCO_3 exchange (AE2) and Na/HCO_3 cotransport (NBCn1) accumulate Cl^- in the CPE (the latter using the inward Na^+ gradient (set up by apical Na^+-K^+-ATPase activity) (Fig. 2.4). At the CSF-facing side, Cl^- moves down its electrochemical gradient into the ventricles. This happens via outward diffusion through Cl^- channels (Johanson et al. 2008; Damkier et al. 2013; Praetorius and Damkier 2017), but also via Na-K-Cl cotransport (NKCC1). The latter is bidirectional, but normally appears to act as an efflux transporter at the CPE (Steffensen et al. 2018). Apical anionic Cl^- effluxes are facilitated by a CSF positivity of ~5 mV.

2.3.4 HCO₃⁻ Transport

The labile anion, HCO_3^-, is both generated and transported by the CP. HCO_3^- transport is more prominent at the BCSFB than BBB. Carbonic anhydrases, present in the cytosol and at both external limiting membranes of CPE, produce 2/3 of the HCO_3^- by hydrating metabolic CO_2; another third is formed by uncatalyzed CO_2

hydration (Maren 1979). Additionally, HCO_3^- is transported into the epithelium by the basolateral cotransporters NCBE and NBCn1, and effluxed via AE2 (Fig. 2.4). The relative importance of these basolateral transporters (in HCO_3^- and fluid transport and pH regulation) awaits experimental elucidation. HCO_3^- transfer into CSF at the BCSFB is an integral part of CSF formation and is decreased markedly after carbonic anhydrase inhibition by acetazolamide (Maren 1988; Johanson 1984). HCO_3^- flux from CP to CSF occurs by apical Na-HCO_3 cotransport activity and by outward diffusion via an anion channel, into the electropositive CSF (Sørensen et al. 1978).

2.3.5 H_2O Flux Across BCSFB

H_2O turnover is great at both the BCSFB and BBB (Brinker et al. 2014). Dynamic turnover of H_2O, though, is not the same as *net* secretion of *fluid*. At the BBB, the great H_2O flux is not associated with high-capacity fluid formation; at the BCSFB, however, the substantial H_2O movement (net) across CPE is associated with a substantial HCO_3 concentration (resulting from net solute flux) that characterizes CSF composition.

Two mechanisms have been described for H_2O flow across the CPE into CSF. In one, H_2O moves down its chemical potential gradient through AQP1 channels on the apical membrane, following the osmotically-active, transcellularly-moved, ions into the ventricles (MacAulay and Zeuthen 2010; Johanson 2015). H_2O is also directly translocated into CSF by coupling with the apically-located, Na-K-Cl co-transporter (NKCC1) (Steffensen et al. 2018). Still, under some conditions, the bidirectional Na-K-Cl co-transporter is able to transport H_2O *into* the CPE by a NKCC1-mediated process (Gregoriades et al. 2019). The precise role of the NKCC1 transporter in the H_2O dynamics of CSF secretion awaits further definition in intact CP preparations with normal blood flow, neural tone and CPE metabolic integrity.

2.3.6   Other Ions/Molecules in CP-CSF Production

While the paragraphs above highlight the role of Na^+, Cl^- and HCO_3^- in CSF formation, K^+ is also secreted into nascent CSF at a stable level of ~3 mmol/L (Spector and Johanson 1989). CSF K^+ is maintained within narrow limits even during hypokalemia and hyperkalemia (Johanson et al. 1974). The importance of CSF K^+ homeostasis for brain function is discussed further in Sect. 2.7.3.1.

The CSF concentrations of the multivalent ions, Ca^{2+}, Zn^{2+}, Mg^{++}, and PO_4^{3-} are also tightly regulated. CP has abundant metal transporters that control ion trafficking into CSF-brain. As with K^+, regulated Ca^{2+} and Mg^{2+} transport across CP is crucial for CNS excitability. Zinc is needed for DNA transcription, carbonic anhydrase activity and a host of other metabolic reactions. Although Zn^{2+} concentrations in CSF are low compared to plasma, the brain needs a bountiful supply. At the CP, 16 transporters from the ZIP and ZNT families regulate Zn^{2+} fluxes between plasma and CSF (Snodgrass and Johanson 2018). Inorganic trivalent phosphate (PO_4^{3-}) concentration in CSF is approximately half that of plasma. An active transporter in the CP apical membrane, PiT2, removes PO_4^{3-} from CSF (Guerreiro et al. 2014).

Trace element concentrations in CSF-brain are also tightly regulated. A plethora of transporters in CP guard the CSF-brain against surges or deficiencies in manganese (Mn^{2+}), copper (Cu^{2+}) and iron (Fe^{2+}). Manganese, an essential trace element, has been extensively investigated by Zheng and colleagues for interactions with Cu^{2+} and Fe^{2+} (Zheng 2005), in the context of CNS toxicity. Mn^{2+} is actively transported by the CP into CSF (Schmitt et al. 2011), but too much exposure to Mn^{2+} can disrupt neurogenesis (Adamson et al. 2018). Excessive brain Mn^{2+} leads to the toxic manganism disorder (e.g., welders overexposed to Mn^{2+} can suffer a Parkinson-like disease). Mn^{2+} overload disrupts Cu^{2+} regulatory transport at the BCSFB, by upregulating the choroidal Cu^{2+} transporter CTR1 and thereby distorting the CP-CSF Cu^{2+} homeostasis (Zheng et al. 2012). Lead exposure also impairs Cu^{2+} transport in CP, leading to brain disorders with an imbalance of trace elements (Zhao et al. 2014). The CP also has a pivotal role in adjusting transport of Fe^{2+}, protecting the brain against the Fe^{2+} deficiency that may cause conditions such as restless legs syndrome (Clardy et al. 2006). The choroidal Fe^{2+}-regulatory hormone hepcidin (Raha-Chowdhury et al. 2015), along with the hepcidin-ferroportin protein associations in the CPE-CSF-ependymal nexus, help to supply sufficient Fe^{2+} for CSF-brain. Together, the CPE and brain capillaries utilize complementary metal transport mechanisms to impart a supportive fluid environment for neuronal networks. Therefore, regulated metal transport activity at these major interfaces insures appropriate neuronal excitability and supplies the metabolic requirements.

2.4 Extrachoroidal Fluid Production in the Brain

In addition to CP, several CNS sites are implicated as fluid-forming regions. Such alternative or complementary sites include brain parenchyma, the BBB and ependyma. These three regions have been less extensively studied than CP for *net* fluid production capability. In view of potential contributions to central fluid dynamics, it is imperative to gain more information about the role of neurons-glia, endothelium-astrocytes, and ependyma in producing fluid. Discussion of current and desired evidence, with regard to non-CP fluid production, will help to build comprehensive models of CNS fluid balance.

2.4.1 *Parenchymal Metabolic H₂O Generation*

Brain H_2O is produced as the result of glucose metabolism. There have been estimates of the rate of metabolic H_2O formation, e.g., for the human brain ~3.3 moles/day (60 ml/day) (Hladky and Barrand 2016). This compares to the much greater human CSF production of ~550 ml/day (Spector et al. 2015b). Thus, metabolic H_2O is a potentially small, but not insignificant, component of H_2O-fluid production in brain. However, it should be noted that unless accompanied by osmolytes, metabolic H_2O may diffuse out of the brain across the BBB. As such, without concurrent osmolyte movement in the direction of H_2O flux, there would consequently be no *net* production of fluid by brain/cord parenchyma. The other product of glucose metabolism, CO_2, is freely diffusible across the BBB.

2.4.2 *Blood-Brain Barrier Fluid Turnover*

In addition to metabolic H_2O production by brain cells, there may be secretion of fluid at the BBB (Cserr 1988). It is generally held that that these two processes account for a minority of brain fluid production (Hladky and Barrand 2016; Spector et al. 2015b). Still, as noted above, it has been claimed that they account for all fluid generation within the CNS, i.e., there is negligible CP production (Oreskovic and Klarica 2010). Due to these wide-ranging points of view, it is essential to weigh all extant evidence for choroidal vs. extra-choroidal fluid production.

With current methodologies, it is not possible to directly assess BBB fluid production. In systemic capillaries there is normally net fluid secretion driven by Starling forces. Thus, fluid secretion, due to the hydrostatic pressure gradient between blood and tissue interstitial fluid (ISF) [or between the capillary glycocalyx and ISF (Adamson et al. 2004)], exceeds the fluid reabsorption, due to the higher oncotic pressure of blood compared to ISF. The resultant net fluid secretion results in peripheral lymph flow. It has been suggested that such filtration also accounts for fluid flow in the brain (Oreskovic and Klarica 2010). The impact of hydrostatic and oncotic pressure on fluid flow, however, also depends upon the hydraulic conductivity (L_p) as shown in the equation:

$$J_v/\mathrm{A} = L_p\,(\Delta p - \sigma\Delta\pi),$$

where J_v is the solvent flow, A is the capillary surface area, Δp and $\Delta\pi$ are the hydrostatic and the colloid osmotic pressure gradients across the capillary wall and σ is the reflection coefficient (Adamson et al. 2004). In cerebral capillaries, L_p is very low compared to systemic capillaries due to occlusion of the paracellular space by endothelial tight junctions and the apparent absence of aquaporin H_2O channels in the cerebral endothelium in vivo (Haj-Yasein et al. 2011). In whole brain for example, Paulson et al. (1977) as well as Fenstermacher and Johnson (1966) estimated L_p values of 2.6 and 0.8×10^{-9} cm s^{-1} cm H$_2$O^{-1} in human and rabbit whole brain, values similar to the 2.0×10^{-9} cm s^{-1} cm H$_2$O^{-1} reported by Fraser et al. (1990) in isolated frog brain microvessels. By contrast, in frog skeletal muscle and mesenteric microvessels, L_p values of 74 and $\sim500 \times 10^{-9}$ cm s^{-1} cm H$_2$O^{-1} have been reported (Curry and Frokjaer-Jensen 1984; Michel et al. 1974; Tucker and Huxley 1990). This difference in L_p would mean that a 35–250 fold higher hydrostatic pressure gradient would be required at the cerebral endothelium to induce the same hydrostatically-driven H$_2$O flow as at a muscle or mesenteric capillary. It should be noted that the human brain (~1.3 kg) secretes ~550 ml of CSF per day (0.4 ml/g/day) while the whole body produces ~4500 ml of lymph per day (~0.06 ml/g/day, assuming 70 kg body weight). That is, fluid flow (per g) is higher, rather than lower, in the brain compared to the rest of the body. In addition, BBB capillaries have a reflection coefficient for ions (the primary blood/tissue osmolytes) and proteins that is close to one; i.e., they are close to impermeable. Unless ions or another osmolyte accompany hydrostatically-driven H$_2$O, an osmotic gradient will be established across the capillary retaining H$_2$O in the vasculature (Hladky and Barrand 2014, 2016).

Investigators have examined alternative mechanisms of BBB fluid secretion, particularly focusing on ion transport. The concept is that BBB has a polarized distribution of ion transporters, similar to CP (Praetorius and Damkier 2017), resulting in a net transport of ions (e.g., Na$^+$ and Cl$^-$) from blood to brain with osmotically-entrained H$_2$O. There is evidence of such a polarized distribution of ion transporters, e.g., Na$^+$-K$^+$-ATPase (Hladky and Barrand 2016; Betz et al. 1980; Mokgokong et al. 2014). It is, however, currently very difficult to measure net ion fluxes across the BBB. While there have been many measurements of blood to brain ion movement in vivo, measuring brain to blood transport is much more difficult and great precision in that measurement is necessary for calculating net transport; see review by Hladky and Barrand (Hladky and Barrand 2016).

An alternate approach is to examine brain endothelial ion and H$_2$O movement in vitro. There have been major advances in such modeling, with in vitro cultures now achieving the high transendothelial resistances found in vivo (Lippmann et al. 2012, 2014). However, whether such systems will be useful in examining brain endothelial ion and fluid transport is still uncertain. It depends on how faithfully they replicate in vivo ion transporter expression and polarity. Note that CPE cultures can transport fluid (Hakvoort et al. 1998).

A complementary hypothesis is that certain solute transporters also transport H$_2$O directly, rather than generating osmotic gradients that entrain H$_2$O movement. This has been best examined for the Na/K/Cl cotransporter NKCC1 (Zeuthen and Macaulay 2012) and the Na$^+$-dependent glucose transporter SGLT1 (Zeuthen et al. 2016). It is estimated that each turnover of NKCC1 (but not the related NKCC2), transports 600 H$_2$O molecules (Zeuthen and Macaulay 2012) and that SGLT1 accounts for two-thirds of passive H$_2$O flow across small intestine (Zeuthen et al. 2016). Whether cotransporters are directly involved in H$_2$O movement at the BBB has not been investigated. Increased NKCC1 and SGLT activity at the BBB has been reported after stroke, a condition causing brain edema (O'Donnell 2014; Vemula et al. 2009). It should be noted, however, that the apparent inability of brain to maintain an osmotic gradient with blood long-term, means that any transporter-associated H_2O movement has to have net solute flux to generate fluid flow across BBB.

2.4.3 Ependymal Wall Fluid Secretion?

The ependymal cells lining the ventricles form a somewhat permeable, wall-like structure that separates internal brain from CSF. Both the ependymal and the CPE are derived from the neuroectoderm suggesting a common fluid-secreting function, or at least some fluid-generating capacity by the ependymal CSF-brain interface in light of the highly-secretory choroidal BCSFB. In addition, the expression of aquaporin 4 channels on the basolateral membrane of ependymal cells as well as periventricular astrocytes prompt consideration of fluid movement into the ventricles. Correspondingly, on the brain's external surface, the relatively-tight glia limitans under the pia mater is a histologic factor intimating possible fluid secretion into cortical SAS.

Pollay and Curl (1967) perfused the isolated Sylvian aqueduct, lined by ependymal cells, and tested (by indicator dilution) for net fluid movement into the channel between the third and fourth ventricles. They reported evidence for a low rate of fluid accretion within the rabbit cerebral aqueduct. This experimental finding awaits corroboration.

Highly-penetrating solute and H_2O movement across the generally-permeable ependyma (where cells are linked by gap, rather than tight, junctions) has been demonstrated in many investigations. What is needed, however, is a systematic study of various ependymal regions (Mitro et al. 2018) to define specific passive vs. active movements of fluid and net ion transport into CSF. To date, the voluminous fluid production consistently demonstrated over decades for the choroidal BCSFB has not been demonstrated for the CSF-brain interface at the ependyma. Even though ependymal cells evidently do not substantially produce fluid, they may participate in regulating paracellular fluid movements into and out of the ventricles. However, it should be noted that ependymal-specific deletion of AQP4 did not affect brain H_2O content in mice (Vindedal et al. 2016).

Interactions between CSF and ISF dynamics are adversely altered in senescence, hydrocephalus and Alzheimer's disease. In all three pathophysiologies, the neurodegeneration is associated with damage to or disappearance of portions of the ependymal wall. Preventing or repairing ependymal damage in CNS diseases/disorders is an important step in stabilizing periventricular regions to assure fluid balance.

2.5 Neurohumoral Modulation of CSF Dynamics

Fine control of CSF pressure is absolutely essential for a sound functioning CNS. Elevated CSF pressure, or ICP, interferes with vascular perfusion of the brain/spinal cord and can impact CSF outflow mechanisms. High ventricular pressure also takes a toll on the ependymal wall and subjacent neurogenic regions around the ventricles. There are both physical and biochemical mechanisms for preventing intracranial hypertension. This section focuses mainly on the latter category, i.e., neurotransmitter and hormonal modulation of CSF production by CP.

2.5.1 Neurotransmitter Regulation of CP Fluid Formation

Several neurotransmitters tonically modulate CSF formation, as well as the blood flow that supports the secretion. The choroidal epithelium and vessels are directly innervated by several systems of nerve fibers; this provides locally-originating tone. Other systems send fibers to the ventricular (ependymal) wall for neurotransmitter release into CSF; thus, over a long distance the neurotransmitter then moves by volume transmission to bind receptors on the apical side of the CPE.

2.5.1.1 Noradrenergic Modulation

Central norepinephrine augmentation inhibits CSF dynamics. Sympathetic (noradrenergic) effects result from both blood flow and epithelial actions. Thus, vasoconstricting catecholamines can curtail CP vascular perfusion by 40–50% and are able to modulate carbonic anhydrase activity in the CPE (Lindvall et al. 1978a). Experimentally presenting biogenic amines to CSF, or electrically stimulating the sympathetic nerve to CP (Lindvall et al. 1978b; Haywood and Vogh 1979), lowers CSF production. Contrariwise, resecting sympathetic fibers from the superior cervical ganglion (SCG) to the lateral CPs increases CSF production. This enhanced CSF flow across the BCSFB results from 'releasing the brake' on CSF formation when sympathetic tone is markedly attenuated (Lindvall et al. 1978b). Electrical activity (neurotransmission) at the preganglionic neurons of the SCG decreases after midnight and remains low until night's end (González Burgos et al. 1994). This factor would cause less stimulation of postganglionic neurons in SCG, attenuate sympathetic tone on CP, and result in increased CSF production at night (Nilsson et al. 1994). There is evidence that at night the brain metabolite clearance by the glymphatic system is augmented (Xie et al. 2013); this, together with increased CSF production, may act in concert.

2.5.1.2   Cholinergic Modulation

Cholinergic transmission to the BCSFB also curtails CSF dynamics. Cholinergic fibers from glossopharyngeal and vagus nerves of several mammals innervate the CP of all four ventricles (Lindvall et al. 1977). Intraventricular or intravenous infusion of carbamylcholine (muscarinic agonist) or acetylcholine reduced CSF production in rabbits up to 55% (Lindvall et al. 1978c). The lowered fluid formation rate, blocked by atropine, was likely an epithelial rather than a vascular effect. Atropine by itself did not affect CSF production, suggesting that normal cholinergic tone at the CP is not strong. The experimental inhibitory cholinergic effect was additive to the adrenergic inhibition of CSF dynamics.

2.5.1.3   Serotoninergic Modulation

Serotonin (5-HT) downregulates CSF production (Nilsson et al. 1992). CP receives 5-HT from indoleaminergic nerves (from Raphe nucleus) locally innervating the epithelium (Napoleone et al. 1982); and from CSF carrying 5-HT released from the supraependymal plexus in the lateral ventricle wall. 5-HT_{2c} (originally called 5-HT_{1c}) serotoninergic receptors, in homodimeric form (Herrick-Davis et al. 2015) densely stud the apical CPE. Autoradiographic (Hartig et al. 1990) and molecular (mRNA) studies agree that CPE has the densest 5-HT_{2c} expression in CNS (Palacios 2016; Pandey et al. 2006). Administration of 5-hydroxytryptophan, to generate 5-HT, decreased CSF production in dogs by 35–40% (Nakamura et al. 1985). A high affinity 5-HT_{2c} ligand, SCH-23390, reduced CSF production up to 50% in rats (Boyson and Alexander 1990). Serotonin likely plays a role in sleep regulation, possibly by modulating CP fluid production (Zeitzer et al. 2002). During the transition from sleep stage 2 to REM, the ventricular CSF serotonin decreases by ~30% in humans; this factor might help increase CSF formation rate and solute flushing (Lindvall et al. 1977). Also, strong circadian rhythms in CP may adjust nocturnal clearance of metabolites from CSF (Myung et al. 2018). In addition, it should be noted that the CP expresses canonical clock genes that have a 24-h rhythm and show sex-dependent regulation (Quintela et al. 2015, 2018). Thus, the CP may exhibit autonomous circadian rhythms.

2.5.2   Hormone Effects on CP Fluid Formation

Ion transporter capacity and choroidal blood flow, predominant factors in CSF formation rate, are modulated by hormones and enzymatic activity. Transport enzymes have a key role: $Na^+\text{-}K^+\text{-ATPase}$ and carbonic anhydrase activities, respectively, promote Na^+ and HCO_3^- flux from CP epithelial cells to CSF. In addition, nitric oxide (NO) synthase activity in CP (Szmydynger-Chodobska et al. 1996), by generating vasodilatory NO, likely maintains basal blood flow (Chodobski et al. 1999; Faraci and Heistad 1992) to support a normal CSF production. NO elevation in response to CNS stressors generally has an overall inhibitory effect on Na^+ fluxes at the CPE (via NHE, Na^+-K^+-ATPase and Na^+/HCO_3^- cotransport).

Disruption of CNS homeostasis hormonally activates the CP-CSF-brain axis. Thus, when CP and brain undergo stressful conditions (e.g., dehydration, hemorrhage, pressure instability, fluid overload), there is activation of complex hormonal systems to restore fluid balance and ICP. A trio of centrally acting hormones— angiotensin II (Ang II), arginine vasopressin (AVP) and atrial natriuretic peptide (ANP)—interactively effect renal-like actions on CP blood flow, epithelial ion transport and fluid turnover (Szmydynger-Chodobska and Chodobski 2005). These adaptive responses help normalize brain extracellular fluid (ECF) volume and CSF pressure. It is clear, then, that these neuropeptides are 'recruited' in CP-CSF when the CNS is under duress.

Ang II, AVP and ANP have variable effects on blood flow, but all inhibit CSF dynamics and suppressing CSF flow helps to reduce ICP. Receptors for these fluid-regulating hormones are present in blood vessels and epithelial cells. The presence of the tight junctions at the CPE allows examination of the importance of localized effects by limiting penetration of peptides between the CSF and blood side of the BCSFB.

2.5.2.1 Angiotensin

Angiotensin II, a potent vasoconstrictor, is a major pressure regulator. The CNS (brain and CP) contains a well-developed renin-angiotensin system (RAS). This enables centrally-derived angiotensin to modulate CSF-brain fluid balance. Angiotensin II is formed by angiotensin converting enzyme (ACE) catalytic action on Ang I (a vasodilator). ACE is present at very high levels in CP (Arregui and Iversen 1978) and ACE inhibition increases CSF formation rate (Vogh and Godman 1989). Thus, the ACE inhibitor quiniprilat (in CSF) prevents Ang I conversion to Ang II, preventing diminished CP blood flow (Vogh and Godman 1989). The presence of ACE at the BCSFB allows for local Ang II generation/release (Maktabi and Faraci 1994) to mediate fluid homeostatic mechanisms. Attenuated CSF formation is observed experimentally with intracerebroventricularly-administered Ang II (Maktabi et al. 1991, 1993a); this is similar to the Ang II-induced inhibition of alveolar fluid clearance dynamics, an effect attributed to lowered epithelial Na^+-K^+-ATPase activity. The primary inhibitory effects of central Ang II on CP-CSF hemodynamics (blood flow) and hydrodynamics (CSF formation) are AT1 receptor-mediated (Chodobski et al. 1994), involving choroidal manufacture (Chodobski et al. 1997) and release of the mediating AVP (Chodobski et al. 1998). Altogether, this speaks to CP uniqueness in the CNS as a hormone-modulating site that regulates ICP.

2.5.2.2   Arginine Vasopressin

Antagonism of AVP V1 receptors in CP does not alter CSF production, suggesting a normally weak vasopressinergic tone at the healthy BCSFB (Maktabi et al. 1993b). However, increasing plasma AVP reduces CP blood flow and CSF production (Faraci et al. 1990). Following systemic hemorrhage, augmented plasma AVP reduces CP hemodynamics and hydrodynamics, and this effect can be eliminated with a V1 receptor antagonist (Maktabi et al. 1993b). Thus, peripheral (systemically)-derived AVP upregulation after hemorrhage likely compensates for whole body blood loss by lowering perfusion to CP. This presumably helps to minimize the lowering of arterial pressure for survival.

Early experiments on dehydration and AVP exposure revealed CPE ultrastructural changes (Schultz et al. 1977). In those conditions, CPE cells shrank, turned dark, and displayed an ultrastructure (Szmydynger-Chodobska and Chodobski 2005) reminiscent of altered fluid transport (Liszczak et al. 1986). Interestingly, salt-loading in rats also increases AVP and V1 receptor mRNA levels in CPE (Zemo and McCabe 2001). Augmented expression of CP Na^+ channels, in response to hypernatremia (Szmydynger-Chodobska et al. 2006), may help adjust BCSFB Na^+ fluxes to attain osmotic balance between plasma and CSF.

Centrally-derived AVP enables the brain to regulate its own fluid balance. Accordingly, in non-hemorrhage and non-dehydration states, AVP derived from the CP-CSF-hypothalamus nexus mediates brain adaptations to elevated ICP. CSF AVP increases as ICP rises in intracranial hypertension (pseudotumor cerebri), intracranial tumors and intracranial hemorrhage and Sorensen et al. postulated that elevated CSF pressure is a stimulus for releasing central AVP (Sorensen and Gjerris 1977). In vitro, CPE cells exposed to AVP turn dark, shrink, have slender (filiform) microvilli, and are less able to transport Cl^- (Johanson et al. 1999a). These dark epithelial cells (5–15% normally) are neuroendocrine-like and usually appear in states of homeostatically-decreased CSF formation and pressure (Johanson et al. 1999b, 2008). Dark cells may reabsorb fluid (Schultz et al. 1977); if so, then drug (hormone)-induced fluid reabsorption by an increasing fraction of total CP cells would result in less *net* CSF formation.

2.5.2.3   Atrial Natriuretic Peptide

ANP is a volume-regulating hormone. In the periphery, ANP stabilizes extracellular (plasma) volume by increasing urinary Na excretion in response to arterial pressure (volume) overload. In the CNS, expanded CSF volume (ICP increase) is compensatorily decreased by reducing Na uptake by the CP-CSF system. Human CSF titers of ANP increase in proportion to ICP increments (Yamasaki et al. 1997; Doczi et al. 1988; Tulassay et al. 1990). This points to ANP as a 'pressure-sensitive' biosignal for a homeostatic feedback system to regulate ICP. The BCSFB is a target organ for ANP, with the CPE having abundant receptors (Herbute et al. 1994;

Johanson et al. 2006), the expression of which changes with CNS fluid shifts and pressure alterations (Grove et al. 1997; Saavedra and Kurihara 1991; Mori et al. 1990). CSF-administered ANP decreases CSF formation rate (Steardo and Nathanson 1987) and ICP (Akdemir et al. 1997). These inhibitory actions of ANP relate to biochemical and ultrastructural changes induced by the hormone at the CPE.

At the epithelial level, ANP exerts intriguing effects on the CP. Both in vitro and in vivo CP exposed to increasing ANP concentrations undergo altered fine structure of the epithelial cells. Similar to the AVP-induced effects, exposure to ANP causes shrinkage of CPE cells, darkening of cytoplasm and narrowing of apical microvilli (reflective of reduced CSF secretion). Such 'dark' cells are considered 'neuroendo-crine-like', since they occur following CP exposure to rising CSF levels of hormonal AVP (Johanson et al. 1999a), ANP (Johanson et al. 2006) and basic fibroblast factor (Johanson et al. 1999b) (all of which decrease CSF formation). These dark neuroendocrine-like cells in CP greatly increase in hydrocephalus (Johanson et al. 2006), in which ICP is elevated and CSF formation is downregulated (Weaver et al. 2004).

Given that ANP is generally antagonistic to AVP release (Gerstberger et al. 1992), it is somewhat surprising that both neuropeptides inhibit CSF flow. However, so important is ICP control for brain viability, that the CNS contains redundant neurohumoral systems to downregulate CSF pressure in various diseases/disorders. ANP and analogs may eventually find a place in the therapeutic armamentarium of agents (Woodard and Rosado 2008) to control hypersecretory CSF dynamics and/or intracranial hypertension. However, as a peptide of considerable molecular weight (~3 kDa), ANP does not appreciably cross the BCSFB and BBB (Marumo et al. 1988). Intranasal delivery may provide an alternate route of delivery.

It is now evident that the CP responds to both systemic and local perturbations in fluid balance and pressure. A fruitful future endeavor would be a systems biology analysis of the BCSFB: integrating the multiple neurohumoral stimuli input, the epithelial processing of signals, and the output of CSF homeostatic responses.

2.6 Pharmacologic and Genetic Manipulation of CSF Production

Restoration of aberrant CP-CSF dynamics in both young and old patients (Stopa et al. 2018) has been a longstanding aspiration of neurologists and neurosurgeons. The limited repertoire of efficacious agents to manage altered CSF pressure, ventriculomegaly and cerebral edema has spurred much interest to implement drugs and strategies. Decades-long usage of corticosteroids has yielded a host of variable effects in trying to reclaim disordered CSF-brain fluid balance; the complexity of interpreting steroid treatments of CSF-brain fluid perturbations precludes their treatment here. Reduction of elevated ICP is essential for brain health. Hormones that curtail CSF formation rate (Ang II, AVP and ANP) have been previously discussed; their application to human CSF imbalances has not yet been refined. Most medicinals for reducing CSF formation are either inhibitors of transport enzymes (Na^+-K^+-ATPase and carbonic anhydrase) or agents that interfere with CP ion transporters. Novel approaches, e.g., inhibitors of 11-OH steroid dehydrogenase (HSD) are in clinical trial; their rationale and promise are discussed below.

2.6.1 Enzyme and Ion Transport Modulators

Carbonic anhydrase inhibitors, with acetazolamide (Diamox) as the class prototype (Maren 1979; Birzis et al. 1958), have been the mainstay (>60 year) of CSF pharmacotherapy for elevated ICP. Acetazolamide lowers ICP (Uldall et al. 2017) by curtailing CSF production rate by ~50%. Experimentally, the effects on CSF dynamics/transport can be assessed by ventriculo-perfusion (indicator dilution method) that has been reliably done for a half century (Cserr 1971; Davson and Segal 1970; Pappenheimer et al. 1961). Some neurosurgery patients requiring acute management of elevated ICP, and pseudotumor cerebri patients needing chronic reduction of intracranial hypertension (Smith and Friedman 2017), have benefited from acetazolamide. However, several undesirable side effects prevent acetazolamide from being an ideal agent to manage CSF dynamics.

Inhibiting CPE Na^+-K^+-ATPase with ouabain reduces CSF formation rate ~50% in animal experiments, only when that cardiac glycoside is placed on the CSF side of the tight junction. This factor, along with CSF K^+ buildup and other potential neurotoxic side effects, make ouabain usage not feasible clinically. Digoxin, another cardiac glycoside inhibitor of the Na^+ pump, is used in some cardiac patients. More lipid-soluble than ouabain, digoxin penetrates the BCSFB and reduces CSF formation rate by ~25% (Allonen et al. 1977). Digitoxin applied on the CSF side nearly halves CSF formation rate (Welch 1963). It should be noted that CSF turnover decreases with age potentially putting elderly patients at neurotoxic risk due to denatured protein retention. This may be exacerbated in digitized elderly patients (e.g., digoxin, digitoxin), due to lowered CSF secretion. Clinically, a combination of acetazolamide and furosemide, a loop diuretic inhibitor of Na-Cl cotransport, was given to reduce the need for shunt placement with post-hemorrhagic hydrocephalus. However, a clinical trial with this combination indicated that children did worse (shunt placement/neurological morbidity) (*Lancet (London, England)*1998). This may have been because of the type of cotransporter present at the CPE (Na-K-Cl rather than Na-Cl cotransport) and its location (apical rather than basolateral). Thus, recent evidence indicates the importance of the CPE apical NKCC1 in CSF secretion. For example, Karimy et al. (2017) have found that intraventricular hemorrhage results in CSF hypersecretion by activating NKCC1 and that this contributes to post-hemorrhagic hydrocephalus. The NKCC1 inhibitor, bumetanide, could reduce the intraventricular hemorrhage-induced CSF hypersecretion and the associated hydrocephalus. However, the bumetanide needs to be given by intracerebroventricular injection to access the CPE apical NKCC1. Such inaccessibility from plasma limits current clinical utility, but it may be possible to develop CSF-penetrating NKCC1 inhibitors.

Guanylyl cyclase is a CP enzyme that generates cGMP in response to ANP (Israel et al. 1988). Elevated cGMP initiates a biochemical cascade that eventually inhibits Na^+-K^+-ATPase and reduces the pumping of Na^+ into ventricular CSF. A promising translational CSF area is the testing of ANP analogs (some in current usage to treat heart/kidney ailments) to suppress CPE Na^+ transport, fluid turnover and ICP.

An enzyme in CP that regulates availability of steroid molecules (e.g., cortisol, a stimulator of CSF formation) is 11β-hydroxysteroid dehydrogenase (11β-HSD). Intracranial hypertensive patients who lose weight, also attain lower levels of 11β-HSD and ICP. An inhibitor of 11β-HSD (type 1), AZD4017, is now in Phase II clinical trial (NCT02017444) to treat idiopathic intracranial hypertension (Markey et al. 2017). Sinclair and investigators hypothesize that AZD4017 will lower the levels of CP cortisol (local, epithelial) (Sinclair et al. 2007) as well as fluid production and CSF pressure in intracranial hypertensive patients.

2.6.2 Manipulating CP Transporters Genetically

Genetic manipulation is an important experimental tool for examining the mechanisms underlying CPE transport and fluid secretion. It is also potentially an important clinical tool but there it carries potentially-sizeable risks and clinical applications are likely on the distant horizon. Still, because of the many possible benefits, it is worthwhile discussing new avenues that are opening up.

Gene deletion or manipulation studies have provided insight into the role of specific transporters at the CP (Keep and Smith 2011). For example, the CP epithelium highly expresses an oligopeptide transporter, Pept2 (Slc15a2), on the apical membrane where it functions to clear oligopeptides/peptidomimetics from the CSF into the epithelium. Gene deletion of Pept2 in mice results in greater entry of systemically-delivered oliopeptides and peptidomimetics (e.g., the antibiotic cefadroxil) into CSF (Kamal et al. 2008). Systemically administered 5-aminolevulinic acid, an endogenous peptidomimetic, also causes greater neurotoxicity in the Pept2 null mice (Hu et al. 2007) while intracerebroventricular injection of L-kyotorphin causes more analgesia because of reduced clearance (Jiang et al. 2009).

Gene manipulation studies have also provided evidence on the importance of CP in fluid secretion (Table 2.1). Within the brain, the H_2O channel AQP1 is primarily expressed on the CP apical membrane (Xu et al. 2017). Gene deletion studies have shown that AQP1 is the primary H_2O channel expressed in the CPE (fivefold decrease in osmotic H_2O permeability in null mice), and loss of AQP1 reduces CSF secretion by 25% (Oshio et al. 2003). In addition, aquaporin-1 KO mice have a smaller ventricular system as well as reduced ventriculomegaly after injection of kaolin to block CSF flow (Wang et al. 2011).

CSF secretion involves the net movement of ions as well as H_2O. Two studies have examined the role of the Na^+-coupled HCO_3^- exchanger, Slc4a10 (NCBE), in

Table 2.1 Studies on genetically manipulating ion transporters or H_2O channels in CP epithelium: effects on CSF secretion or hydrocephalus

Target	KO/ siRNA	Effect on CSF secretion/ hydrocephalus	Refs.
Direct			
Aquaporin 1	KO	Reduced CSF production (mouse)	Oshio et al. (2003)
Aquaporin 1	KO	Reduced ventricle size and reduced kaolin-induced hydrocephalus (mouse)	Wang et al. (2011)
Sodium-dependent phosphate cotransporter, Slc20a2	KO	Spontaneous hydrocephalus (mouse)	Wallingford et al. (2017)
Sodium-coupled bicarbonate exchanger, Slc4a10	KO	Reduced ventricle size (mouse)	Jacobs et al. (2008)
Sodium-coupled bicarbonate exchanger, Slc4a10	siRNA	Reduced CSF secretion and hydrocephalus after germinal matrix hemorrhage (rat)	Li et al. (2018)
Sodium bicarbonate cotransporter, Slc4a5	KO	Reduced ventricle size and lower intracranial pressure (mouse)	Kao et al. (2011)
Sodium bicarbonate cotransporter, Slc4a5	KO	No significant reduction in ventricle size (mouse)	Groger et al. (2012)
Indirect			
Toll-like receptor 4	KO	Reduced CSF hypersecretion after IVH by inhibiting SPAK-NKCC1 (Slc12A2) complex (rat)	Karimy et al. (2017)
Ste20-type stress kinase (SPAK)	AS-ODN	Reduced CSF hypersecretion after IVH by inhibiting NKCC1 phosphorylation (rat)	Karimy et al. (2017)
Liver X receptor α and β	KO	Reduced ventricle size. Associated with reduced CP aquaporin 1 and carbonic anhydrase IX (mouse)	Dai et al. (2016)

The transporters/channels listed are present in CP epithelium but some may also be expressed in other areas of the brain stressing the need for future cell-specific manipulations

AS-ODN antisense oligodeoxyribonucleotide, *IVH* intraventricular, *KO* knockout, *siRNA, NKCC1* sodium potassium chloride cotransporter 1; small interfering RNA, *Slc* solute carrier, SPAK, Ste20-type stress kinase

CSF secretion by genetic manipulation. *Slc4a10* is present on the basolateral membrane of the CPE. Jacobs et al. (2008) found that Slc4a10 null mice have a markedly reduced ventricle size. Li et al. (2018) determined that depleting *Slc4a10* via intracerebroventricular injection of a small interfering RNA (siRNA) decreased the hydrocephalus induced by a germinal matrix hemorrhage in neonatal rats. Both studies suggest an important role of *Slc4a10* in CSF secretion.

The Na^+-HCO_3^- cotransporter, *Slc4a5* (NBCe2), is present at the apical membrane of CPE (Christensen et al. 2018). The effects of deleting this transporter on CSF dynamics have been equivocal. One group found that null mice have reduced ventricle size and lower intracranial pressure (Kao et al. 2011), while another group didn't find a significant effect on ventricular size (Groger et al. 2012). These differences may reflect alternate methods for inducing the gene deletion (Christensen et al. 2013), but there is also the need for direct measurements of CSF secretion to clarify the genetic findings.

This is also the case for the Na^+-dependent phosphate cotransporter, *Slc20a2*. This transporter is present at the CP (epithelium and endothelium), and loss of the transporter in mice results in spontaneous hydrocephalus (Wallingford et al. 2017). However, this transporter is not solely present in the CP and there needs to be studies examining whether the null mouse has increased CSF secretion.

There have also been studies using genetic manipulations that have indirectly provided evidence on the role of CP H_2O-ion movement in CSF secretion. Karimy et al. (2017) found that intraventricular hemorrhage (IVH) in rats causes CSF hypersecretion by stimulating the Na-K-Cl cotransporter, NKCC1 (*Slc12A2*), in CP. Deleting toll-like receptor 4 or reducing Ste20-type stress kinase (SPAK) with an antisense oligodeoxyribonucleotide in rats both prevented *Slc12A2* activation after IVH, and reduced IVH-induced CSF hypersecretion. Examining the effects of deleting both Liver X Receptor α and β in mice, Dai et al. (2016) found reduced ventricle size that was associated with lower expression of AQP1 and carbonic anhydrase IX in the CP, two important proteins in CSF secretion.

These genetic manipulation studies provide insight into the role of the CP in CSF secretion. There is, however, a need for further studies. Global genetic deletion may affect other cell types operative in CNS fluid dynamics, and there may also be adaptive responses to ablation. There is a need for studies using cell-specific knockouts that can be induced during development.

2.7 Choroid Plexus-CSF Streaming in Relation to Solute Exchange with Brain

2.7.1 Overview of Transependymal Solute Distribution

Substances secreted by the CPs flow through the ventricles down to the basal cisterns (Fig. 2.5). These CSF-borne solutes contact the expansive surface area of the ependyma bordering the ventricular system (Johanson et al. 2011b; Del Bigio 1995). By active transport and paracellular diffusion, substances move from CSF into ependymal cells or penetrate the underlying ISF and glia. The penetrating materials include vitamins, protein stabilizers, immunologic modulators, growth factors/hormones, and osmotic solutes. Simultaneously, in the opposite direction, cerebral catabolites and protein waste diffuse into ventricular CSF for removal by bulk flow and active reabsorption at the BCSFB (CP plus arachnoid). Such bidirectional transependymal fluxes contribute to maintaining brain ISF composition.

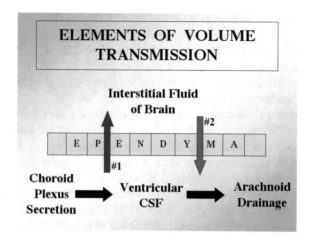

Fig. 2.5 Volume transmission of CSF: Flow of CSF is mainly convective, originating as CP secretion that distributes solutes and a few blood cells throughout the ventriculo-subarachnoid space (SAS) nexus. The hydrostatic pressure head from active CP secretion is an impetus to 'push' CSF widely throughout the CNS; inhibiting CSF formation by acetazolamide slows down the more distal glymphatic clearance. Before draining at arachnoidal sites into lymphatic/venous fluids, the CSF exchanges solutes between ventricles/SAS and brain interstitial fluid (ISF). Vitamin C, cystatin C, transthyretin, among many solutes from CP, diffuse into ISF (arrow #1) across the permeable ependyma. Concurrently, surplus K, proteins, amino acids and catabolic organic anions diffuse (arrow #2) from ISF into CSF for clearance. Acting simultaneously as a supply and excretory system, CSF volume transmission mediates solute/cell distribution along diverse extracellular pathways (Fig. 2.7). Endocrine signaling by hormone transit, from CP to CSF to hypothalamus, is enabled by CSF bulk flow. Adapted from Johanson et al. (2008)

2.7.2 Solute Influx: CSF to Brain

Early tracer studies in CSF physiology demonstrated ependymal interface permeation by high molecular weight substances including inulin (~5 kDa) and albumin (~67 kDa) (Rall 1968). In contrast to the tight junctions linking the endothelial cells of the BBB and the epithelial cells of the BCSFB, the adult (but not fetal) ependymal cells are linked by gap junctions that, unlike tight junctions, are leaky and thus permit paracellular diffusion. An exception is some areas of the third ventricle regions which do have diffusion-restricting tight junctions. Transependymal penetrability impacts cerebral metabolism by allowing the entry of many compounds from CSF to brain.

2.7.2.1 Micronutrient Vitamins

Neurons have high concentrations of the vitamin ascorbate, up to ~10 mM. The BBB does not normally express SVCT2, the active transporter of vitamin C, but SVCT2 is highly expressed in the CPE and actively transports ascorbate into CSF to a level

3–4× that of plasma (Spector and Johanson 2014). CSF ascorbate can readily diffuse across the ependyma and then distribute to neurons that accumulate ascorbate via SVCT2 (Fig. 2.2). This crucial pathway, from blood to CP to CSF to ependyma to brain, is maintained even when plasma vitamin C falls. SVCT2 at the BCSFB is a prototype for molecules that reach CNS preponderantly via the CP-CSF pathway and not brain capillaries.

Folate is another H_2O-soluble vitamin that extensively uses the CP en route to penetrating the CSF-brain nexus. Folate has multiple functions, including in DNA synthesis (Spector and Johanson 2014). Folate deficiency can lead to neural tube defects and folate supplementation during pregnancy curtails the occurrence of hydrocephalus and spina bifida. Experiments reveal that 10-formyl-tetrahydrofolate-dehydrogenase (FDH) is a key regulator for folate availability and metabolic interconversion (Naz et al. 2016). FDH deficiency decreases DNA methylation, which may be problematic in some forms of hydrocephalus. Experimental strategies to normalize CSF folate in hydrocephalus include maternal supplementation with specific folate metabolites (Cains et al. 2009).

2.7.2.2   Transthyretin

Proteins secreted by CP into CSF gain access to brain ISF by diffusing across ependymal interfaces throughout the ventricular system. Transthyretin (TTR) is the hallmark molecule in CP protein metabolism, accounting for up to 40% of CPE protein synthesis. TTR has a stabilizing role in attenuating β-amyloid protein misfolding in brain interstitium, and also facilitates iodide transport into brain across the BCSFB.

Other carrier proteins manufactured/secreted by the CP include transferrin and ceruloplasmin, regulators of Fe^{2+} and Cu^{2+}, respectively. Controlled distribution of Fe^{2+} and Cu^{2+} among plasma, CP-CSF and brain helps to maintain oxidative balance in CNS (Andersen et al. 2014). Due to interactive transport of Fe^{2+}, Cu^{2+} and Mn^{2+} across the BCSFB and ependyma, mutations in various carrier proteins and injuries such as stroke (Xiang et al. 2017) cause complex distortions in trace element distribution/metabolism in the CP-CSF-brain nexus.

2.7.2.3   Interferon-γ

There is a growing understanding of the role of the CP in leukocyte entry into brain in inflammatory conditions (Engelhardt and Ransohoff 2012; Schwartz and Baruch 2014). For example, interferon-γ (IFN-γ) receptor activity in CP controls entry of $CD4^+$ T cells and monocyte-derived macrophages into CSF, a process integral to central immunologic repair (Kunis et al. 2013). IFN-γ presents in both plasma and CSF. Certain cytokine responses to central injury/inflammation involve a feedback mechanism, via brain-CSF and CP, whereby the latter is modulated by brain-derived IFN-γ to stimulate leucocyte movement into CSF. Thus, following movement across the BCSFB, the CD4[+] T cells and macrophages distribute across ependyma to the neuronal injury site for immuno-reparation. When this IFN-γ regulation of CP is inefficient or inoperative, the consequent imbalanced trafficking of specific leucocytes from CP-to-CSF-to brain may result in exacerbated neuroinflammatory/ degenerative states (Schwartz and Deczkowska 2016).

2.7.2.4 Cystatin C

Cystatin C is synthesized and secreted by the CP in response to brain injury such as in stroke and trauma. Transcript for cystatin C, a cysteine protease inhibitor, was evaluated in CP by Aldred et al. (1995). This regulatory protein in CSF is available for distribution across the CSF-brain (ependymal) interface. Cystatin C upregulates after seizures and ischemia, with its concentration increasing in hippocampus (Palm et al. 1995; Pirttila et al. 2005). A choroidal supply source of cystatin C presumably complements astrocytic synthesis in the injured brain. Cystatin C has a putative role in pruning/reorganizing neuronal networks damaged by physical injury or diminished blood flow. Neurogenesis and migration in periventricular/hippocampal regions is also positively associated with cystatin C availability (Pirttila et al. 2005).

2.7.2.5 Insulin-Like Growth Factors

Insulin-like growth factor-II (IGF-II), manufactured by the CPE, is upregulated after physical and biochemical insults to the CNS. Normally IGF-II and its binding protein IGFBP-2 are present mainly in mesenchymal support structures such as CP (Ocrant and Parmelee 1992). Following traumatic injury to the cerebral cortex, the CP expression of IGF-II and IGFBP-2 are augmented (Walter et al. 1999). This homeostatic response to a trauma distant from the CP, is presumably initiated by the flow of 'injury signals' (e.g., cytokines) to the BCSFB epithelium. Thus, during the first week (acute phase) following a penetrating injury of rat brain, transcripts for IGF-II and its binding protein increase in CP; and their respective proteins rise in CSF. IGF-II protein immunoreactivity, but not mRNA, increases at the site of injury. These phenomena fit the interpretation that soon after cortical trauma, the CSF carries newly-manufactured IGF-II from CP to the injury locus for healing. This IGF-II homeostatic action via CSF volume distribution is facilitated by the permeable ependyma (and pia-glia) allowing distribution of CP-synthesized hormone into wounded brain for repair.

IGF-I is actively transported from plasma to CSF across the CP. Neurons require IGF-1 for energy regulation, growth and for mediating cognitive processes (Carro et al. 2006a). Reduced peripheral IGF-I may contribute to neurological conditions such as Alzheimer's disease. Gradual systemic delivery of IGF-I in mice increases levels of synaptic proteins and enhances cognitive performance (Carro et al. 2006b). IGF-1 nuclear translocation in hippocampus is induced by TTR (Vieira et al. 2015).

Clearly, IGF-1 and TTR secretion by CP into CSF enable many neuroprotectant and trophic mechanisms.

2.7.2.6 Brain-Derived Neurotrophic Factor

Multiple investigators and cell preparations have demonstrated CP synthesis and secretion of BDNF into CSF/medium (Timmusk et al. 1995; Huang et al. 2014; Borlongan et al. 2004). Lateral ventricular injection of BDNF is neuroprotective (Sharma and Johanson 2007), and BDNF is one of many neurotrophins and growth factors secreted by CP that modulates neurogenesis and recovery from CNS insults (Johanson et al. 2011c). A dipeptide mimetic (GSB-106), that reproduces the homodimeric structure of BDNF, stimulates synaptogenesis (Gudasheva et al. 2018).

CSF levels of BDNF are reduced in mild cognitive impairment (Martin-de-Pablos et al. 2018), Alzheimer's disease (Du et al. 2018) and certain psychiatric disorders. Huntington's disease is exacerbated by BDNF depletion, thereby prompting consideration to use encapsulated cells or viral vectors in CSF to deliver neurotrophins such as BDNF (Savolainen et al. 2018). Motor neuron levels of BDNF mRNA are experimentally downregulated by toxic CSF from amyotrophic lateral sclerosis (ALS) patients (Shruthi et al. 2017); this adverse effect can be countered with supplemental BNDF. While CSF levels do not always reflect brain BDNF mRNA (Lanz et al. 2012), boosting CSF BDNF levels may benefit neurodegenerative diseases such as ALS.

There are exciting prospects for pharmacologically stimulating CP secretion of BDNF and other neurotrophins. Copolymer-1 is a neuroprotectant, immunomodulatory peptide used in multiple sclerosis. Copolymer-1 stimulates CP expression of BDNF, neurotrophin-3 and insulin-like growth factor-1 and this increased peptide expression correlates with enhanced neurogenesis (Cruz et al. 2018). Other CP-expressed peptides/proteins have recently been discovered: DRR1, a stress-induced protein that augments stress resilience (Masana et al. 2018), and prosaposin, a secretory neurotrophic factor and regulator of lysosomal enzymes (Nabeka et al. 2017). The continual discovery of new peptides to strengthen the homeostatic capabilities of CP enhances the prospects for translational neuromedicine (Johanson 2017).

2.7.2.7 Osmotic Solutes

Severe hyperosmolality carries high morbidity to the CNS; homeostatic mechanisms are essential for rapidly restoring isoosmolal balance between brain and hypertonic plasma. Net CSF-to-brain ion flux, e.g., Na^+ and Cl^-, takes place across the ependyma in response to plasma hyperosmolality (Pullen et al. 1987). Hyperosmotic blood draws H_2O out of the CNS. As the brain interstitial fluid volume consequently contracts, CSF compensatorily flows into the cerebral extracellular space to stabilize/

restore lost volume. CSF derived from CP is rich in Na^+ and Cl^-, as well as H_2O (99%). Accordingly, the BCSFB homeostatically (likely by the Na-K-Cl cotransporter) provides the osmotic solute and H_2O to replace the brain fluid withdrawn across the BBB by rising plasma osmolality. AVP, a H_2O-regulating hormone that modulates Na-K-Cl cotransport, is expressed in CP epithelium (Chodobski et al. 1997). Salt loading in the plasma upregulates AVP mRNA and its cognate V1 receptor in CP (Zemo and McCabe 2001) as well as hypothalamus; this hormonal response putatively alters ion-H_2O transport to attain osmolar balance between CNS and the periphery. Volume regulation, then, is one of many CP transport functions supporting the brain.

2.7.3 Solute Flux: Brain to CSF

Because the brain does not have the lymphatic system found in other tissues, it needs disposal mechanisms to eliminate acute ion excesses, unneeded catabolites, and proteinaeous injury debris. The ventricular CSF, and its resident CP tissues, are convenient repositories for collecting and eliminating waste materials. Therefore, the transfer dynamics allows CSF to mediate functions 'bidirectionally', across the ependyma, as a quasi-lymphatic clearance system as well as the nutrient supply source previously addressed. This transependymal versatility of the CP-CSF-brain nexus is amplified further by the functions of the paravascular/glymphatic system (see Sect. 2.9).

2.7.3.1 K^+, Ca^{++}, Creatinine and Amino Acid Spillover into CSF

CSF K^+ concentration, ~3 mmol/L, is lower than that in plasma. Although not a major component of CSF osmolality (~1/100th), K^+ is crucial for stabilizing CNS excitability, and CSF K^+ concentration is maintained constant during variations in plasma concentration. CSF K^+ is kept low by removal via the Na^+-K^+ pump and by 'buffering' from the CPE Na-K-Cl cotransport ion exchanges with CSF. Excessive brain-CSF K^+, e.g., from a high rate of neuronal firing (seizures) or uremia (Hise and Johanson 1979), is removed by efflux into blood by CP as well as BBB.

Similar to K^+, extracellular Ca^{2+} concentration is crucial for neuronal excitability, and the concentration of Ca^{2+} in CSF and brain ISF is maintained constant during hypercalcemia (Jones and Keep 1988). Normal CSF Ca^{2+} concentration is about 50% of plasma levels where half is bound to plasma proteins. Regulation of CSF Ca^{2+} appears to involve alterations in Ca^{2+} transport at the CPE apical as opposed to the basolateral membrane (Murphy et al. 1989).

Creatinine, a byproduct of brain creatine metabolism, diffuses across ependyma into CSF. Creatinine concentration in human CSF is about 50–70 μmol/L (Deignan et al. 2010; Johanson et al. 2018). Elevations in CSF creatinine have been associated with seizure initiation (Hiramatsu et al. 1988; Tachikawa and Hosoya 2011). The

BCSFB, not BBB, expresses transporters such as Oct3/SLC22A3 that actively remove (Anderson and Heisey 1975) excessive creatinine from CSF (Hosoya and Tachikawa 2011). Such creatinine removal from CSF by CP improves seizure control.

Almost all amino acid concentrations in CSF are just a fraction of their corresponding plasma levels (an exception being glutamine). The CPE expresses multiple transporters that clear amino acids from the ventricular fluid. For example, glutamate concentrations in brain ISF are finely controlled, and any harmful excess of cerebral extracellular glutamate can diffuse into CSF for active transport elimination at the choroidal BCSFB. There are three forms of the Na^+-dependent glutamate transporters in the CPE apical membrane: GLAST1a, GLAST1c and GLT1b that clear glutamate from CSF (Lee et al. 2012). In addition, an apical cystine-glutamate antiporter activity promotes glutathione (an anti-oxidant) formation in CP (Albrecht et al. 2010). Thus, the shuttling of glutamate from brain to CSF to CP mediates two actions: active removal of excess CNS glutamate from CSF to blood, and the formation of cell-protecting glutathione from the CP cycling of cystine and glutamate. Ethanol intoxication and thiamine toxicity impair glutamate clearance, BCSFB permeability and glutathione production (Nixon 2008). This CP breakdown in handling glutamate contributes to disorders such as Wernicke's periventricular encephalopathy.

2.7.3.2   Organic Anion and Cation Catabolites

Certain catabolites, such as products of biogenic amine metabolism, are released by cortical parenchyma into ECF-CSF (Min-Chu et al. 1981; Majchrzak et al. 1980). There, breakdown products such as 5-hydroxyindoleacetic acid (5-HIAA) from serotonin, and homovanillic acid (HVA) from dopamine, move along extracellular routes to the ependymal interface, and thence into CSF. Then, CSF bulk flow carries 5-HIAA and HVA to CP tissues (Huang and Wajda 1977) for active removal from the ventricles (Cserr and VanDyke 1971). Mitochondrial/energy failure in CP reduces HVA reabsorptive transport, causing CSF HVA concentration to rise (Batllori et al. 2018). In these energy-depleted patients (Mitro et al. 2018), the alternative clearance pathway for 5-HIAA and HVA, i.e., venous-lymphatic convection, evidently does not fully compensate for the defective CP organic anion removal capacity.

Of interest to pharmacologists and toxicologists are the many CPE and arachnoid (Zhang et al. 2018) transporters that remove drugs, catabolites and toxic substances from CSF (Fig. 2.6). The apical membrane at the BCSFB contains Oat and Oatp organic anion transporters, as well as the Oct2 and Oct3 organic cation transporters (Gibbs and Thomas 2005); these transport systems help to cleanse the CSF from foreign compounds (xenobiotics) and harmful products of metabolism. In addition, a basolateral ABC transporter, Mrp1, clears toxic organic compounds out of the CPE into the interstitial zone and thence to choroidal capillaries. An interesting difference

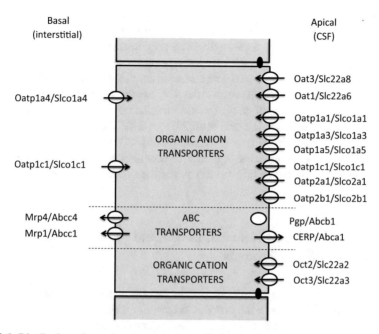

Fig. 2.6 Distribution of organic anion transporter (Oat), organic anion transporting polypeptide (Oatp), organic cation transporter (Oct) and ATP-binding cassette (ABC) transporter family members at the rodent CP epithelium. For the Oat, Oatp and Oct transporters, the protein and solute carrier (Slc) names are given. For the evidence on the distribution please see reviews Farthing and Sweet (2014), Stieger and Gao (2015) and Hartz and Bauer (2011). The wide range of organic anion transporters probably reflects the importance of such systems for regulating/removing products of brain metabolism and neuropeptides/hormones. Note that as well as Oct2/3, the CP also expresses the organic anion/carnitine transporter (Oct1/2), Slc22a4 and Slc22a5, but the subcellular location(s) remain unknown (Farthing and Sweet 2014). The location of p-glycoprotein (Pgp) in the epithelium is probably sub-apical, but there is controversy. *MRP* multidrug resistance-associated protein, *CERP* cholesterol efflux regulatory protein

between the BCSFB and the BBB is the localization of p-glycoprotein (Pgp; Fig. 2.6). In cerebral endothelial cells, Pgp is on the blood-facing membrane where it clears xenobiotics back into blood. Pgp is not present at the corresponding basolateral membrane of the CPE.

2.7.3.3 Denatured and Toxic Proteins

The CSF sink mechanism for excretory dispersal is threefold: (1) CSF bulk flow (sweeping effect of the CSF stream), (2) active removal of useless products by the arachnoid membrane (Zhang et al. 2018); (3) and active transport of molecules across CP epithelial cells at the BCSFB. Denatured proteins from distorted brain metabolism, and excessive leakage of plasma proteinaceous material across BBB, need removal. The flow of CSF, normally very low in protein concentration, acts as an effective hydrodynamic drain on unnecessary proteins. CSF streaming distally to sites of absorption from the SAS mediates dispersive clearance of macromolecules. Once in CSF, certain proteins are also actively removed and discarded by a 3-step transcytotic process: adsorptive apical endocytosis, protein transport through cytoplasm, and finally exocytosis across the CP basolateral membrane (Balin and Broadwell 1988). Clearance of neurotoxic proteins remains a major challenge in neurodegenerative diseases (Boland et al. 2018). CP-CSF formation and sink action diminish in Alzheimer's disease (Silverberg et al. 2001), contributing to protein accumulation. A key homeostasis question is whether CP protein adsorptive mechanisms (for removal) can be boosted to counter the decreased CSF turnover.

2.8 Choroid Plexus Generation of Fluid for Manifold Distribution Channels

For nearly a century, the conventional view has been that CP-formed CSF flows down the ventricular axis to the cisterna magna and other basal cisterns. From the cisterns, the CSF distributes along many SAS pathways that surround the brain and spinal cord. However, one group that disputes a major role of CP in CSF formation also challenges the *net flow* of CSF down the ventriculo-subarachnoid nexus to venous/lymphatic drainage sites (Oreskovic and Klarica 2014). Because the CSF origin and its flow routes are lynchpins in the thinking about CSF dynamics, we deem it important to recapitulate and re-emphasize the substantial evidence for CSF origin, circulation (flow) and egress.

2.8.1 Ventricular CSF Flow Across Velae/Foramina into Basal Cisterns and Nasal Lymph

The CPs can be regarded as the origin or 'headwaters' of a major fluid production locus within the CNS. In adult humans, the CSF generated by the four CPs in their respective ventricles moves quickly into various reaches of the subarachnoid system. The relatively large subarachnoid cisterns at the base of the brain (Fig. 2.7) collect CSF as it percolates down the neuraxis. The most proximal points of CSF origin are the two lateral ventricles. Even though there is gentle bidirectional movement of CSF within the ventricular system, responding dynamically to ever-changing hydrostatic pressure gradients (from rhythmic cardio-respiratory pulses and postural changes), the overall (net) movement of CSF is down the ventriculo-subarachnoid axis for clearance out of the CNS. From the basal cisterns, CSF moves down through the cribriform plate into the nasal submucosa-cervical lymphatics, and also slowly up over the cortical hemispheres for penetration of the Virchow-Robin spaces as the gate to brain perivascular flow.

NET CSF MOVEMENT FROM THE VENTRICULAR SYSTEM
OUT TO THE SUBARACHNOID SPACES AND DRAINAGE SITES

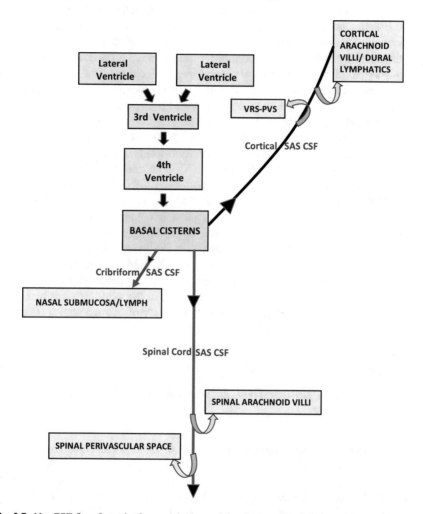

Fig. 2.7 Net CSF flow from the four cerebral ventricles down to the SAS pathways and drainage sites into lymph/venous blood: CSF formed by the CPs moves by convection to the basal cisterns, and thence outward to three major regions: (1) the cribriform plate-olfactory nerve distribution to the nasal submucosa-cervical lymphatics, (2) the cortical SAS CSF containing Virchow-Robin spaces (VRS) as an entry point to the perivascular system (PVS), and (3) the spinal cord SAS with a perivascular system to receive materials from the subarachnoidal CSF for delivery to the spinal parenchyma. CSF flowing up over the cortical hemispheres has access to the VRS, the arachnoid villi (normally a minor CSF outflow site into the venous sinuses), and perhaps also the lymphatic vessels in the dura mater. Dural lymph drainage runs to the cervical lymph glands. Additional anatomic/functional information is needed to clarify the dynamics of CSF flow into spinal PVS, and subsequent drainage into lymphatic regions surrounding the spinal cord and its exiting nerves. The CSF flow pathways depicted in this figure (for healthy humans) can be substantially altered in disorders such as hydrocephalus and Alzheimer's disease

Animal experimentation reveals that tracer ^{14}C-sucrose injected into a lateral ventricle rapidly accesses the basal cisterns (Fenstermacher et al. 1997). Within minutes, the ^{14}C-sucrose marker reaches the aqueduct and fourth ventricle, as well as velae (interpositum and superior medullary); the velae are extensions of the sub-arachnoid space (SAS) into regions adjacent to the third and fourth ventricles. Accordingly, by 5–10 min post-administration to an adult rat lateral ventricle, the ^{14}C-sucrose reaches the major basal cisterns (Fig. 2.7) either by the velar route or via the aqueduct-fourth ventricle-foramina nexus. Significantly, for glymphatic fluid modeling, the ^{14}C-sucrose penetrates (for >3 h) pial perivascular spaces of arteries/arterioles (Ghersi-Egea et al. 1996) by accessing cortex from SAS CSF.

Another set of convincing experiments, for CSF flow from cisternal SAS to nasal-cervical lymph, has been conducted by the Johnston group in Toronto. Following injection of ^{125}I-human serum albumin (HSA) in an adult rat lateral ventricle, the CSF-borne HSA tracer drains down to the olfactory turbinates (cribriform plate-to-nasal SAS-lymph), achieving a peak concentration there at 30 min (Nagra et al. 2006). Moreover, yellow microfil dye injected into the cisterna magna of adult monkeys distributes throughout the SAS, eventually draining into the olfactory submucosa (Johnston et al. 2005). Clearly, radio-tracers and dyes injected into ventricles or cisterna magna CSF distribute widely (and sequentially) through the ventriculo-SAS axis to drainage sites in lymph and venous blood. This long-appreciated CSF flow-through system, originating in the ventricles, has been char-acterized by numerous laboratories utilizing a variety of methodologies (Bradbury et al. 1981).

2.8.2 CSF Distribution from Basal Cisterns to Cortical/Spinal SAS Sites

CSF flow routes are anatomically complex, allowing very extensive penetration of CSF to outer reaches of the CNS. The basal cisterns (large pockets of SAS CSF at the brain base) contain a pool of CSF, dynamically available for distribution up over the brain (cortical SAS) or down to the spinal cord (spinal SAS). Basal cisterns (e.g., cisterna magna), contiguous with the proximal ventricular CSF, can be viewed hydrodynamically as a hub, a center from which CSF distribution radiates outward. Diseases or injuries affecting the basal cisterns and adjacent regions may substan-tially impact distribution of CP-derived substances to the most distal areas of CSF distribution.

2.8.2.1 Cisternal CSF Flow to Cortical Virchow-Robin Spaces

Intimately connected to questions on sites of CNS fluid production is the impact of CSF secretion by CP on perivascular fluid and solute flow within the brain parenchyma. There is longstanding evidence that large-cavity CSF acts as a sink to remove waste products/potential toxins from brain (Oldendorf and Davson 1967). This occurs different ways; one is by molecular waste diffusion across the ependyma into ventricular CSF, the other across the glia limitans-pia mater to SAS CSF (Oldendorf and Davson 1967). However, another recently proposed mechanism is a brain glymphatic system (Iliff et al. 2012). This includes a perivascular flow of SAS CSF into the brain along penetrating arterioles, astrocyte AQP-4-mediated fluid flow within cerebral gray matter and a putative perivenous outflow of fluid back to the SAS. The glymphatic system provides a fluid flow dynamic for an efficient 'inside-the-brain sink' to remove toxins internally within neural-glial tissue (Iliff et al. 2012). The egress of glymphatic fluid, from brain to SAS CSF and beyond, is under investigation; the dural lymphatic vessels likely play a role in CSF movement to the cervical lymph glands (Aspelund et al. 2015). In any event, the glymphatic system ultimately relies on sink action of CSF flow (thus basically upon CP-CSF production), eventually draining into cervical lymph.

2.8.2.2 Cisternal CSF Flow to the Spinal Cord Perivascular Space

Volume transmission of CSF to the perivascular spaces of the spinal cord is drawing attention as an unrecognized flow pathway for CNS fluids needing more investigation. Tracers injected into the spinal SAS were found in white and gray matter to define perivascular spaces around arterioles and venules, but not spinal capillaries (Lam et al. 2017). By 5 min post-injection, the tracer was also found in the spinal cord ECF and in blood vessels (suggesting trans-vascular clearance). This finding indicates that the rat spinal SAS CSF has rapid access to many compartments in the cord. That report (Nagra et al. 2006) provides an anatomical/ultrastructural substrate for how M2 macrophage trafficking across CP, and subsequent CSF volume transmission, help to heal spinal cord injuries distant from the BCSFB (Shechter et al. 2013); and how ascorbate transported across CP (but not brain/cord capillaries) can support cellular elements of the cord that require ample vitamin C (Zarebkohan et al. 2009).

2.9 Putative Functional Implications of CP-CSF for the Glymphatic System

A fascinating question is: how does CP secretion of CSF relate to 'distant' brain/cord parenchymal hydrodynamics and solute transport kinetics? The CP secretory machine, together with CSF pulsations (waveforms resulting from cardiac systole and respiration), help to drive CSF to the basal cisternal subarachnoid spaces. Consequently, the CSF volume and pressure pulses, generated by the CP tissues and pulsating peripheral organs, provide the essential hydrodynamic underpinning to support the glymphatic system.

CSF entering the Virchow-Robin spaces and perivascular channels is proposed eventually to penetrate the astrocytic feet-AQP4 complex (inside the BBB) to gain access to brain interstitium. Hypothetically, this trans-astrocyte-aquaporin flux into brain interstitium provides a mechanism to deliver nutrients (from CP), and to carry away waste material (such as Aβ) for eventual elimination in downstream cervical lymph.

An hemodynamic/hydrodynamic analysis of possible driving forces that create CSF movement is pertinent here. Although cardiac systolic force helps to drive capillary fluid across permeable capillaries (the hydrostatic pressure factor) into ISF (as in muscle), this force from heart pumping cannot significantly drive fluid across BBB tight junctions (low hydraulic conductance) into the interstitium (Nakada and Kwee 2019). Alternatively, by a more circuitous route to the brain ISF, the ample fluid secretion (active) across CP into CSF creates a hydrostatic pressure head to drive fluid down the ventriculo-SAS system; fluid movement down the ventricles is also assisted by the beating ependymal cilia. Moreover, the physical structure of CP villi with pulsating arterioles (from systole) generates a CSF pulse wave (Bering Jr 1955) for transmission through the choroidal-ventriculo-subarachnoid system. Cooperatively, the CP secretory/pulsating factors, together with the ependymal ciliary propulsive effect, likely generate a combined forward force to drive CSF into the subarachnoidal Virchow-Robin spaces for access to brain ISF.

CSF flow from CP is biochemically as well as biophysically significant. Intimately connected to the issue of CNS fluid production sites, and CSF percolation pathways, is the potential impact of CP homeostatic secretion on CSF-borne solute conveyance into the brain interior. Because BBB does not transport vitamin C and other substances, it is a significant that CP-derived substances such as vitamin C, transthyretin and IGF-1 likely penetrate deep into brain by the CSF-perivascular flow system (starting at the Virchow-Robin spaces).

The proposed brain glymphatic system model (Iliff et al. 2012; Iliff and Nedergaard 2013) provides for a brisk 'flux' (fluid throughput) enabling an efficient sink to remove toxins, such as β-amyloid. Unless there is a direct connection between the venous perivascular system and a non-CSF drainage site (such as the dural membrane), the downstream clearance of toxins by the glymphatic system ultimately relies on the sink action of CSF flow [and thus ultimately the upstream CSF production driving the 'third circulation' originally proposed by Harvey Cushing (Black 1999)].

One of many roles of CP is to regulate CSF composition, e.g., the K^+ (Johanson et al. 1974), Mg^{2+} (Reed and Yen 1978) and Ca^{2+} (Jones and Keep 1988) concentrations. It should be noted that the proposed rapid perivascular movement of CSF into brain (near Virchow-Robin spaces) stresses the importance for such CSF regulation. Failure of CP to regulate CSF ions and micronutrients (Spector and Johanson 2014) quickly would destabilize brain parenchyma and synapses.

Another substantial function of CP-CSF is to mediate an endocrine-like task of distributing solutes over relatively long distances between the secreting cells (CPE) and targets (e.g., hypothalamic nuclei). Volume transmission of CSF was originally ascribed largely to periventricular regions (Konsman et al. 2000), e.g., neurogenesis (Mudò et al. 2009). However, now with the developing glymphatic paradigm, CSF distribution modeling may need marked adjustment to accommodate the factor of CSF circulating within the cerebral parenchyma. This could have striking implications for understanding brain-wide information transfer by signaling molecules. The emerging role of the perivascular glymphatics in overall CSF circulation also affords new facets of delivery for neuroendocrine agents and medicinals to the brain interior.

There is still controversy about proposed histological/anatomical elements for the glymphatic system- and how they relate to flow mechanisms (Abbott et al. 2018; Smith and Verkman 2018). Areas of lively debate include the role of astrocytes and AQP4 in parenchymal fluid flow vs. diffusion; whether brain capillaries have a vascular space that participates in CSF flow through the parenchyma; and the nature (even existence) of perivenular glymphatic drainage pathways. Although bulk CSF flow occurs in the interstitium of white matter, it very well may not in gray matter.

Another major issue in the glymphatics debate is the driving force for perivascular CSF flow. Is it ICP? The ability of both acetazolamide-induced reduction (Lundgaard et al. 2017) of ICP and neurosurgical cisternostomy (Cherian et al. 2016) to curtail CSF perivascular flow inside brain, and the clearance of lactate, strongly implicate an ICP influence on glymphatic efficiency. On the other hand, the possible linkage between intracranial hypertension (pseudotumor cerebri) and glymphatic stagnation (Bezerra et al. 2017) suggests that substantially-elevated ICP may slow down CSF outflow from within the brain. Accordingly, an optimal ICP range (and CSF hydrostatic pressure gradient) may be essential for healthy glymphatic fluid dynamics.

2.10 Conclusions

In the current era of increasingly-powerful analytical tools for pin-pointing transport and metabolic functions, by way of CP gene ablations and disease transcriptomics, abundant opportunities are arising to manipulate the BCSFB for therapeutic benefit. A long-term goal in the clinical CSF field has been to control more effectively the fluid dynamics in hydrocephalus, intracranial hypertension and cerebral edema. As more light is shed on CP H_2O and ion movement, e.g., via the newly-characterized HCO_3 cotransporters, prospects should improve in regard to fine tuning CSF pH and fluid production.

The main locus of CSF formation in CP, and the flow from this secretory epithelium down the ventriculo-subarachnoid axis, have long been recognized as essential factors in CSF dynamics. As new BCSFB transport data emerge, this CSF dynamics model is strengthened. Whether designated as CSF flow, movement or circulation, it seems likely that the CNS contains orderly, circumscribed-pathways for moving a *net* volume of fluid towards/into the SAS for clearance into lymphatic/venous egress sites. The alternate theory, that CSF is formed mainly at the neural capillaries due to profuse H_2O exchange across BBB, lacks supporting evidence for substantial, *net* fluid movement across BBB into the interstitium (including acetazolamide-sensitive Na^+ and HCO_3^- transport). These criteria, however, have already been fulfilled for CP as the preponderant *fluid* generator.

New secretory/excretory roles are continuously being revealed for the CP-CSF system. One of the most intriguing novel findings is the circadian modulation of transport phenomena at the BCSFB. The CP itself expresses clock genes that may engender a 'cell autonomous' circadian rhythm. Information about diurnal rhythms in hormone secretion, metabolite clearance from CSF by CP, and even CSF formation itself, will almost most certainly have application to glymphatic system modeling, as glymphatic flow increases in the brain nocturnally.

The importance of CP-generated CSF and intraventricular pressure for glymphatic function is just starting to gain appreciation. Attenuation of CSF volume and ICP, by acetazolamide and by cisternostomy, presumably points to essential fluid output from CP impacting CSF flow through the perivascular spaces. This working hypothesis is worthy of further testing, due to the potential significance of CP-secreted hormones, vitamins, trace elements and growth factors in accessing the brain interior by the perivascular system. Clearly one of the main challenges (and opportunities) in the area of CNS fluids/glymphatics, is to elucidate how the CP-CSF secretions penetrate the depths of the brain.

Finally, as more laboratories analyze BCSFB and BBB in the same experimentation, it will become expedient in the barrier field to merge transport/fluid data into more efficacious modeling of phenomena for brain health and disease (Johanson and Johanson 2016). In addition, the CSF-brain interface, especially in the cortical SAS, promises to be an exciting region of interest for advancing the CSF-glymphatics topic. Thus, transport modeling for the CSF-brain system is entering an era in which data for all CNS transport interfaces need to be integrated for optimal insight and therapeutic outcome.

Acknowledgments This study was supported by grants to RFK: NS-093399, and NS-106746 from the National Institutes of Health (NIH). NIH funding to CEJ, as NS-27601 and NIA-AG027910, helped to produce some of the information and concepts in this chapter.

Conflicts of Interest The authors have no conflicts of interest to declare.

References

Abbott NJ, Pizzo ME, Preston JE, Janigro D, Thorne RG (2018) The role of brain barriers in fluid movement in the CNS: is there a 'glymphatic' system? Acta Neuropathol 135(3):387–407

Adamson RH, Lenz JF, Zhang X, Adamson GN, Weinbaum S, Curry FE (2004) Oncotic pressures opposing filtration across non-fenestrated rat microvessels. J Physiol 557(Pt 3):889–907

Adamson SX, Shen X, Jiang W, Lai V, Wang X, Shannahan JH et al (2018) Subchronic manganese exposure impairs neurogenesis in the adult rat hippocampus. Toxicol Sci 163(2):592–608

Agnati LF, Bjelke B, Fuxe K (1995) Volume versus wiring transmission in the brain: a new theoretical frame for neuropsychopharmacology. Med Res Rev 15(1):33–45

Akdemir G, Luer MS, Dujovny M, Misra M (1997) Intraventricular atrial natriuretic peptide for acute intracranial hypertension. Neurol Res 19(5):515–520

Albrecht P, Lewerenz J, Dittmer S, Noack R, Maher P, Methner A (2010) Mechanisms of oxidative glutamate toxicity: the glutamate/cystine antiporter system xc- as a neuroprotective drug target. CNS Neurol Disord Drug Targets 9(3):373–382

Aldred AR, Brack CM, Schreiber G (1995) The cerebral expression of plasma protein genes in different species. Comp Biochem Physiol B Biochem Mol Biol 111(1):1–15

Allonen H, Anderson KE, Iisalo E, Kanto J, Stromblad LG, Wettrell G (1977) Passage of digoxin into cerebrospinal fluid in man. Acta Pharmacol Toxicol 41(3):193–202

Andersen HH, Johnsen KB, Moos T (2014) Iron deposits in the chronically inflamed central nervous system and contributes to neurodegeneration. Cell Mol Life Sci 71(9):1607–1622

Anderson DK, Heisey SR (1975) Creatinine, potassium, and calcium flux from chicken cerebro-spinal fluid. Am J Phys 228(2):415–419

Anonymous (1998) International randomised controlled trial of acetazolamide and furosemide in posthaemorrhagic ventricular dilatation in infancy. International PHVD Drug Trial Group. Lancet (London, England) 352(9126):433–440

Arregui A, Iversen LL (1978) Angiotensin-converting enzyme: presence of high activity in choroid plexus of mammalian brain. Eur J Pharmacol 52(1):147–150

Aspelund A, Antila S, Proulx ST, Karlsen TV, Karaman S, Detmar M et al (2015) A dural lymphatic vascular system that drains brain interstitial fluid and macromolecules. J Exp Med 212(7):991–999

Balin BJ, Broadwell RD (1988) Transcytosis of protein through the mammalian cerebral epithelium and endothelium. I. Choroid plexus and the blood-cerebrospinal fluid barrier. J Neurocytol 17 (6):809–826

Batllori M, Molero-Luis M, Ormazabal A, Montero R, Sierra C, Ribes A et al (2018) Cerebrospinal fluid monoamines, pterins, and folate in patients with mitochondrial diseases: systematic review and hospital experience. J Inherit Metab Dis 41:1147–1158

Bergen AA, Kaing S, ten Brink JB, Gorgels TG, Janssen SF, Bank NB (2015) Gene expression and functional annotation of human choroid plexus epithelium failure in Alzheimer's disease. BMC Genomics 16:956

Bering EA Jr (1955) Choroid plexus and arterial pulsation of cerebrospinal fluid; demonstration of the choroid plexuses as a cerebrospinal fluid pump. AMA Arch Neurol Psychiatry 73 (2):165–172

Betz AL, Firth JA, Goldstein GW (1980) Polarity of the blood-brain barrier: distribution of enzymes between the luminal and antiluminal membranes of brain capillary endothelial cells. Brain Res 192(1):17–28

Bezerra MLS, Ferreira ACAF, de Oliveira-Souza R (2017) Pseudotumor cerebri and glymphatic dysfunction. Front Neurol 8:734

Birzis L, Carter CH, Maren TH (1958) Effects of acetazolamide on CSF pressure and electrolytes in hydrocephalus. Neurology 8(7):522–528

Black PM (1999) Harvey Cushing at the Peter Bent Brigham Hospital. Neurosurgery 45 (5):990–1001

Boland B, Yu WH, Corti O, Mollereau B, Henriques A, Bezard E et al (2018) Promoting the clearance of neurotoxic proteins in neurodegenerative disorders of ageing. Nat Rev Drug Discov 17:660–688

Borlongan CV, Skinner SJ, Geaney M, Vasconcellos AV, Elliott RB, Emerich DF (2004) Intrace-rebral transplantation of porcine choroid plexus provides structural and functional neuroprotection in a rodent model of stroke. Stroke 35(9):2206–2210

Boyson SJ, Alexander A (1990) Net production of cerebrospinal fluid is decreased by SCH-23390. Ann Neurol 27(6):631–635

Bradbury MW, Cserr HF, Westrop RJ (1981) Drainage of cerebral interstitial fluid into deep cervical lymph of the rabbit. Am J Phys 240(4):F329–F336

Brinker T, Stopa E, Morrison J, Klinge P (2014) A new look at cerebrospinal fluid circulation. Fluids Barriers CNS 11:10

Cains S, Shepherd A, Nabiuni M, Owen-Lynch PJ, Miyan J (2009) Addressing a folate imbalance in fetal cerebrospinal fluid can decrease the incidence of congenital hydrocephalus. J Neuropathol Exp Neurol 68(4):404–416

Carro E, Trejo JL, Spuch C, Bohl D, Heard JM, Torres-Aleman I (2006a) Blockade of the insulin-like growth factor I receptor in the choroid plexus originates Alzheimer's-like neuropathology in rodents: new cues into the human disease? Neurobiol Aging 27(11):1618–1631

Carro E, Trejo JL, Gerber A, Loetscher H, Torrado J, Metzger F et al (2006b) Therapeutic actions of insulin-like growth factor I on APP/PS2 mice with severe brain amyloidosis. Neurobiol Aging 27(9):1250–1257

Cherian I, Bernardo A, Grasso G (2016) Cisternostomy for traumatic brain injury: pathophysiologic mechanisms and surgical technical notes. World Neurosurg 89:51–57

Chodobski A, Szmydynger-Chodobska J, Vannorsdall MD, Epstein MH, Johanson CE (1994) AT1 receptor subtype mediates the inhibitory effect of central angiotensin II on cerebrospinal fluid formation in the rat. Regul Pept 53(2):123–129

Chodobski A, Loh YP, Corsetti S, Szmydynger-Chodobska J, Johanson CE, Lim YP et al (1997) The presence of arginine vasopressin and its mRNA in rat choroid plexus epithelium. Brain Res Mol Brain Res 48(1):67–72

Chodobski A, Szmydynger-Chodobska J, Johanson CE (1998) Vasopressin mediates the inhibitory effect of central angiotensin II on cerebrospinal fluid formation. Eur J Pharmacol 347 (2–3):205–209

Chodobski A, Szmydynger-Chodobska J, Johanson CE (1999) Angiotensin II regulates choroid plexus blood flow by interacting with the sympathetic nervous system and nitric oxide. Brain Res 816(2):518–526

Christensen HL, Nguyen AT, Pedersen FD, Damkier HH (2013) Na(+) dependent acid-base transporters in the choroid plexus; insights from slc4 and slc9 gene deletion studies. Front Physiol 4:304

Christensen HL, Barbuskaite D, Rojek A, Malte H, Christensen IB, Fuchtbauer AC et al (2018) The choroid plexus sodium-bicarbonate cotransporter NBCe2 regulates mouse cerebrospinal fluid pH. J Physiol 596:4709–4728

Clardy SL, Wang X, Boyer PJ, Earley CJ, Allen RP, Connor JR (2006) Is ferroportin-hepcidin signaling altered in restless legs syndrome? J Neurol Sci 247(2):173–179

Cruz Y, García EE, Gálvez JV, Arias-Santiago SV, Carvajal HG, Silva-García R et al (2018) Release of interleukin-10 and neurotrophic factors in the choroid plexus: possible inductors of neurogenesis following copolymer-1 immunization after cerebral ischemia. Neural Regen Res 13(10):1743–1752

Cserr HF (1971) Physiology of the choroid plexus. Physiol Rev 51(2):273–311

Cserr HF (1988) Role of secretion and bulk flow of brain interstitial fluid in brain volume regulation. Ann N Y Acad Sci 529:9–20

Cserr HF, VanDyke DH (1971) 5-Hydroxyindoleacetic acid accumulation by isolated choroid plexus. Am J Phys 220(3):718–723

Curry FE, Frokjaer-Jensen J (1984) Water flow across the walls of single muscle capillaries in the frog, *Rana pipiens*. J Physiol 350:293–307

Dai YB, Wu WF, Huang B, Miao YF, Nadarshina S, Warner M et al (2016) Liver X receptors regulate cerebrospinal fluid production. Mol Psychiatry 21(6):844–856

Damkier HH, Brown PD, Praetorius J (2013) Cerebrospinal fluid secretion by the choroid plexus. Physiol Rev 93(4):1847–1892

Davson H, Segal MB (1970) The effects of some inhibitors and accelerators of sodium transport on the turnover of 22Na in the cerebrospinal fluid and the brain. J Physiol 209(1):131–153

Deignan JL, De Deyn PP, Cederbaum SD, Fuchshuber A, Roth B, Gsell W et al (2010) Guanidino compound levels in blood, cerebrospinal fluid, and post-mortem brain material of patients with argininemia. Mol Genet Metab 100(Suppl 1):S31–S36

Del Bigio MR (1995) The ependyma: a protective barrier between brain and cerebrospinal fluid. Glia 14(1):1–13

Doczi T, Joo F, Vecsernyes M, Bodosi M (1988) Increased concentration of atrial natriuretic factor in the cerebrospinal fluid of patients with aneurysmal subarachnoid hemorrhage and raised intracranial pressure. Neurosurgery 23(1):16–19

Du Y, Wu HT, Qin XY, Cao C, Liu Y, Cao ZZ et al (2018) Postmortem brain, cerebrospinal fluid, and blood neurotrophic factor levels in Alzheimer's disease: a systematic review and meta-analysis. J Mol Neurosci 65(3):289–300

Engelhardt B, Ransohoff RM (2012) Capture, crawl, cross: the T cell code to breach the blood-brain barriers. Trends Immunol 33(12):579–589

Faraci FM, Heistad DD (1992) Does basal production of nitric oxide contribute to regulation of brain-fluid balance? Am J Phys 262(2 Pt 2):H340–H344

Faraci FM, Mayhan WG, Heistad DD (1990) Effect of vasopressin on production of cerebrospinal fluid: possible role of vasopressin (V1)-receptors. Am J Phys 258(1 Pt 2):R94–R98

Farthing CA, Sweet DH (2014) Expression and function of organic cation and anion transporters (SLC22 family) in the CNS. Curr Pharm Des 20(10):1472–1486

Fenstermacher JD, Johnson JA (1966) Filtration and reflection coefficients of the rabbit blood-brain barrier. Am J Phys 211(2):341–346

Fenstermacher JD, Ghersi-Egea JF, Finnegan W, Chen JL (1997) The rapid flow of cerebrospinal fluid from ventricles to cisterns via subarachnoid velae in the normal rat. Acta Neurochir Suppl 70:285–287

Fraser PA, Dallas AD, Davies S (1990) Measurement of filtration coefficient in single cerebral microvessels of the frog. J Physiol 423:343–361

Gath U, Hakvoort A, Wegener J, Decker S, Galla HJ (1997) Porcine choroid plexus cells in culture: expression of polarized phenotype, maintenance of barrier properties and apical secretion of CSF-components. Eur J Cell Biol 74(1):68–78

Gerstberger R, Schütz H, Luther-Dyroff D, Keil R, Simon E (1992) Inhibition of vasopressin and aldosterone release by atrial natriuretic peptide in conscious rabbits. Exp Physiol 77(4):587–600

Ghersi-Egea JF, Finnegan W, Chen JL, Fenstermacher JD (1996) Rapid distribution of intraventricularly administered sucrose into cerebrospinal fluid cisterns via subarachnoid velae in rat. Neuroscience 75(4):1271–1288

Gibbs JE, Thomas SA (2005) Choroid plexus and drug therapy for AIDS encephalopathy. In: Zheng W, Chodobski A (eds) The blood-cerebrospinal fluid barrier. CRC/Taylor & Francis Group, Boca Raton, FL, pp 391–411

González Burgos GR, Rosenstein RE, Cardinali DP (1994) Daily changes in presynaptic cholinergic activity of rat sympathetic superior cervical ganglion. Brain Res 636(2):181–186

Gregoriades JMC, Madaris A, Alvarez FJ, Alvarez-Leefmans FJ (2019) Genetic and pharmacological inactivation of apical Na+-K+-2Cl- cotransporter 1 in choroid plexus epithelial cells reveals the physiological function of the cotransporter. Am J Physiol Cell Physiol 316(4):C525–CC44

Groger N, Vitzthum H, Frohlich H, Kruger M, Ehmke H, Braun T et al (2012) Targeted mutation of SLC4A5 induces arterial hypertension and renal metabolic acidosis. Hum Mol Genet 21 (5):1025–1036

Grove KL, Goncalves J, Picard S, Thibault G, Deschepper CF (1997) Comparison of ANP binding and sensitivity in brains from hypertensive and normotensive rats. Am J Phys 272(4 Pt 2): R1344–R1353

Gudasheva TA, Povarnina PY, Antipova TA, Seredenin SB (2018) Dipeptide mimetic of the BDNF GSB-106 with antidepressant-like activity stimulates synaptogenesis. Dokl Biochem Biophys 481(1):225–227

Guerreiro PM, Bataille AM, Parker SL, Renfro JL (2014) Active removal of inorganic phosphate from cerebrospinal fluid by the choroid plexus. Am J Physiol Renal Physiol 306(11):F1275–F1284

Haj-Yasein NN, Vindedal GF, Eilert-Olsen M, Gundersen GA, Skare O, Laake P et al (2011) Glial-conditional deletion of aquaporin-4 (Aqp4) reduces blood-brain water uptake and confers barrier function on perivascular astrocyte endfeet. Proc Natl Acad Sci U S A 108 (43):17815–17820

Hakvoort A, Haselbach M, Wegener J, Hoheisel D, Galla HJ (1998) The polarity of choroid plexus epithelial cells in vitro is improved in serum-free medium. J Neurochem 71(3):1141–1150

Hartig PR, Hoffman BJ, Kaufman MJ, Hirata F (1990) The 5-HT1C receptor. Ann N Y Acad Sci 600:149–166. discussion 66–67

Hartz AM, Bauer B (2011) ABC transporters in the CNS – an inventory. Curr Pharm Biotechnol 12 (4):656–673

Haywood JR, Vogh BP (1979) Some measurements of autonomic nervous system influence on production of cerebrospinal fluid in the cat. J Pharmacol Exp Ther 208(2):341–346

Herbute S, Oliver J, Davet J, Viso M, Ballard RW, Gharib C et al (1994) ANP binding sites are increased in choroid plexus of SLS-1 rats after 9 days of spaceflight. Aviat Space Environ Med 65(2):134–138

Herrick-Davis K, Grinde E, Lindsley T, Teitler M, Mancia F, Cowan A et al (2015) Native serotonin 5-HT2C receptors are expressed as homodimers on the apical surface of choroid plexus epithelial cells. Mol Pharmacol 87(4):660–673

Hiramatsu M, Edamatsu R, Fujikawa N, Shirasu A, Yamamoto M, Suzuki S et al (1988) Measurement during convulsions of guanidino compound levels in cerebrospinal fluid collected with a catheter inserted into the cisterna magna of rabbits. Brain Res 455(1):38–42

Hise MA, Johanson CE (1979) The sink action of cerebrospinal fluid in uremia. Eur Neurol 18 (5):328–337

Hladky SB, Barrand MA (2014) Mechanisms of fluid movement into, through and out of the brain: evaluation of the evidence. Fluids Barriers CNS 11(1):26

Hladky SB, Barrand MA (2016) Fluid and ion transfer across the blood-brain and blood-cerebrospinal fluid barriers; a comparative account of mechanisms and roles. Fluids Barriers CNS 13(1):19

Hosoya K, Tachikawa M (2011) Roles of organic anion/cation transporters at the blood-brain and blood-cerebrospinal fluid barriers involving uremic toxins. Clin Exp Nephrol 15(4):478–485

Hu Y, Shen H, Keep RF, Smith DE (2007) Peptide transporter 2 (PEPT2) expression in brain protects against 5-aminolevulinic acid neurotoxicity. J Neurochem 103(5):2058–2065

Huang JT, Wajda IJ (1977) The effects of morphine on the accumulation of homovanillic and 5-hydroxyindoleacetic acids in the choroid plexus of rats. Br J Pharmacol 60(3):363–367

Huang SL, Wang J, He XJ, Li ZF, Pu JN, Shi W (2014) Secretion of BDNF and GDNF from free and encapsulated choroid plexus epithelial cells. Neurosci Lett 566:42–45

Iliff JJ, Nedergaard M (2013) Is there a cerebral lymphatic system? Stroke 44(6 Suppl 1):S93–S95

Iliff JJ, Wang M, Liao Y, Plogg BA, Peng W, Gundersen GA et al (2012) A paravascular pathway facilitates CSF flow through the brain parenchyma and the clearance of interstitial solutes, including amyloid beta. Sci Transl Med 4(147):147ra11

Israel A, Garrido MR, Barbella Y, Becemberg I (1988) Rat atrial natriuretic peptide (99-126) stimulates guanylate cyclase activity in rat subfornical organ and choroid plexus. Brain Res Bull 20(2):253–256

Jacobs S, Ruusuvuori E, Sipila ST, Haapanen A, Damkier HH, Kurth I et al (2008) Mice with targeted Slc4a10 gene disruption have small brain ventricles and show reduced neuronal excitability. Proc Natl Acad Sci U S A 105(1):311–316

Janssen SF, van der Spek SJ, Ten Brink JB, Essing AH, Gorgels TG, van der Spek PJ et al (2013) Gene expression and functional annotation of the human and mouse choroid plexus epithelium. PLoS One 8(12):e83345

Jiang H, Hu Y, Keep RF, Smith DE (2009) Enhanced antinociceptive response to intracerebroventricular kyotorphin in Pept2 null mice. J Neurochem 109(5):1536–1543

Johanson CE (1984) Differential effects of acetazolamide, benzolamide and systemic acidosis on hydrogen and bicarbonate gradients across the apical and basolateral membranes of the choroid plexus. J Pharmacol Exp Ther 231(3):502–511

Johanson CE (1988) The choroid plexus-arachnoid-cerebrospinal fluid system. In: Boulton A, Baker G, Walz W (eds) Neuromethods: neuronal microenvironment- electrolytes and water spaces, vol 9. Humana, Clifton, NJ, pp 33–104

Johanson CE (2015) Fluid-forming functions of the choroid plexus: what is the role of aquaporin-1? In: Dorovini-Zis K (ed) The blood-brain barrier in health and disease, vol 1: Morphology, biology and immune function. CRC, Boca Raton, pp 140–171

Johanson CE (2017) Choroid plexus–cerebrospinal fluid transport dynamics: support of brain health and a role in neurotherapeutics. In: Conn PM (ed) Conn's translational neuroscience. Elsevier, Academic, pp 233–261

Johanson C, Johanson N (2016) Merging transport data for choroid plexus with blood-brain barrier to model CNS homeostasis and disease more effectively. CNS Neurol Disord Drug Targets 15 (9):1151–1180

Johanson C, Woodbury D (1974) Changes in CSF flow and extracellular space in the developing rat. In: Vernadakis A, Weiner N (eds) Drugs and the developing brain. Plenum, New York, pp 281–287

Johanson CE, Reed DJ, Woodbury DM (1974) Active transport of sodium and potassium by the choroid plexus of the rat. J Physiol 241(2):359–372

Johanson CE, Preston JE, Chodobski A, Stopa EG, Szmydynger-Chodobska J, McMillan PN (1999a) AVP V1 receptor-mediated decrease in cl- efflux and increase in dark cell number in choroid plexus epithelium. Am J Phys 276(1 Pt 1):C82–C90

Johanson CE, Szmydynger-Chodobska J, Chodobski A, Baird A, McMillan P, Stopa EG (1999b) Altered formation and bulk absorption of cerebrospinal fluid in FGF-2-induced hydrocephalus. Am J Phys 277(1 Pt 2):R263–R271

Johanson CE, Donahue JE, Spangenberger A, Stopa EG, Duncan JA, Sharma HS (2006) Atrial natriuretic peptide: its putative role in modulating the choroid plexus-CSF system for intracranial pressure regulation. Acta Neurochir Suppl 96:451–456

Johanson CE, Duncan JA III, Klinge PM, Brinker T, Stopa EG, Silverberg GD (2008) Multiplicity of cerebrospinal fluid functions: new challenges in health and disease. Cerebrospinal Fluid Res 5:10

Johanson CE, Stopa E, McMillan PN (2011a) The blood-cerebrospinal fluid barrier: structure and functional significance. Methods Mol Biol 686:101–131

Johanson CE, Stopa E, McMillan PN, Roth DR, Funk J, Krinke G (2011b) The distributional nexus of choroid plexus to CSF, ependyma and brain: toxicologic/pathologic phenomena, periventricular destabilization and lesion spread. Toxicol Pathol 39(1):186–212

Johanson C, Stopa E, Baird A, Sharma H (2011c) Traumatic brain injury and recovery mechanisms: peptide modulation of periventricular neurogenic regions by the choroid plexus-CSF nexus. J Neural Transm (Vienna) 118(1):115–133

Johanson CE, Stopa EG, Daiello L, de la Monte S, Keane M, Ott B (2018) Disrupted blood-CSF barrier to urea and creatinine in mild cognitive impairment and Alzheimer's disease. J Alzheimers Dis Parkinsonism 8:2

Johnston M, Zakharov A, Koh L, Armstrong D (2005) Subarachnoid injection of microfil reveals connections between cerebrospinal fluid and nasal lymphatics in the non-human primate. Neuropathol Appl Neurobiol 31(6):632–640

Jones HC, Keep RF (1988) Brain fluid calcium concentration and response to acute hypercalcaemia during development in the rat. J Physiol 402:579–593

Kamal MA, Keep RF, Smith DE (2008) Role and relevance of PEPT2 in drug disposition, dynamics, and toxicity. Drug Metab Pharmacokinet 23(4):236–242

Kao L, Kurtz LM, Shao X, Papadopoulos MC, Liu L, Bok D et al (2011) Severe neurologic impairment in mice with targeted disruption of the electrogenic sodium bicarbonate cotransporter NBCe2 (Slc4a5 gene). J Biol Chem 286(37):32563–32574

Karimy JK, Zhang J, Kurland DB, Theriault BC, Duran D, Stokum JA et al (2017) Inflammation-dependent cerebrospinal fluid hypersecretion by the choroid plexus epithelium in posthemorrhagic hydrocephalus. Nat Med 23(8):997–1003

Keep RF, Jones HC (1990) A morphometric study on the development of the lateral ventricle choroid plexus, choroid plexus capillaries and ventricular ependyma in the rat. Brain Res 56 (1):47–53

Keep RF, Smith DE (2011) Choroid plexus transport: gene deletion studies. Fluids Barriers CNS 8 (1):26

Konsman JP, Tridon V, Dantzer R (2000) Diffusion and action of intracerebroventricularly injected interleukin-1 in the CNS. Neuroscience 101(4):957–967

Kunis G, Baruch K, Rosenzweig N, Kertser A, Miller O, Berkutzki T et al (2013) IFN-gamma-dependent activation of the brain's choroid plexus for CNS immune surveillance and repair. Brain 136(Pt 11):3427–3440

Lam MA, Hemley SJ, Najafi E, Vella NGF, Bilston LE, Stoodley MA (2017) The ultrastructure of spinal cord perivascular spaces: implications for the circulation of cerebrospinal fluid. Sci Rep 7 (1):12924

Lanz TA, Bove SE, Pilsmaker CD, Mariga A, Drummond EM, Cadelina GW et al (2012) Robust changes in expression of brain-derived neurotrophic factor (BDNF) mRNA and protein across the brain do not translate to detectable changes in BDNF levels in CSF or plasma. Biomarkers 17(6):524–531

Lee A, Anderson AR, Rayfield AJ, Stevens MG, Poronnik P, Meabon JS et al (2012) Localisation of novel forms of glutamate transporters and the cystine-glutamate antiporter in the choroid plexus: implications for CSF glutamate homeostasis. J Chem Neuroanat 43(1):64–75

Li Q, Ding Y, Krafft P, Wan W, Yan F, Wu G et al (2018) Targeting germinal matrix hemorrhage-induced overexpression of sodium-coupled bicarbonate exchanger reduces posthemorrhagic hydrocephalus formation in neonatal rats. J Am Heart Assoc 7(3):31

Lindvall M, Edvinsson L, Owman C (1977) Histochemical study on regional differences in the cholinergic nerve supply of the choroid plexus from various laboratory animals. Exp Neurol 55 (1):152–159

Lindvall M, Edvinsson L, Owman C (1978a) Histochemical, ultrastructural, and functional evidence for a neurogenic control of CSF production from the choroid plexus. Adv Neurol 20:111–120

Lindvall M, Edvinsson L, Owman C (1978b) Sympathetic nervous control of cerebrospinal fluid production from the choroid plexus. Science (New York, NY) 201(4351):176–178

Lindvall M, Edvinsson L, Owman C (1978c) Reduced cerebrospinal fluid formation through cholinergic mechanisms. Neurosci Lett 10(3):311–316

Lippmann ES, Azarin SM, Kay JE, Nessler RA, Wilson HK, Al-Ahmad A et al (2012) Derivation of blood-brain barrier endothelial cells from human pluripotent stem cells. Nat Biotechnol 30 (8):783–791

Lippmann ES, Al-Ahmad A, Azarin SM, Palecek SP, Shusta EV (2014) A retinoic acid-enhanced, multicellular human blood-brain barrier model derived from stem cell sources. Sci Rep 4:4160

Liszczak TM, Black PM, Foley L (1986) Arginine vasopressin causes morphological changes suggestive of fluid transport in rat choroid plexus epithelium. Cell Tissue Res 246(2):379–385

Lundgaard I, Lu ML, Yang E, Peng W, Mestre H, Hitomi E et al (2017) Glymphatic clearance controls state-dependent changes in brain lactate concentration. J Cereb Blood Flow Metab 37 (6):2112–2124

MacAulay N, Zeuthen T (2010) Water transport between CNS compartments: contributions of aquaporins and cotransporters. Neuroscience 168(4):941–956

Majchrzak H, Kmieciak-Kołada K, Herman Z, Wencel T (1980) Homovanillic (HVA) and 5-hydroxyindoleacetic (5-HIAA) acid concentration in the cerebrospinal fluid of patients with supratentorial tumors and symptoms of intracranial hypertension (preliminary report). Neurol Neurochir Pol 14(1):87–90

Maktabi MA, Faraci FM (1994) Endogenous angiotensin II inhibits production of cerebrospinal fluid during posthypoxemic reoxygenation in the rabbit. Stroke 25(7):1489–1493. discussion 94

Maktabi MA, Heistad DD, Faraci FM (1991) Effects of central and intravascular angiotensin I and II on the choroid plexus. Am J Phys 261(5 Pt 2):R1126–R1132

Maktabi MA, Stachovic GC, Faraci FM (1993a) Angiotensin II decreases the rate of production of cerebrospinal fluid. Brain Res 606(1):44–49

Maktabi MA, Elbokl FF, Faraci FM, Todd MM (1993b) Halothane decreases the rate of production of cerebrospinal fluid. Possible role of vasopressin V1 receptors. Anesthesiology 78(1):72–82

Maren TH (1979) Effect of varying CO2 equilibria on rates of HCO3- formation in cerebrospinal fluid. J Appl Physiol Respir Environ Exerc Physiol 47(3):471–477

Maren TH (1988) The kinetics of HCO3- synthesis related to fluid secretion, pH control, and CO2 elimination. Annu Rev Physiol 50:695–717

Markey KA, Ottridge R, Mitchell JL, Rick C, Woolley R, Ives N et al (2017) Assessing the efficacy and safety of an 11beta-hydroxysteroid dehydrogenase type 1 inhibitor (AZD4017) in the idiopathic intracranial hypertension drug trial, IIH:DT: clinical methods and design for a phase II randomized controlled trial. JMIR Res Protoc 6(9):e181

Martin-de-Pablos A, Córdoba-Fernández A, Fernández-Espejo E (2018) Analysis of neurotrophic and antioxidant factors related to midbrain dopamine neuronal loss and brain inflammation in the cerebrospinal fluid of the elderly. Exp Gerontol 110:54–60

Marumo F, Masuda T, Masaki Y, Ando K (1988) The presence of atrial natriuretic peptide in canine cerebrospinal fluid and its possible origin in the brain. J Endocrinol 119(1):127–131

Masana M, Westerholz S, Kretzschmar A, Treccani G, Liebl C, Santarelli S et al (2018) Expression and glucocorticoid-dependent regulation of the stress-inducible protein DRR1 in the mouse adult brain. Brain Struct Funct 223:4039–4052

Michel CC, Mason JC, Curry FE, Tooke JE, Hunter PJ (1974) A development of the Landis technique for measuring the filtration coefficient of individual capillaries in the frog mesentery. Q J Exp Physiol Cogn Med Sci 59(4):283–309

Min-Chu L, Mackenna BR, Watt JA (1981) Evidence for the contribution by the cerebral cortex to 5-hydroxyindoleacetic acid found in the cerebrospinal fluid of cats. Brain Res 209(1):235–239

Mitro A, Lorencova M, Kutna V, Polak S (2018) Labelling of individual ependymal areas in lateral ventricles of human brain: ependymal tables. Bratisl Lek Listy 119(5):265–271

Mokgokong R, Wang S, Taylor CJ, Barrand MA, Hladky SB (2014) Ion transporters in brain endothelial cells that contribute to formation of brain interstitial fluid. Pflugers Arch 466 (5):887–901

Mori K, Tsutsumi K, Kurihara M, Kawaguchi T, Niwa M (1990) Alteration of atrial natriuretic peptide receptors in the choroid plexus of rats with induced or congenital hydrocephalus. Childs Nerv Syst 6(4):190–193

Mudò G, Bonomo A, Di Liberto V, Frinchi M, Fuxe K, Belluardo N (2009) The FGF-2/FGFRs neurotrophic system promotes neurogenesis in the adult brain. J Neural Transm (Vienna) 116 (8):995–1005

Murphy VA, Johanson CE (1989a) Acidosis, acetazolamide, and amiloride: effects on 22Na transfer across the blood-brain and blood-CSF barriers. J Neurochem 52(4):1058–1063

Murphy VA, Johanson CE (1989b) Alteration of sodium transport by the choroid plexus with amiloride. Biochim Biophys Acta 979(2):187–192

Murphy VA, Smith QR, Rapoport SI (1989) Uptake and concentrations of calcium in rat choroid plexus during chronic hypo- and hypercalcemia. Brain Res 484(1–2):65–70

Myung J, Wu D, Simonneaux V, Lane TJ (2018) Strong circadian rhythms in the choroid plexus: implications for sleep-independent brain metabolite clearance. J Exp Neurosci 12:1179069518783762

Nabeka H, Saito S, Li X, Shimokawa T, Khan MSI, Yamamiya K et al (2017) Interneurons secrete prosaposin, a neurotrophic factor, to attenuate kainic acid-induced neurotoxicity. IBRO Rep 3:17–32

Nagra G, Koh L, Zakharov A, Armstrong D, Johnston M (2006) Quantification of cerebrospinal fluid transport across the cribriform plate into lymphatics in rats. Am J Physiol Regul Integr Comp Physiol 291(5):R1383–R1389

Nakada T, Kwee IL (2019) Fluid dynamics inside the brain barrier: current concept of interstitial flow, glymphatic flow, and cerebrospinal fluid circulation in the brain. Neuroscientist 25 (2):155–166

Nakamura S, Maeda K, Sasaki J, Tsubokawa T (1985) Serotonergic effect on cerebrospinal fluid production. No to Shinkei = Brain Nerve 37(3):237–242

Napoleone P, Sancesario G, Amenta F (1982) Indoleaminergic innervation of rat choroid plexus: a fluorescence histochemical study. Neurosci Lett 34(2):143–147

Naz N, Jimenez AR, Sanjuan-Vilaplana A, Gurney M, Miyan J (2016) Neonatal hydrocephalus is a result of a block in folate handling and metabolism involving 10-formyltetrahydrofolate dehydrogenase. J Neurochem 138(4):610–623

Nilsson C, Lindvall-Axelsson M, Owman C (1992) Neuroendocrine regulatory mechanisms in the choroid plexus-cerebrospinal fluid system. Brain Res Brain Res Rev 17(2):109–138

Nilsson C, Stahlberg F, Gideon P, Thomsen C, Henriksen O (1994) The nocturnal increase in human cerebrospinal fluid production is inhibited by a beta 1-receptor antagonist. Am J Phys 267(6 Pt 2):R1445–R1448

Nixon PF (2008) Glutamate export at the choroid plexus in health, thiamin deficiency, and ethanol intoxication: review and hypothesis. Alcohol Clin Exp Res 32(8):1339–1349

O'Donnell ME (2014) Blood-brain barrier Na transporters in ischemic stroke. Advances in pharmacology (San Diego). CAL 71:113–146

Ocrant I, Parmelee JT (1992) Immunofluorescent cytometry and electron microscopic immunolocalization of insulin-like growth factor (IGF)-II receptors in infant rat choroid plexus. Mol Cell Neurosci 3(4):354–359

Oldendorf WH, Davson H (1967) Brain extracellular space and the sink action of cerebrospinal fluid. Measurement of rabbit brain extracellular space using sucrose labeled with carbon 14. Arch Neurol 17(2):196–205

Oreskovic D, Klarica M (2010) The formation of cerebrospinal fluid: nearly a hundred years of interpretations and misinterpretations. Brain Res Rev 64(2):241–262

Oreskovic D, Klarica M (2014) A new look at cerebrospinal fluid movement. Fluids Barriers CNS 11:16

Oshio K, Song Y, Verkman AS, Manley GT (2003) Aquaporin-1 deletion reduces osmotic water permeability and cerebrospinal fluid production. Acta Neurochir Suppl 86:525–528

Palacios JM (2016) Serotonin receptors in brain revisited. Brain Res 1645:46–49

Palm DE, Knuckey NW, Primiano MJ, Spangenberger AG, Johanson CE (1995) Cystatin C, a protease inhibitor, in degenerating rat hippocampal neurons following transient forebrain ischemia. Brain Res 691(1–2):1–8

Pandey GN, Dwivedi Y, Ren X, Rizavi HS, Faludi G, Sarosi A et al (2006) Regional distribution and relative abundance of serotonin(2c) receptors in human brain: effect of suicide. Neurochem Res 31(2):167–176

Pappenheimer J, Heisey S, Jordan E (1961) Active transport of diodrast and phenolsulfonphthalein from cerebrospinal fluid to blood. Am J Phys 200:1–10

Parmelee JT, Johanson CE (1989) Development of potassium transport capability by choroid plexus of infant rats. Am J Phys 256(3 Pt 2):R786–R791

Paulson OB, Hertz MM, Bolwig TG, Lassen NA (1977) Filtration and diffusion of water across the blood-brain barrier in man. Microvasc Res 13(1):113–124

Pirttila TJ, Lukasiuk K, Hakansson K, Grubb A, Abrahamson M, Pitkanen A (2005) Cystatin C modulates neurodegeneration and neurogenesis following status epilepticus in mouse. Neurobiol Dis 20(2):241–253

Pollay M, Curl F (1967) Secretion of cerebrospinal fluid by the ventricular ependyma of the rabbit. Am J Phys 213(4):1031–1038

Praetorius J, Damkier HH (2017) Transport across the choroid plexus epithelium. Am J Physiol 312 (6):C673–CC86

Praetorius J, Nielsen S (2006) Distribution of sodium transporters and aquaporin-1 in the human choroid plexus. Am J Physiol 291(1):C59–C67

Pullen RG, DePasquale M, Cserr HF (1987) Bulk flow of cerebrospinal fluid into brain in response to acute hyperosmolality. Am J Phys 253(3 Pt 2):F538–F545

Quintela T, Sousa C, Patriarca FM, Goncalves I, Santos CR (2015) Gender associated circadian oscillations of the clock genes in rat choroid plexus. Brain Struct Funct 220(3):1251–1262

Quintela T, Albuquerque T, Lundkvist G, Carmine Belin A, Talhada D, Gonçalves I et al (2018) The choroid plexus harbors a circadian oscillator modulated by estrogens. Chronobiol Int 35 (2):270–279

Raha-Chowdhury R, Raha AA, Forostyak S, Zhao JW, Stott SR, Bomford A (2015) Expression and cellular localization of hepcidin mRNA and protein in normal rat brain. BMC Neurosci 16:24

Rall DP (1968) Transport through the ependymal linings. Prog Brain Res 29:159–172

Reed DJ, Yen MH (1978) The role of the cat choroid plexus in regulating cerebrospinal fluid magnesium. J Physiol 281:477–485

Saavedra JM, Kurihara M (1991) Autoradiography of atrial natriuretic peptide (ANP) receptors in the rat brain. Can J Physiol Pharmacol 69(10):1567–1575

Savolainen M, Emerich D, Kordower JH (2018) Disease modification through trophic factor delivery. Methods Mol Biol 1780:525–547

Schmitt C, Strazielle N, Richaud P, Bouron A, Ghersi-Egea JF (2011) Active transport at the blood-CSF barrier contributes to manganese influx into the brain. J Neurochem 117(4):747–756

Schultz WJ, Brownfield MS, Kozlowski GP (1977) The hypothalamo-choroidal tract. II. Ultrastructural response of the choroid plexus to vasopressin. Cell Tissue Res 178 (1):129–141

Schwartz M, Baruch K (2014) The resolution of neuroinflammation in neurodegeneration: leukocyte recruitment via the choroid plexus. EMBO J 33(1):7–22

Schwartz M, Deczkowska A (2016) Neurological disease as a failure of brain-immune crosstalk: the multiple faces of neuroinflammation. Trends Immunol 37(10):668–679

Sharma HS, Johanson CE (2007) Intracerebroventricularly administered neurotrophins attenuate blood cerebrospinal fluid barrier breakdown and brain pathology following whole-body hyperthermia: an experimental study in the rat using biochemical and morphological approaches. Ann N Y Acad Sci 1122:112–129

Shechter R, Miller O, Yovel G, Rosenzweig N, London A, Ruckh J et al (2013) Recruitment of beneficial M2 macrophages to injured spinal cord is orchestrated by remote brain choroid plexus. Immunity 38(3):555–569

Shruthi S, Sumitha R, Varghese AM, Ashok S, Chandrasekhar Sagar BK, Sathyaprabha TN et al (2017) Brain-derived neurotrophic factor facilitates functional recovery from ALS-cerebral spinal fluid-induced neurodegenerative changes in the NSC-34 motor neuron cell line. Neurodegener Dis 17(1):44–58

Silverberg GD, Heit G, Huhn S, Jaffe RA, Chang SD, Bronte-Stewart H et al (2001) The cerebrospinal fluid production rate is reduced in dementia of the Alzheimer's type. Neurology 57(10):1763–1766

Sinclair AJ, Onyimba CU, Khosla P, Vijapurapu N, Tomlinson JW, Burdon MA et al (2007) Corticosteroids, 11beta-hydroxysteroid dehydrogenase isozymes and the rabbit choroid plexus. J Neuroendocrinol 19(8):614–620

Smith SV, Friedman DI (2017) The idiopathic intracranial hypertension treatment trial: a review of the outcomes. Headache 57(8):1303–1310

Smith AJ, Verkman AS (2018) The "glymphatic" mechanism for solute clearance in Alzheimer's disease: game changer or unproven speculation? FASEB J 32(2):543–551

Snodgrass SR, Johanson CE (2018) Choroid plexus: source of cerebrospinal fluid and regulator of brain development and function. In: Cinalli G (ed) Pediatric hydrocephalus. Springer, Berlin, pp 1–36

Sorensen SC, Gjerris F (1977) Adaptation of intraventricular pressure to acute changes in brain volume. Exp Eye Res 25(Suppl):387–390

Sørensen E, Olesen J, Rask-Madsen J, Rask-Andersen H (1978) The electrical potential difference and impedance between CSF and blood in unanesthetized man. Scand J Clin Lab Invest 38 (3):203–207

Spector R, Johanson CE (1989) The mammalian choroid plexus. Sci Am 261(5):68–74

Spector R, Johanson CE (2014) The nexus of vitamin homeostasis and DNA synthesis and modification in mammalian brain. Mol Brain 7:3

Spector R, Keep RF, Snodgrass SR, Smith QR, Johanson CE (2015a) A balanced view of choroid plexus structure and function: focus on adult humans. Exp Neurol 267:78–86

Spector R, Snodgrass SR, Johanson CE (2015b) A balanced view of the cerebrospinal fluid composition and functions: focus on adult humans. Exp Neurol 273:57–68

Steardo L, Nathanson JA (1987) Brain barrier tissues: end organs for atriopeptins. Science (New York, NY) 235(4787):470–473

Steffensen AB, Oernbo EK, Stoica A, Gerkau NJ, Barbuskaite D, Tritsaris K et al (2018) Cotransporter-mediated water transport underlying cerebrospinal fluid formation. Nat Commun 9(1):2167

Stieger B, Gao B (2015) Drug transporters in the central nervous system. Clin Pharmacokinet 54 (3):225–242

Stopa EG, Tanis KQ, Miller MC, Nikonova EV, Podtelezhnikov AA, Finney EM et al (2018) Comparative transcriptomics of choroid plexus in Alzheimer's disease, frontotemporal dementia and Huntington's disease: implications for CSF homeostasis. Fluids Barriers CNS 15(1):18

Szmydynger-Chodobska J, Chodobski A (2005) Peptide-mediated regulation of CSF formation and blood flow to the choroid plexus. In: Zheng W, Chodobski A (eds) The blood-cerebrospinal fluid barrier. CRC/Francis & Taylor Group, Boca Raton, FL, pp 101–117

Szmydynger-Chodobska J, Monfils PR, Lin AY, Rahman MP, Johanson CE, Chodobski A (1996) NADPH-diaphorase histochemistry of rat choroid plexus blood vessels and epithelium. Neurosci Lett 208(3):179–182

Szmydynger-Chodobska J, Chung I, Chodobski A (2006) Chronic hypernatremia increases the expression of vasopressin and voltage-gated Na channels in the rat choroid plexus. Neuroendocrinology 84(5):339–345

Tachikawa M, Hosoya K (2011) Transport characteristics of guanidino compounds at the blood-brain barrier and blood-cerebrospinal fluid barrier: relevance to neural disorders. Fluids Barriers CNS 8(1):13

Timmusk T, Mudò G, Metsis M, Belluardo N (1995) Expression of mRNAs for neurotrophins and their receptors in the rat choroid plexus and dura mater. Neuroreport 6(15):1997–2000

Tucker VL, Huxley VH (1990) Evidence for cholinergic regulation of microvessel hydraulic conductance during tissue hypoxia. Circ Res 66(2):517–524

Tulassay T, Khoor A, Bald M, Ritvay J, Szabo A, Rascher W (1990) Cerebrospinal fluid concentrations of atrial natriuretic peptide in children. Acta Paediatr Hung 30(2):201–207

Uchida Y, Zhang Z, Tachikawa M, Terasaki T (2015) Quantitative targeted absolute proteomics of rat blood-cerebrospinal fluid barrier transporters: comparison with a human specimen. J Neurochem 134(6):1104–1115

Uldall M, Botfield H, Jansen-Olesen I, Sinclair A, Jensen R (2017) Acetazolamide lowers intracranial pressure and modulates the cerebrospinal fluid secretion pathway in healthy rats. Neurosci Lett 645:33–39

Vemula S, Roder KE, Yang T, Bhat GJ, Thekkumkara TJ, Abbruscato TJ (2009) A functional role for sodium-dependent glucose transport across the blood-brain barrier during oxygen glucose deprivation. J Pharmacol Exp Ther 328(2):487–495

Vieira M, Gomes JR, Saraiva MJ (2015) Transthyretin induces insulin-like growth factor I nuclear translocation regulating its levels in the hippocampus. Mol Neurobiol 51(3):1468–1479

Vindedal GF, Thoren AE, Jensen V, Klungland A, Zhang Y, Holtzman MJ et al (2016) Removal of aquaporin-4 from glial and ependymal membranes causes brain water accumulation. Mol Cell Neurosci 77:47–52

Vogh BP, Godman DR (1989) Effects of inhibition of angiotensin converting enzyme and carbonic anhydrase on fluid production by ciliary process, choroid plexus, and pancreas. J Ocul Pharmacol 5(4):303–311

Wallingford MC, Chia JJ, Leaf EM, Borgeia S, Chavkin NW, Sawangmake C et al (2017) SLC20A2 deficiency in mice leads to elevated phosphate levels in cerebrospinal fluid and glymphatic pathway-associated arteriolar calcification, and recapitulates human idiopathic basal ganglia calcification. Brain Pathol 27(1):64–76

Walter HJ, Berry M, Hill DJ, Cwyfan-Hughes S, Holly JM, Logan A (1999) Distinct sites of insulin-like growth factor (IGF)-II expression and localization in lesioned rat brain: possible roles of IGF binding proteins (IGFBPs) in the mediation of IGF-II activity. Endocrinology 140 (1):520–532

Wang D, Nykanen M, Yang N, Winlaw D, North K, Verkman AS et al (2011) Altered cellular localization of aquaporin-1 in experimental hydrocephalus in mice and reduced ventriculomegaly in aquaporin-1 deficiency. Mol Cell Neurosci 46(1):318–324

Weaver C, McMillan P, Duncan JA, Stopa E, Johanson C (2004) Hydrocephalus disorders: their biophysical and neuroendocrine impact on the choroid plexus epithelium. In: Hertz L (ed) Non-neuronal cells of the nervous system: function and dysfunction, vol 31. Elsevier, Amsterdam, pp 269–293

Welch K (1963) Secretion of cerebrospinal fluid by choroid plexus of the rabbit. Am J Phys 205:617–624

Woodard GE, Rosado JA (2008) Natriuretic peptides in vascular physiology and pathology. Int Rev Cell Mol Biol 268:59–93

Xiang J, Routhe LJ, Wilkinson DA, Hua Y, Moos T, Xi G et al (2017) The choroid plexus as a site of damage in hemorrhagic and ischemic stroke and its role in responding to injury. Fluids Barriers CNS 14(1):8

Xie L, Kang H, Xu Q, Chen MJ, Liao Y, Thiyagarajan M et al (2013) Sleep drives metabolite clearance from the adult brain. Science (New York, NY) 342(6156):373–377

Xu M, Xiao M, Li S, Yang B (2017) Aquaporins in nervous system. Adv Exp Med Biol 969:81–103

Yamasaki H, Sugino M, Ohsawa N (1997) Possible regulation of intracranial pressure by human atrial natriuretic peptide in cerebrospinal fluid. Eur Neurol 38(2):88–93

Zarebkohan A, Javan M, Satarian L, Ahmadiani A (2009) Effect of chronic administration of morphine on the gene expression level of sodium-dependent vitamin C transporters in rat hippocampus and lumbar spinal cord. J Mol Neurosci 38(3):236–242

Zeitzer JM, Maidment NT, Behnke EJ, Ackerson LC, Fried I, Engel J et al (2002) Ultradian sleep-cycle variation of serotonin in the human lateral ventricle. Neurology 59(8):1272–1274

Zemo DA, McCabe JT (2001) Salt-loading increases vasopressin and vasopressin 1b receptor mRNA in the hypothalamus and choroid plexus. Neuropeptides 35(3–4):181–188

Zeuthen T, Macaulay N (2012) Cotransport of water by Na(+)-K(+)-2Cl(−) cotransporters expressed in Xenopus oocytes: NKCC1 versus NKCC2. J Physiol 590(5):1139–1154

Zeuthen T, Gorraitz E, Her K, Wright EM, Loo DD (2016) Structural and functional significance of water permeation through cotransporters. Proc Natl Acad Sci U S A 113(44):E6887–E6e94

Zhang Z, Tachikawa M, Uchida Y, Terasaki T (2018) Drug clearance from cerebrospinal fluid mediated by organic anion transporters 1 (Slc22a6) and 3 (Slc22a8) at arachnoid membrane of rats. Mol Pharm 15(3):911–922

Zhao H, Yang H, Yan L, Jiang S, Xue L, Guan W et al (2014) Effects of lead exposure on copper and copper transporters in choroid plexus of rats. Zhonghua Lao Dong Wei Sheng Zhi Ye Bing Za Zhi 32(11):819–822

Zheng W (2005) Blood-CSF barrier in iron regulation and manganese-induced parkinsonism. In: Zheng W, Chodobski A (eds) The blood-cerebrospinal fluid barrier. CRC/Taylor and Francis, Boca Raton, FL

Zheng G, Chen J, Zheng W (2012) Relative contribution of CTR1 and DMT1 in copper transport by the blood-CSF barrier: implication in manganese-induced neurotoxicity. Toxicol Appl Pharmacol 260(3):285–293

Chapter 3
A Frog Model for CSF Secretion

Donald D. F. Loo and Ernest M. Wright

Abstract Model systems ranging from the frog skin to the fish gall bladder have played an indisputable role in advancing our understanding of epithelial physiology. Apart from the curiosity and talents of scientists such as Hans Ussing and Jared Diamond, a critical factor was the simplicity and viability of the epithelia chosen for study. Almost 50 years ago, we chose the frog choroid plexus to unravel the mysteries of cerebrospinal fluid secretion using the then state-of-the art physiological, biochemical and biophysical tools. Here we summarize ion transport across the frog choroidal epithelium and our interpretation of the transport mechanisms involved. The challenge now in the era of functional genomics is to identify the transporters and channels responsible, and to extend our understanding to CSF secretion in man.

3.1 Introduction

There is compelling evidence that the choroid plexus is responsible for a substantial fraction of the cerebrospinal fluid (CSF) secretion, approximately 600 ml/day in man, and that the epithelium plays an important role in the regulation of CSF composition (Wright 1978). However, the complexity of brain anatomy in mammals has restricted the direct study of the mechanisms of solute and water transport across the choroidal epithelium, both in vivo and in vitro. We instead have used a model system, the amphibian choroid plexus. The structure of the frog choroid plexus, the rate of CSF secretion and CSF composition, are similar to those in man (Wright 1978). It is a simple task to isolate the frog IV ventricle choroid plexus and mount it in Ussing chambers separating the ventricular and vascular compartments, and then measure ion, non-electrolyte and water fluxes **across** the epithelium and **across** the apical membrane. Furthermore, it is straightforward to impale the epithelium mounted in such chambers with microelectrodes and measure transmembrane

D. D. F. Loo · E. M. Wright (✉)
Department of Physiology, David Geffen School of Medicine at UCLA, Los Angeles, CA, USA
e-mail: ewright@mednet.ucla.edu

© The American Physiological Society 2020
J. Praetorius et al. (eds.), *Role of the Choroid Plexus in Health and Disease*,
Physiology in Health and Disease, https://doi.org/10.1007/978-1-0716-0536-3_3

voltages and electrical resistances, conduct an equivalent circuit analysis, and determine the intracellular activities of ions (Na^+, K^+, and Cl^-). Ion channels in the apical membrane may also be studied by patch-clamp techniques. Finally, a major advantage of amphibian in vitro preparations, unlike mammalian ones, is that they are viable for many hours.

Our task here is to quantitate the rates of ion and water transport across the epithelium through the paracellular and cellular pathways, and across the apical and basolateral membranes, and identify the mechanisms involved. All of our earlier work was carried out prior to the molecular biology revolution, and so the current challenge is to determine which of the hundreds of genes coding for membrane proteins are responsible for CSF secretion (Hediger et al. 2013). It is not a trivial task to determine the functional significance of ion channels and transporters, as one has to determine their density and turnover number under physiological conditions. Here we review the mechanism of ion and water transport across the amphibian choroid plexus to provide a background for the understanding of CSF secretion in man.

3.2 The Frog Choroid Plexus

Figure 3.1a shows a drawing of a mid-line sagittal section of the frog brain with the posterior choroid plexus forming the roof of the IV ventricle. Essentially, this is the same anatomy as in higher vertebrates, but the enlargement of the cerebrum and cerebellum in higher vertebrates obscures the IV ventricle choroid plexus. The tedious dissection of this plexus in mammals combined with its low viability has discouraged investigators until now, and so a quantitative description of the transport mechanisms is lacking.

A continuous layer of cuboidal epithelial cells covers the highly folded ventricular surface of the plexus resting upon a very thin supporting vascular tissue (Figs. 3.1b and 3.1c). The choroid plexuses account for only 3% of the total brain weight, but they contain 35% of the brain blood volume. It is estimated that the blood flow to the choroid plexuses is tenfold higher than that to the brain parenchyma and this reflects the high metabolic demand of the epithelium.

3.3 Trans-Epithelial Transport

The amphibian posterior choroid plexus is readily dissected from the brain and mounted as a vertical flat sheet between Lucite half-chambers (Ussing chambers) separating two saline solutions, referred to as the ventricular (CSF) and serosal (vascular) solutions. The transport characteristics and electrical properties of the epithelium have been studied using standard methods (Wright 1972a, b, c). In some experiments, the plexus was mounted horizontally to measure fluxes across the

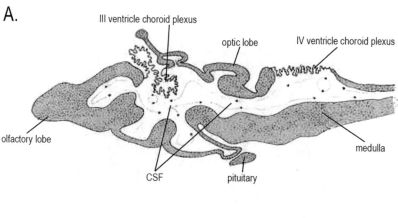

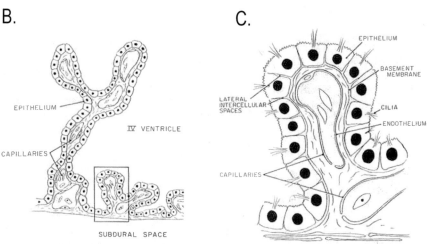

Fig. 3.1 (**a**) A drawing of a mid-line sagittal section of the frog brain showing the location of the choroid plexus in the III and IV ventricle. The choroid plexus in the lateral ventricles is not shown. (**b**) Drawing of cross-section of frog choroid plexus showing folded epithelium resting on a thin stroma of vascular tissue and a basement membrane, and (**c**) Expanded drawing of the box in (**b**) showing the ciliated cuboidal epithelium (Wright 1972a, b, c). Each epithelial cell (20 × 20 × 20 μm) is joined to the neighboring cell by tight junctions. Cross sections of the amphibian posterior choroid plexus were drawn from histological sections. Each cell has 3–40 motile cilia beating at an average frequency of 7/s (see also Nelson and Wright 1974). The fenestrated capillaries contain nucleated red blood cells

apical membrane and fluid secretion. The epithelium was impaled with microelectrodes to measure the intracellular membrane potential, intracellular ion activities, and the equivalent circuit parameters for the epithelium (Wright 1974; Wright et al. 1977; Zeuthen and Wright 1978, 1981; Saito and Wright 1984, 1987).

In the steady state the unidirectional fluxes across the cellular pathway are related to the unidirectional fluxes across the apical and basolateral membranes (Fig. 3.4b) by the expressions (Wright 1974)

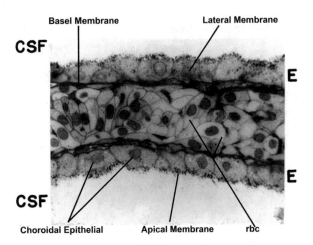

Fig. 3.2 Autoradiography of [³H]-ouabain binding to the frog choroid plexus. The isolated plexus was exposed to 5×10^{7} M [³H]-ouabain in frog saline for 2.5 h, and then washed free of unbound ouabain before freezing at $-155\,°C$, freeze-drying, fixing in osmium tetroxide vapor, and sectioning at 1 μm. The sections were coated with a 2-μm film of emulsion and exposed for 1 week at 4 °C before photographic development. The section shows a fold of the plexus with the cuboidal epithelial cells resting on a basement membrane and a fenestrated capillary packed with nucleated red blood cells. The width of the fold is 50 μm. Image take from the study by Quinton et al. (1973)

$$J_{net} = J_{vs} - J_{sv} = J_{vc} - J_{cv} = J_{cs} - J_{sc},$$

$$J_{vs} = (J_{vc} \times J_{cs})/(J_{cv} + J_{cs}), \text{and}$$

$$J_{sv} = (J_{sc} \times J_{cv})/(J_{cv} + J_{cs}),$$

where J_{net} are the net fluxes across the epithelium, and J_{vs}, J_{sv}, J_{vc}, J_{cv}, J_{cs} and J_{sv} are the unidirectional ventricle to serosal, serosal to ventricle, ventricle to cell, cell to ventricle, cell to serosal, and serosal to cell fluxes respectively.

It is then possible to estimate the remaining unidirectional fluxes by knowing the magnitude of the paracellular unidirectional fluxes, the net fluxes for Na⁺, K⁺ and Cl⁻, and the unidirectional influxes for these ions across the apical membrane, (Fig. 3.4b). Clearly, the accuracy of these estimates is subject to some uncertainty related to the magnitude of the paracellular fluxes [Fig. 3.4b; (Fig. 2, in Saito and Wright 1984)].

An early finding was a low electrical potential difference and low electrical resistance across the frog plexus. The potential difference between identical solutions was less than 1 mV with the CSF positive, and this was similar to that observed for the choroid plexus isolated from sharks, rays and mammals (see Wright 1972a, b, c). The electrical resistance was 200 ohms cm², expressed as the area of the window between the half-chambers, but when corrected for the actual area of the epithelium, the resistance is 25 ohms cm² (see below). Thus >95% of the conductance of the epithelium is due to current flow between the cells, the paracellular pathway (Zeuthen and Wright 1981; Saito and Wright 1984).

Table 3.1 Unidirectional fluxes of ions across the frog choroid plexus (Wright 1972a, b, c, 1977)

	n-mol cm^{-1} h^{-1}		
	J_{sv}	J_{vs}	J_{net}
Sodium	580 ± 40(22)	380 ± 40(17)	+200
Chloride	540 ± 20(36)	420 ± 20(36)	+120
Potassium	18 ± 2(6)	24 ± 2(5)	−6
Calcium	17 ± 3(6)	8 ± 2(7)	+7
Glycodiazine	160 ± 22(7)	100 ± 6(6)	+60

The ventricular and serosal compartments contained frog saline buffered with bicarbonate or glycodiazine, and fluxes were measured using radioactive tracers. Positive net fluxes refer to secretion into the ventricular fluid. The fluxes are normalized to the area of the epithelium, taken here as a fivefold increase due to tissue folding over the window area of the Ussing chamber

The unidirectional ion fluxes across the plexus were measured using radioactive tracers (Table 3.1). The unidirectional fluxes are high relative to the net transport, and this simply reflects the high passive permeability of the paracellular pathway. There was a net secretion of Na$^+$ and Cl$^-$ towards the ventricular compartment, but the secretion of chloride was only 60% of that for sodium in bicarbonate buffered saline (NaHCO$_3$ 25 mM). The net sodium secretion decreased to the level of chloride in the absence of bicarbonate (not shown). When 25 mM sodium bicarbonate was replaced with 25 mM sodium glycodiazine, a non-volatile buffer (pK 5.9), the net secretion of glycodiazine accounts for the deficit between sodium and chloride secretion (Wright 1977). Both the net secretions of sodium, chloride and glyodiazine were blocked by the addition of ouabain to the CSF side of the plexus (see below). We conclude that the frog choroid plexus secretes a mixture of NaCl and NaHCO$_3$ from the blood into the CSF (see Fig. 3.4). The secretion of sodium is accompanied by a relatively small absorption of potassium (Table 3.1) both at normal (2 mM) and higher (10 mM) KCl concentrations (Wright 1972a, b, c). There was no significant net Ca^{2+} transport across the epithelium in either direction.

3.4   The Apical Membrane Na$^+$/K$^+$-ATPase Pump

Ouabain added to the ventricular solution, but not the serosal solution, eliminated net sodium secretion (Wright 1972a, b, c). This hinted that the sodium/potassium pump was on the apical membrane of the epithelium, in contradistinction to that for most other epithelia. The conjecture was confirmed by [^{3}H]-ouabain binding and autoradiography (Quinton et al. 1973). The time course of binding closely follows the time course of the inhibition of active sodium transport, and specific binding to the apical surface was 40-fold higher than the vascular surface. Autoradiography showed the specific binding occurred only on the apical membrane, and none on the basolateral membrane (Fig. 3.2). We estimated that there are about 10^4 pumps per μm^2 of apical membrane, considerably higher than in red blood cells (1 per μm^2), but is not unexpected given the secretory role of the epithelium.

In most epithelia the Na^+/K^+-pump is located on the basolateral membrane and, this together with complex anatomy of epithelial tissues, has made it difficult to study pump kinetics. In the frog choroid plexus mounted in Ussing chambers, the apical membrane is fully exposed to the ventricular solution. This permits direct measurement of the properties of the pump in the intact epithelium using radioactive $^{42}K^+$ uptakes and $^{22}Na^+$ effluxes, and electrical recordings (Zeuthen and Wright 1978, 1981; Saito and Wright 1982). A valuable tool in these studies is the sensitivity of the apical pump to ventricular ouabain; ventricular ouabain binds rapidly to the apical pump with kinetics similar to those reported for single cells, e.g. red blood cells. The maximum binding was found to be 4.5×10^{-11} moles/plexus with a K_B of 8×10^{-7} M. Estimates of the On and Off rates indicate that at a concentration of 0.1 mM, ouabain 2.25×10^{-11} moles bind within 10 s, i.e. an amount sufficient to inhibit the pump by more than 75%. The rapid exposure of the apical membrane to 0.1 mM ouabain was then used to determine if the epithelial pump is electrogenic.

Single and double-barreled microelectrodes were used to measure the electrical potential across the apical membrane (E^{vc}) and the intracellular potassium activity (K_c) (Zeuthen and Wright 1978, 1981). When 0.1 mM ouabain was applied to the ventricular solution, there was an abrupt depolarization of the epithelial cell, 10 mV in 10 s, and then the potential gradually dropped as potassium is lost from the cells (Fig. 3.3). The abrupt depolarization is evidence for an electrogenic Na^+/K^+ pump, and the magnitude is precisely that is expected from a pumping ratio of 3 Na^+ to 2 K^+. It can be shown that the sudden arrest of the Na^+/K^+-pump should bring about a depolarization of RT/F ln r ($= 25$ mV $\times$ ln r), where R is the gas constant, T the absolute temperature and F Faradays constant, and r the ratio between Na^+ and K^+ transport. For r equal to 1.5, the expected depolarization is 10.4 mV is close to the observed value of 10.2 mV (see also Thomas 1972),

The kinetics of the pump were further investigated by measuring the initial rate of ^{42}K uptake into the plexus from the ventricular solution, and the ^{22}Na efflux from the epithelium into the ventricular solution (Zeuthen and Wright 1981; Saito and Wright 1982). The K^+ uptake across the apical membrane was inhibited 85% by ventricular ouabain (K_i 1×10^{-7} M), with a further 10% on addition of ethacrynic acid. The rate (J^K) increased as a sigmoidal functional of the ventricular K^+ (K_v) concentration, i.e. $J^K = J^K_{max}/(1 + K^K_m/K_v)^2$, where K^K_m is the apparent affinity for ventricular K^+ (1.5 mM). The rate of K^+ uptake into the plexus from the serosal solution was only 6% of that across the apical membrane, and this was presumably into the supporting cells (see Fig. 3.1b).

The pump component of the $^{22}Na^+$ efflux from the epithelium into the ventricular solution across the apical membrane was extracted from the activation of the Na^+ efflux by ventricular K^+ (there was no significant serosal K^+ activation of Na^+ efflux

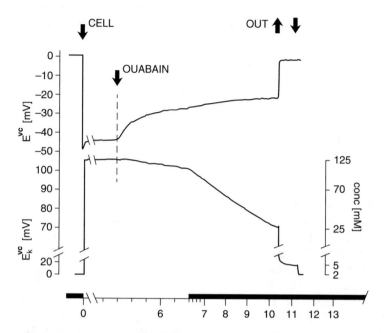

Fig. 3.3 Effects of ouabain on the intracellular electrical potential E^{vc}, the intracellular potassium chemical potential E_K, and apparent intracellular K^+ concentration (mM). All electrical potentials are referred to the apical solution. At the first arrow a double-barreled K^+-electrode was advanced into the epithelium, and after 5.5 min ouabain was added to the ventricular solution (final concentration 1×10^{-4} M) (Zeuthen and Wright 1981)

into the serosal solution). The concentration dependence of the Na^+ efflux on ventricular K^+ was sigmoidal with similar kinetics to K^+ uptake (K_m 1.1 mM and Hill coefficient of 2), and the K^+-stimulation was mimicked by Rb^+ (1.0), Tl^+ (1.0), Cs^+ (0.03), NH_4^+ (0.04), but not by Li^+, as expected for Na^+/K^+-ATPases. The intracellular Na^+ (Na_c) dependence of the Na^+ efflux across the apical membrane was determined by equilibrating the tissue in the cold with 2–45 mM Na^+ in K^+-free solutions, and then measuring the Na^+ efflux stimulated by ventricular K^+ at room temperature. The Na^+ pump rate (J^{Na}) was a sigmoid function of the intracellular Na^+ concentration, i.e. $J^{Na} = J^{Na}_{max}/(1 + K_m^{Na}/Na_c)^3$ with an apparent affinity of 7 mM. The total pump rate J^P across the apical membrane of the choroid plexus is therefore a function of the intracellular Na^+ and ventricular K^+, with a 3/2 coupling of Na^+ to K^+ transport.

3.5   The K^+ Efflux Across the Ventricular Membrane

K^+ that enters the epithelium via the Na^+/K^+-pump, returns to the ventricular solution, with only a minor fraction ($<<4\%$) entering the serosal solution. Several questions are raised about what constitutes the return pathway from the cell into the ventricular compartment, is it electrodiffusive through channel or electroneutral mediated by cotransporters?

The conductance of the ventricular cell membrane is higher than that of the serosal membrane by a factor of ten, and furthermore, the passive permeability for K^+ is much larger than that for other ions. This suggests that the K^+ that is pumped into the cell may leave the cell again at least in part passively across the ventricular membrane through channels (Zeuthen and Wright 1981).

The first channel identified in the apical membrane of the Necturus choroid plexus by the patch clamp technique was a 200 pS Ca^{++} activated maxi-K^+ channel (Christensen and Zeuthen 1987, Brown et al. 1988). However, whole-cell patch clamp experiments showed that these channels account for less than 25% of the total conductance, at a normal resting potential (-60 mV) and normal intracellular K^+ and Ca^{++} concentrations (Loo et al. 1988). The contribution of the maxi-K^+ channel to K^+ efflux across the apical membrane is likely to be low, but this may increase with depolarization of the apical membrane potential and a rise in intracellular Ca^{++}. A second type of channel in the apical membrane, 90 pS, voltage-independent, inward rectifying K^+, may contribute to the K^+ efflux, but the channel has yet to be established. Other possibilities include an apical KCl cotransporter (see below and Steffensen and Zeuthen 2020).

3.6 Na^+ Entry into the Epithelium

Na^+ pumping out of the epithelium across the apical membrane, 3.6 µmoles cm^{-2} h^{-1}, requires a balanced Na^+ entry into the cell, as much as 2.6 µmoles cm^{-2} h^{-1} to account for the difference between the pump rate and the rate of Na^+ secretion. A minimum of 1 µmoles cm^{-2} h^{-1} must enter the cell across the basolateral membrane to account for the rate of Na^+ secretion across the plexus. The driving force for basolateral Na^+ entry is the sum of the electrical and chemical potentials across the membrane, -117 mV ($E_m - 45$ mV, and E_{Na} 72 mV). However, it is unlikely that entry occurs through Na^+ channels and it is more likely to occur via transporters such Na^+/Cl^- and Na^+/HCO_3^- cotransporters to drive the uphill accumulation of Cl^- and HCO_3^- (see Fig. 3.4c).

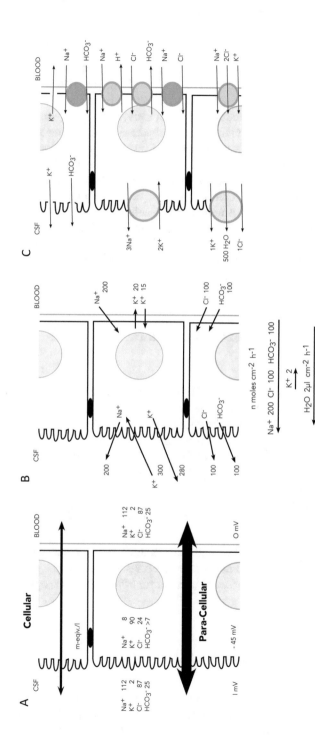

Fig. 3.4 Cartoon showing the pathways for ions and water transport across the frog choroid plexus. (**a**) There is net secretion of Na⁺, Cl⁻, HCO₃⁻ and water, and a small reabsorption of K⁺. Fluid, Na⁺, Cl⁻, and HCO₃⁻ secretion is blocked by the presence of ouabain in the CSF. There are high paracellular fluxes of Na⁺, K⁺, Cl⁻ and HCO₃⁻ between the epithelial cells (through leaky tight junctions). The Na⁺ paracellular fluxes are three- to fourfold higher than the rate of Na⁺ secretion across the cellular pathway. The apparent intracellular ion concentrations were determined using Na⁺, K⁺, and Cl⁻ microelectrodes. K⁺, Cl⁻ and HCO₃⁻ are accumulated within the epithelium above that predicted by the membrane potential, and Na⁺ is lower than predicted (−45 mV). (**b**) A summary of the determined rates of Na⁺, Cl⁻, HCO₃⁻, K⁺ and water transport across the cells of the choroid plexus epithelium, and the unidirectional fluxes across the apical and basolateral membrane. (**c**) The transporters and channels that may account for the rates of ion transport across the cell membranes (see text). There is clear experimental evidence for the apical 3Na⁺/2K⁺-pump and the K⁺/Cl⁻/H₂O transporter, and the basolateral Na⁺/Cl⁻ and Na⁺/HCO₃⁻ cotransporters, but there is no direct evidence for the basolateral Cl⁻/HCO₃⁻ exchanger and the Na⁺/K⁺/2Cl⁻ cotransporters (see text)

3.7   Transport of HCO_3^-

Na^+ secretion across the choroid plexus depends on the presence of HCO_3^-. In HCO_3^--free solutions, Na^+ transport is dramatically reduced (Wright 1972a, b, c, 1977). The presence of HCO_3^- in the serosal solution increased intracellular $[Na^+]$, and the activity of the Na^+/K^+ pump. Besides basolateral entry, HCO_3^- may also be generated intracellularly from CO_2 by the carbonic anhydrase catalyzed reaction: $CO_2 + H_2O \leftrightarrow H^+ + HCO_3^-$. This reaction is catalyzed by carbonic anhydrase at $37°$ C, but the enzyme is not required at room temperature and this explains why sodium secretion in frogs is not sensitive to Diamox, the carbonic anhydrase inhibitor (Wright 1972a, b, c). HCO_3^- exit across the apical membrane into the CSF is conductive, possibility through a non-selective anion channel (Saito and Wright 1984). In frog choroid plexus, an apical bicarbonate conductance is stimulated by cAMP (Saito and Wright 1983). Bicarbonate could also exit the apical membrane via an electrogenic Na^+/HCO_3^- transporter, but we have no evidence for this in frogs.

3.8 Cl^- Transport Pathways in the Basolateral and Apical Membranes

In the amphibian choroid plexus, there is a net Cl^- flux from blood to CSF (Wright 1972a, b, c). In bicarbonate-buffered saline, the brush-border membrane potential was -43 mV, and intracellular Cl^- activity (measured by Cl^- sensitive microelectrodes) was 24 mM, twice the predicted equilibrium activity (12 mM). Cl^- accumulation was dependent on Na^+ but insensitive of HCO_3^- (Saito and Wright 1987). This led us to the conclusion that a NaCl cotransporter mediates Cl^- accumulation across the basolateral membrane.

Cell-attached patches in the apical membrane of Necturus CP revealed 3 Cl^- channels (Christensen et al. 1989): (1) a large ~375 PS channels, rarely observed; (2) a 28 pS channel that also conducts HCO_3^-, and not simulated by cAMP; and (3) 7 pS channel stimulated by cAMP in 28% of the patches. It was concluded that the current through these channels could not account for transepithelial steady state Cl^- exit. A KCl cotransporter was proposed to account for efflux across the apical membrane (Zeuthen 1994).

3.9   Regulation of Ion Transport

Bicarbonate plays a central role in the regulation of Na^+ secretion: The rate of net secretion across the plexus is stimulated by the presence of bicarbonate; the apical $3Na^+/2 K^+$-pump is stimulated 100% by HCO_3^- with a K_m of 7 mM; and HCO_3^-

accounts for 40% of the anions accompanying Na^+ secretion (Wright 1972a, b, c, 1977). Furthermore, secretion is increased up to 300% by cAMP, either directly using dibutryl-cAMP, or indirectly by blocking phosphodiesterases (e.g., Theophylline) or stimulating adenylate cyclase (Forskolin, PGE1, or ACTH) (Saito and Wright 1983). Furosemide, apical or basolateral, partially inhibited the response to cAMP, but bumetanide did not. Both the basal and cAMP -stimulated secretions were dependent on the presence of Na^+ and HCO_3^-, but were insensitive to Cl^-. In the frog, the only hormone that stimulated secretion was ACTH, but we also found that the β-adrenergic agonist, isoproterenol, also increased secretion and this was blocked by propranolol.

The mode of action of cAMP was elucidated by an equivalent circuit analysis (Saito and Wright 1984). Using theophylline or IBMX to increase intracellular cAMP, we found that this produced a depolarization of the apical membrane along with a large increase in conductance, and was Cl^--independent (Wright 1974). There was no effect of cAMP on the electrical properties of the basolateral membrane.

3.10   Inhibitors

An important component of the armamentarium used to parse out the mechanisms of epithelial transport is transport inhibitors, and is also true for studies of the frog choroid plexus. As we have already discussed, the cardiac glycoside ouabain, 10^{-8} to 10^{-4} M, has been valuable to dissecting the role of apical Na^+/K^+-pumps in Na^+ and fluid secretion. Ouabain also inhibits the accumulation of trace anions, e.g. iodide, amines including choline and LSD, and PAH (Wright 1972a, b, c, 1974; Ehrlich and Wright 1982), but this is a secondary effect linked to dissipation of the Na^+ gradient that drives secondary active transport, Na^+/cotransport.

Inhibitors of carbonic anhydrase, Diamox, acetazolamide and ethoxyzolamide, had minimum effects up to 1 mM, and is likely due to the fact that our experiments were conducted at 22° C where the hydration of CO_2 is independent to the enzyme. Amiloride (1 μM), an inhibitor of the ENaC channel and Na^+/H^+ exchangers, did not influence transepithelial Na^+ fluxes (Wright 1972a, b, c), and furosemide (0.6 mM) on the vascular side had only modest effects on intracellular chloride accumulation (Saito and Wright 1987). Bicarbonate secretion in the resting state was promptly inhibited by 1 mM furosemide, while the more potent inhibitor of K^+/Cl^- cotransporters, bumetanide, had little effect (Saito and Wright 1983). In the absence of knowledge about the kinetics and specificity of the inhibitors in the frog, it is problematic to interpret these effects, or lack thereof, in terms of transport mechanisms.

3.11   CSF Secretion by Frog Choroid Plexus

In the absence of osmotic gradients the rate of fluid secretion was 2 $\mu l/cm^2/h$, and was blocked by ouabain in the ventricular fluid (Wright et al. 1977). This is comparable to estimates of CSF secretion in mammals, 0.1–0.4 up/min/mg at 37 °C. Collection of freshly secreted fluid under oil demonstrated a higher osmolality and sodium concentration relative to the serosal bathing solution (254 vs. 205 milliosmolar, and 125 vs. 110 m-equiv/l). The chloride was not significantly different from that in saline, (85 m-equivl/l), thus infers that the anion deficit, 40 m-equiv/l, is accounted for by HCO_3^-. Pre-exposure of the ventricular surface of the plexus to 0.1 mM ouabain reduced the osmolality and Na^+ concentration of the reduced CSF secretion. This leads to the conclusion that fluid secretion across the frog choroid plexus is intimately linked to the secretion of Na^+, Cl^- and HCO_3^-. The link is indirect in that the apical $3Na^+/2 K^+$ – pump is responsible for Na^+ secretion and the intracellular accumulation of K^+ ($E_K > E_m$). The K^+ that is pumped into the epithelium is largely matched by an efflux back across the apical membrane into the CSF. In part, this K^+ efflux occurs through apical K^+-channels, and in part by a KCl/water cotransporter (1 K^+, 1 Cl^-, 500 H_2O) (Zeuthen 1991, 1994). Zeuthen and Steffensen review the mechanisms of water transport further in the accompanying chapter.

3.12   Other Transport Systems

Apart from ion transport and CSF secretion reviewed above, our studies have provided insights in the role of frog choroid plexus in regulation of the composition of the cerebrospinal fluid. These include the transport of non-electrolytes, sugars, amino acids, amines, and anions (Wright 1972a, b, c, 1974; Ehrlich and Wright 1982; Prather and Wright 1970; Wright and Prather 1970). These have confirmed that many of the molecules are accumulated in the mammalian choroid plexus, but there is no significant net transport of amino acids, sugars or amines across the epithelium. However, there is net transport of trace anions, I^-, Br^-, NO_3^-, ReO_4^- and ClO_4^-, from CSF to blood (Wright 1974). The permeability of the paracellular pathway is similar to Na^+ and Cl^-. The trace anions are accumulated across the apical membrane by a Na^+-dependent, ouabain-sensitive mechanism, with properties very similar to the cloned Na^+/iodide cotransporter NIS (Eskandari et al. 1997). The physiological importance of this choroid plexus anion transport is unknown, but it does explain the perchlorate sensitive I^- uptake into the human choroid plexus in vivo.

3.13 Summary

What we know about ion transport across the frog choroid plexus is summarized in Fig. 3.4. Panel A shows the composition of the solutions bathing the CSF and vascular faces of the in vitro preparations of the choroid plexus, the intracellular ion concentrations, and the relative magnitude of the unidirectional fluxes across the cellular and paracellular pathways, e.g. the Na^+ unidirectional fluxes through the tight junctions is 3–4-times higher than the rate of Na^+ secretion. Panel B summarizes the net fluxes across the epithelium and the unidirectional ion fluxes across the apical and basolateral membranes. The membrane transporters and channels that may account for these fluxes are illustrated in Panel C. As discussed above, the density and turnover number of the apical $3Na^+/2 K^+$-pump accounts for the coupled efflux of Na^+ and the influx of K^+, and the efflux of K^+ is distributed between apical K^+-channels, and the apical $K^+/Cl^-/H_2O$ cotransporter. Both the outward K^+ and Cl^- gradients drive the efflux of coupled K^+ and Cl^-, but we have no direct estimate of the rates relative to the electrodiffusive K^+ efflux. If the cotransporter accounts for >90% of the K^+ efflux, >250 nmoles $cm^{-2} h^{-1}$, then the cotransport could account for >125,000 nmoles $cm^{-2} h^{-1}$ of water secretion, or 2.3 µl $cm^{-2} h^{-1}$, i.e., close to the observed rate of fluid secretion. We have evidence that basolateral Na^+/Cl^- and Na^+/HCO_3^- cotransporters account for Cl^- and HCO_3^- accumulation, but none for Na^+/H^+ exchangers or $Na^+/K^+/2Cl^-$ cotransporters.

All of our data was collected prior to the molecular biology revolution, and now with the identification of hundreds of genes coding of pumps, channels and transporters (Hediger et al. 2013), it should now be possible to establish the molecular identity of the membrane proteins, as has been attempted for mammalian epithelia (Damkier et al. 2010; Praetorius and Damkier 2017). This is a non-trivial task as we have learned in our studies of intestinal and renal epithelia. First, gene expression analysis assumes that mRNA levels are directly proportional to protein levels, and such assays do not indicate where proteins are distributed in the cell: in sheep intestine the levels of SGLT1 protein expression can vary by 200-fold with little or no change in mRNA levels (see Wright et al. 2011). Second, antibodies raised against short stretches of amino acids sequences of cloned membrane proteins are often used to identify the level and distribution of transporters and channels in epithelial cells using Western Blots and/or immunohistochemistry (see Wright et al. 2011). In using antibodies, particularly commercial antibodies (*caveat emptor*), the specificity of such antibodies are not often considered, and given the currently practiced techniques, it is well-nigh impossible to quantitate the level of protein expression. These two factors, specificity and quantification, have resulted in much confusion in the literature about the proteins expressed in choroid plexus from one species to another. Obviously, for a channel, the density of protein expression, the conductance and open probability are required to draw meaningful conclusions about the role of a given channel in ion fluxes across a cell membrane. Likewise, for transporters, the density and turnover number determine their role in the transport of molecules or ions into and out of cells. How can we solve questions about what membrane proteins are responsible for transport into and out of cells under physiological conditions? One approach is to selectivity knock down the expression of targeted proteins using selective gene knockout, RNA_i or Cre; quantitate the level of protein expression; and determine the change in functional activity in the epithelium. This functional genomics approach does pose a significant challenge since it requires a rare combination of skills in molecular biology, biochemistry and physiology.

Acknowledgments We are very appreciative of all the invaluable contributions made by our colleagues, Peter Brown, Paul Quinton, Yoshitaka Saito, John Tormey and Thomas Zeuthen; the technical assistance of Patricia Ilg; illustrations by Wendy Ravenhill, and the funding from the US National Institutes of Health.

References

Brown PD, Loo DD, Wright EM (1988) Ca^{2+}-activated K^+ channels in the apical membrane of Necturus choroid plexus. J Membr Biol 105(3):207–219

Christensen O, Zeuthen T (1987) Maxi K^+ channels in leaky epithelia are regulated by intracellular Ca^{2+}, pH and membrane potential. Pflugers Arch 408(3):249–259

Christensen O, Simon M, Randlev T (1989) Anion channels in a leaky epithelium. A patch-clamp study of choroid plexus. Pflugers Arch 415(1):37–46

Damkier HH, Brown PD, Praetorius J (2010) Epithelial pathways in choroid plexus electrolyte transport. Physiology (Bethesda) 25(4):239–249

Ehrlich BE, Wright EM (1982) Choline and PAH transport across blood-CSF barriers: the effect of lithium. Brain Res 250(2):245–249

Eskandari S, Loo DD, Dai G, Levy O, Wright EM, Carrasco N (1997) Thyroid Na^+/I^- symporter. Mechanism, stoichiometry, and specificity. J Biol Chem 272(43):27230–27238

Hediger MA, Clemencon B, Burrier RE, Bruford EA (2013) The ABCs of membrane transporters in health and disease (SLC series): introduction. Mol Asp Med 34(2–3):95–107

Loo DD, Brown PD, Wright EM (1988) Ca^{2+}-activated K^+ currents in Necturus choroid plexus. J Membr Biol 105(3):221–231

Nelson DJ, Wright EM (1974) The distribution, activity, and function of the cilia in the frog brain. J Physiol 243(1):63–78

Praetorius J, Damkier HH (2017) Transport across the choroid plexus epithelium. Am J Physiol Cell Physiol 312(6):C673–C686

Prather JW, Wright EM (1970) Molecular and kinetic parameters of sugar transport across the frog choroid plexus. J Membr Biol 2(1):150–172

Quinton PM, Wright EM, Tormey JM (1973) Localization of sodium pumps in the choroid plexus epithelium. J Cell Biol 58(3):724–730

Saito Y, Wright EM (1982) Kinetics of the sodium pump in the frog choroid plexus. J Physiol 328:229–243

Saito Y, Wright EM (1983) Bicarbonate transport across the frog choroid plexus and its control by cyclic nucleotides. J Physiol 336:635–648

Saito Y, Wright EM (1984) Regulation of bicarbonate transport across the brush border membrane of the bull-frog choroid plexus. J Physiol 350:327–342

Saito Y, Wright EM (1987) Regulation of intracellular chloride in bullfrog choroid plexus. Brain Res 417(2):267–272

Steffensen AB, Zeuthen T (2020) Cotransport of water in the choroid plexus epithelium. From amphibians to mammals. In: Praetorius J, Blazer-Yost B, Damkier H (eds) Role of the choroid plexus in health and disease. Springer, Heidelberg

Thomas RC (1972) Electrogenic sodium pump in nerve and muscle cells. Physiol Rev 52 (3):563–594

Wright EM (1972a) Accumulation and transport of amino acids by the frog choroid plexus. Brain Res 44(1):207–219

Wright EM (1972b) Active transport of lysergic acid diethylamide. Nature 240(5375):53–54

Wright EM (1972c) Mechanisms of ion transport across the choroid plexus. J Physiol 226 (2):545–571

Wright EM (1974) Active transport of iodide and other anions across the choroid plexus. J Physiol 240(3):535–566

Wright EM (1977) Effect of bicarbonate and other buffers on choroid plexus Na+/K+pump. Biochim Biophys Acta 468(3):486–489

Wright EM (1978) Transport processes in the formation of the cerebrospinal fluid. Rev Physiol Biochem Pharmacol 83:3–34

Wright EM, Prather JW (1970) The permeability of the frog choroid plexus to nonelectrolytes. J Membr Biol 2(1):127–149

Wright EM, Wiedner G, Rumrich G (1977) Fluid secretion by the frog choroid plexus. Exp Eye Res 25(Suppl):149–155

Wright EM, Loo DDF, Hirayama BA (2011) Biology of human sodium glucose transporters. Physiol Rev 91:733–794

Zeuthen T (1991) Water permeability of ventricular cell membrane in choroid plexus epithelium from *Necturus maculosus*. J Physiol 444:133–151

Zeuthen T (1994) Cotransport of K^+, Cl^- and H2O by membrane proteins from choroid plexus epithelium of *Necturus maculosus*. J Physiol 478(Pt 2):203–219

Zeuthen T, Wright EM (1978) An electrogenic Na^+/K^+ pump in the choroid plexus. Biochim Biophys Acta 511(3):517–522

Zeuthen T, Wright EM (1981) Epithelial potassium transport: tracer and electrophysiological studies in choroid plexus. J Membr Biol 60(2):105–128

Chapter 4
Cotransport of Water in the Choroid Plexus Epithelium: From Amphibians to Mammals

Annette B. Steffensen and Thomas Zeuthen

Abstract Transport by the choroid plexus epithelium is central for cerebrospinal fluid formation. The final steps take place across the brush border membranes, and the availability of these membranes allows detailed studies of relevant transport mechanisms. Here we review data from amphibians and relate them to mammals. Osmotic water permeabilities have been determined in amphibians with high precision by means of microelectrodes. No significant unstirred layers were observed: this accords with vesicle studies of both amphibian and mammalian membranes. Simple osmotic mechanisms do not explain fluid transport in epithelia adequately; the water permeabilities are not large enough. Furthermore, osmotic models cannot describe fluid transport against large osmotic gradients of up to half plasma osmolarity. Cotransporter-mediated water transport has been demonstrated in luminal membranes of the choroid plexus epithelium: in amphibia a K^+/Cl^- cotransporter and in mammals a $Na^+/K^+/2Cl^-$ cotransporter. These cotransporters are water permeable and the water flux can be energized by the cotransport of ions. We suggest that a hyperosmolar compartment within the cotransport-protein is central to this coupling. The water fluxes can proceed against significant osmotic gradients. Models of the choroid plexus epithelium employing cotransporter-mediated water transport give a quantitative description of the properties of the living tissue.

4.1 Introduction

Fluid transport by the choroid plexus epithelium (CPE) is interesting from a clinical, a physiological, and a molecular point of view. A major fraction of the cerebrospinal fluid (CSF) is secreted into the ventricles of the brain by the CPE, in humans about 500 ml a day, for some recent reviews see Damkier et al. (2013), Spector et al. (2015), Praetorius and Damkier (2017). The rate of secretion must be under strict

A. B. Steffensen · T. Zeuthen (✉)
Department of Neuroscience, The Panum Institute, University of Copenhagen, Copenhagen (København N), Denmark
e-mail: tzeuthen@sund.ku.dk

© The American Physiological Society 2020
J. Praetorius et al. (eds.), *Role of the Choroid Plexus in Health and Disease*,
Physiology in Health and Disease, https://doi.org/10.1007/978-1-0716-0536-3_4

control. If, for example, production exceeds that of the efflux mechanisms, the intracranial pressure will increase leading to potentially life-threatening conditions. Clearly, it would be important to be able to modulate cerebrospinal fluid production, but, unfortunately, the physiological mechanism of fluid transport in this epithelium is not understood very well. A general picture of the ion transport mechanisms is emerging, but the way the secretion of water is coupled to that of ions is not clear. The choroid plexus epithelium is particularly well suited for the study of the molecular mechanisms involved in the transport processes. This epithelium is one of the few in which the final steps in the fluid production takes place across a membrane not covered by connective tissues, another example is the airway epithelium. The exit membrane, which faces the ventricles, is directly accessible to electrophysiological experiments and the individual membrane proteins involved in the transport can be identified and probed by specific drugs (see Fig. 4.1).

Amphibians have several experimental advantages. The tissues are easy to handle and remain viable for hours even at room temperature, which makes them ideal for electrophysiological investigations. This is probably why the fundamentals of several physiological mechanisms were unraveled initially in cold-blooded animals: nerve conduction and muscle contraction are good examples. The use of choroid plexus from amphibians for the study of epithelial ion and water transport was pioneered by Wright and coworkers (Wright et al. 1977; Wright 1972) and has been central for the study of epithelial transport mechanisms. It is harder to do experiments on excised mammalian epithelia; they do not last so long. Nevertheless, the mammalian organs are larger, which facilitates studies of how the whole organ works and biochemical experiments such as molecular identification of transporters. In many respects, the results from amphibians and mammals complement each other. Here we primarily review what is known about fluid transport in the amphibian choroid plexus. In the concluding section, we will bridge the data obtained in amphibians with those from mammals. The specifics of ion transport will be the focus of a separate review in this book (Wright and Loo, this book).

4.1.1   Anatomy and Direction of Transport

The CPE is composed of a single layer of cuboidal cells joined together at their apical end by tight junctions. The apical, or brush border membrane faces the lumen of the ventricles, while the basolateral membranes of the cells face the serosal compartment which contains the capillaries. Thus, the basolateral membranes define the lateral intercellular spaces. In most epithelia, such as the small intestine and the kidney proximal tubule, the direction of salt and water transport is in the direction from the luminal towards the basolateral membrane; these epithelia are named 'forwards facing'. In the CPE, however, the direction of transport is opposite. It proceeds from the basolateral to the apical membrane; such epithelia are called 'backwards facing'. Apparently, the direction and mechanisms of salt and water transport is not linked to the micro-anatomy of the epithelial cells, see Fig. 4.1. Transport is closely

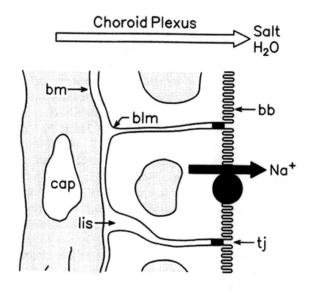

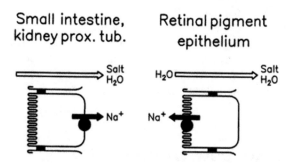

Fig. 4.1 Ultrastructure, ATPases, and the direction of salt and water transport. The choroid plexus epithelium (top) is characterized by a microvillus brush border membrane (bb), tight junctions (tj), a basolateral membrane (blm), and lateral intercellular spaces (lis). The cell layer rests on a leaky basement membrane (bm) and connective tissue, which contains fenestrated capillaries (cap). The Na^+/K^+-ATPase (filled symbols) links the active transport of Na^+ to the consumption of metabolic energy. The protein is found primarily in the brush border or ventricular membrane. Salt and water (cerebrospinal fluid) is transported in the direction from blood into the ventricle; i.e. across the basolateral membrane through the cell and finally across the brush border membrane into the ventricle. It is interesting to compare the epithelial cell from choroid plexus with those from other epithelia. The epithelial cell from small intestine, kidney proximal tubule and gall-bladder are alike as illustrated in the lower left panel. For these cells the direction of fluid transport is opposite to that of the choroid plexus: from brush border towards basolateral membrane. The lower right panel shows a cell from the retinal pigment epithelium. For this type of epithelia the direction of active Na^+ transport is opposite to that of the fluid transport. Apparently, ultrastructure and direction of active Na^+ transport do not have direct roles for the direction of salt and water transport in leaky epithelia. As reviewed elsewhere, the location of cotransporters such as the KCC and the NKCC1 do correlate to the direction on fluid transport. Adapted from Zeuthen (1992, 1996)

linked to the specific types of proteins that occupy the two ends of the epithelial cells. In fact, epithelial cells are polarized in the sense that the population of transport proteins in the apical membrane differs from that of the basolateral membrane. In the CPE the types of transport proteins that occupy the apical membrane are similar to those that occupy the basolateral membranes of the forwards facing epithelia, and vice versa. For example, in the CPE, the luminal membrane contains the Na^+/K^+-ATPase, while in the small intestine this membrane protein is located exclusively to the basolateral membrane facing the blood side (Fig. 4.1). The Na^+/K^+-ATPase is the link between the metabolic energy and ion transport. Thus, one might think that the direction of the active Na^+ transport by the ATPase would coincide with the overall direction of the epithelial NaCl and water transport, as is the case for the CPE and the epithelium small intestine. But this is not the general rule. In the retinal pigment epithelium and in the airway epithelium the direction of active Na^+ transport is opposite to that of the salt and water transport, for references see Zeuthen and Stein (1994).

Thus, neither the location of the Na^+/K^+-ATPase nor the microanatomy gives any idea about the mechanism of coupling between salt and water transport. Instead, as discussed below, the distribution of another class of membrane protein, the cotransporters, seems to play an important role. The CPE is an excellent choice for the study of membrane proteins in the ventricular membrane. The membrane is easily accessible from the ventricular solution, it has only short microvilli with lengths of about 1 µm, and the solution abutting the membrane is even stirred by cilia emerging from some of the cells (Nelson and Wright 1974). As can be ascertained with ion-selective microelectrodes, the concentrations and osmolarity of the solution secreted by the epithelium are maintained right up to the ventricular membrane: there are no significant unstirred layers.

4.2   A Note on Ion Transport in the CPE

In the following we give a brief overview of ion transport with focus upon those mechanisms most relevant for water transport: the Na^+/K^+-ATPase and the associated leakage pathway for K^+. Both mechanisms are located in the ventricular membrane (Fig. 4.2a). The ATPase is directly involved in the generation of osmotic gradients qua its role in controlling the intracellular K^+ concentration and the Na^+ concentration in the secretion. The leak is shared between a K^+/Cl^- cotransporter (KCC) and channel mediated transport (Fig. 4.2b). Importantly, transport in the KCC gives rise to water transport (Zeuthen 2010) while the channel-mediated transport does not. A more complete review of ion transport is given elsewhere in this book (Wright and Loo, this book).

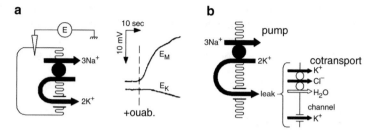

Fig. 4.2 Pumps and leaks in the ventricular membrane. (**a**) The Na^+/K^+-ATPase is electrogenic. The intracellular electrical potential (E_m) and the chemical potential for K^+ (E_K) are recorded from an epithelial cell of the bullfrog choroid plexus by means of a double barreled microelectrode. The activity of the ventricular Na^+/K^+-ATPase gives rise to a current since it translocate $3Na^+$ for each $2K^+$, subsequently, the K^+ leaks out of the cell again (black arrows). Application of the specific inhibitor ouabain stops the pump abruptly. This is seen as a depolarization in E_M of about $10\,mV$ and a leak of K^+ from the cell apparent as a linear decrease in E_K, redrawn from Zeuthen and Wright (1981). (**b**) The K^+ ions pumped into the cell by the ventricular Na^+/K^+-ATPase leak out of the cell again predominantly via the ventricular membrane. This can take place via K^+ channels and KCl cotransporters. The latter can cotransport water

4.2.1 The Na^+/K^+-ATPase

The ventricular localization of the Na^+/K^+-ATPase can be demonstrated from the effects of ouabain. When this specific inhibitor is applied to the ventricular solution, an abrupt depolarization of the epithelial cell of $10\,mV$ ($10.2\,mV$ to be precise) *and* a gradual loss of K^+ from the cells are initiated (Fig. 4.2a). This was observed by means of intracellular microelectrodes in the choroid plexus from Bullfrog (Zeuthen and Wright 1978, 1981). The depolarization is good evidence for an electrogenic pump, and the magnitude is precisely what would be expected from a pumping ratio (r) of 3 Na^+ to 2 K^+.

4.2.2 The KCC

K^+ and Cl^- are cotransported in the ventricular membrane of the salamander (*Necturus maculosus*) choroid plexus. This was first shown in experiments in which Cl^- ions were returned to tissues depleted of Cl^-. This caused a rapid influx of Cl^- which was accompanied by a numerical equal influx of K^+. The influxes of K^+ and Cl^- were *not* associated with any significant changes in electrical potential and were abolished by the cotransporter-specific inhibitor furosemide (Zeuthen 1991a, b, 1994). This was taken as good evidence for the existence of an electroneutral KCl cotransporter. The involvement of Na^+ was of minor importance as tested in several lines of experiments. For example, the furosemide sensitive influx of Na was maximally one quarter of that of K^+. In view of the stoichiometry of a supposed $Na^+/K^+/2Cl^-$ cotransporter the influx of Na^+ should have equaled that of

K^+. The distinction is important since a $Na^+/K^+/2Cl^-$ cotransporter is found in the ventricular membrane of the mammalian CPE (Fig. 4.6). Our data from the amphibian tissue seem to exclude this possibility, although a final verdict will have to await a more precise investigation, as for example by means of specific antibodies.

4.2.3   The K^+ Channels

The conductance of the ventricular cell membrane is higher than that of the serosal membrane by a factor of about ten. Furthermore, the cells have an electro-diffusive permeability for K^+ which is much larger than that for other ions. This shows that K^+ pumped into the cell may leave the cell across the ventricular membrane at least in part by electro-diffusion (Zeuthen and Wright 1978, 1981). Patch clamp techniques has enabled a dissection of this permeability into its channel components. The first channel to be investigated was a so called K^+ maxi-channel with a conductance of 200 pS (Christensen and Zeuthen 1987). The density of the channels was low, and only *if* all the channels were open the channels could explain the K^+ permeability of the membrane. However, the channels were practically closed under physiological conditions. At Ca^{2+} concentrations of 1 μM, *pH* of 7, and membrane potentials of −90 mV, the probability for being open was as low as 10^{-5}. Other patch clamp studies support this notion (Loo et al. 1988; Brown et al. 1988). Summing up, overall channel conductance is determined by several factors: The density of channels in the membrane and the opening probability of the channel, the latter being controlled by membrane potential, intracellular *pH*, and intracellular Ca^{2+}. These factors should be taken in account in order to estimate the significance of the given channel in the given physiological situation. A rough estimate obtained from the patch clamp data indicates that only 10 to 50% of the K^+ exit was mediated by channels under physiological conditions (Fig. 4.2b).

4.2.4   The Ratio Between Channel-Mediated and Cotransport of K^+

An estimate of the ratio between channel-mediated and cotransport of K^+ can be obtained from arresting the Na^+/K^+-ATPase abruptly with ouabain and observe the rate of K^+ loss into the ventricular solution (Fig. 4.2a). The rate will be a function of the K^+ concentration in the ventricular solution: channel-mediated transport will be much less sensitive to increases in ventricular K^+ concentrations than cotransport-mediated transport. In Bullfrog choroid plexus we found that a fivefold increase in the concentration of K^+ from 2 to 10 mM reduced the rate of K^+ loss by about 75% (Zeuthen and Wright 1981). In case of an entirely channel mediated exit of K^+, the rate should have been reduced by only about 15% as estimated from the Goldman equation. In contrast, the rate of cotransport is a linear function of the concentrations in the ventricular solution and the transport rate would be much more sensitive to the ventricular concentration of K^+ (Zeuthen 2010). Taken at face value, the data from (Zeuthen and Wright 1981) suggest that three quarters of the exit was mediated by cotransport and one quarter by electro-diffusion. This estimate is in good agreement with that obtained from the patch clamp data reviewed above and by Wright and Loo (this book).

The discussions above comply with the fact that the major job of channels is to maintain intracellular potentials, not to mediate large fluxes, while that of cotransporters is to move large number of ions without disturbing the cell too much. The importance of the KCC cotransporter emphasizes its role for water transport outlined in the following.

4.3   Fluid Transport by the Choroid Plexus Epithelium

Epithelial water transport is vital to whole body water homeostasis. Specifically, it is central for water homeostasis of the brain as well as a wide variety of other organs such as the eye, the inner ear and gallbladder. In man, the kidney proximal tubule reabsorbs more than 150 l each day and the small intestine reabsorbs up to 10 l in order to maintain the body's content of about 50 l. Despite the various specific functions of epithelia, the mechanism by which they transport water is remarkably similar. In the following, we briefly review some of the general properties of epithelial fluid transport and relate them to the choroid plexus epithelium.

4.3.1   General Properties of Epithelial Fluid Transport

Epithelial fluid transport is a sum of two components: an active and a passive one. The active, or vital, component of water transport (J_o) is coupled to the active transport of ions within the epithelial cell, and therefore depends on an intact cellular metabolism. If the tissues are poisoned by cyanide or deprived of glucose, the vital component of water transport is abolished. The passive component is determined by the passive water permeability of the epithelium (L_P) and is driven by external transepithelial osmotic gradients (Δosm). For detailed references see (Zeuthen and Stein 1994). L_P is not affected by poisoning of the tissue. This bimodal transport has been demonstrated for a wide range of epithelia both from amphibians and mammals and can be expressed as $J_V = J_o + L_P \times \Delta osm$ (Fig. 4.3). Interestingly, the combined fluid transport (J_V) can proceed against large opposing osmotic gradients. For a wide range of adverse transepithelial osmolarities, the active component of water transport more than compensate for the osmotic back flux. The relations between water transport and osmotic gradients have given rise to a number of important discussions: What determines the osmotic water permeability L_P? What is the mechanism

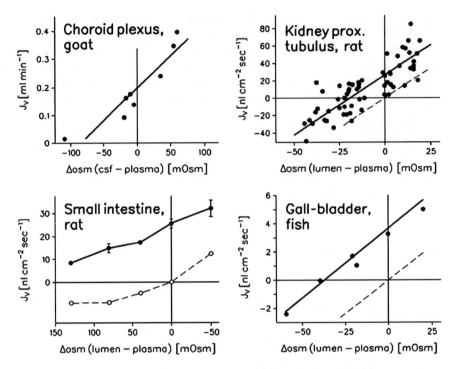

Fig. 4.3 The relation between transepithelial fluid transport (J_v) and transepithelial osmotic gradient (Δosm). For goat, the rate of cerebrospinal fluid production (J_v) is a linear function of the difference in osmolarity between the ventricles and the blood plasma (Δosm). When the ventricle is perfused by a hypertonic solution, fluid transport is increased; if the perfusate is hypotonic it is reduced. Importantly, secretion persists within a wide range of opposing osmotic gradients. In fact, it takes an opposing osmotic gradient of around 100 mOsm to arrest transport. Data from goat choroid plexus in vivo are plotted from Table 3 in Heisey et al. (1962); similar data were obtained in cat (Hochwald et al. 1974) and rabbit (Welch et al. 1966). Data from rat kidney proximal tubule are from Green et al. (1991). For this epithelium it takes 25 mOsm to arrest transport. Data from rat small intestine in vitro are from Parsons and Wingate (1961). Here it requires more than 150 mOsm to arrest transport. Data from fish gallbladder are from Diamond (1962) where it takes around 40 mOsm to arrest transport. The effect of arresting metabolism was tested by adding cyanid, in the intestine by removing glucose. In that case the relation between J_v and Δosm is given by the broken line

behind the vital component J_o? What enables the epithelium to transport against large opposing osmotic gradients?

4.3.2   The Osmotic Water Permeability

The osmotic water permeability L_P of the intact epithelium is a central parameter for any model of epithelial water transport. It is given by the permeability of the brush border (or apical) membrane in series with that of the basolateral membrane, the paracellular pathway can be ignored in this respect due to its very low cross sections area. In the following we will discuss how the water permeabilities can be determined and the errors involved.

The permeabilities of the apical and the basolateral membranes can be estimated with good precision by the water permeability of the plasma membrane determined from vesicle preparations in combination with estimates of the true size of the folded membranes found in the tissues. In general, the water permeabilities of cell membranes from mammals range from 0.4 cm s^{-1} (basolateral membranes) to 0.04 cm s^{-1} (apical membranes); the latter being similar to that of the red blood cell. For amphibians the estimates are ten times lower. The size of the cellular membranes can be determined from stereological analysis, which suggests a folding factor of 10 for apical membranes and 20 to 30 for basolateral membranes; for references see Zeuthen (1992, 1996). In rabbit proximal tubule cells the L_p derived from vesicles and stereological analysis matches the L_p of the intact tissue within a factor of two; for amphibians this estimate based upon vesicles and stereological analysis is actually smaller than that of the intact tissue. It would appear that there is no reason to suspect that the L_p determined from the intact epithelium severely underestimate of the true value of the cell membrane L_p. For a review of the data reviewed above see Zeuthen (1992).

The water permeability of the ventricular (apical) membrane of choroid plexus from *Necturus* has been measured with high precision by ion and volume sensitive microelectrodes (Zeuthen 1991a, 1994). By this method unstirred layers effects and other errors can be probed directly and corrected for. Consider, for example, the case where the L_P for the ventricular membrane of the *Necturus* choroid plexus is measured during a shift in the osmolarity of the external bathing solution implemented by addition of NaCl. In this case the change in ion concentrations inside the cell, just outside the cell and 5, 10, 15, 20, 30, and 40 μm away from the cell can be monitored at a resolution of 0.5 s. There is no significant difference in the time of onset of the change in concentration between any of these measurements, which means that any significant unstirred layer effects can be ruled out. Errors from the non-instantaneous change of the external solution due to convection amount to 50% of the ideal, maximal change. Errors from changes in intracellular osmolarity during the osmotic challenge were about 15%. Fortunately, these two errors are monitored directly by the microelectrodes and can easily be corrected for. Incidentally, there were no indications of the water flow sweeping away osmolytes from the surface. In summary, it is safe to conclude that the estimate of the L_P of the apical membrane of the *Necturus* choroid plexus is underestimated by about a factor two in agreement with the observations in *Necturus* gallbladder by Persson and Spring (1982) and Cotton et al. (1989).

A measure of the water permeability of the basolateral membranes of the *Necturus* choroid plexus cell can also be obtained by microelectrodes. Consider the case where the osmolarity of the ventricular solution is increased by 100 mOsm and a steady state osmotic flow is established through the cell. The intracellular osmolarity was measured to increase by about 35 mOsm, which means that the gradient imposed across the apical membrane is 65 mOsm and that across the basolateral membrane 35 mOsm (Zeuthen 1991a). Consequently, the water permeability of the basolateral membrane must be to be twice that of the apical membrane. The precise values of these water permeabilities are important parameters in the models of fluid transport in the choroid plexus discussed below.

It has been suggested that the L_p of the whole epithelium is underestimated by orders of magnitude due to unstirred layer effects (Diamond 1979). On this model unstirred layers cause the osmolarity of the solutions surrounding the cells to be different from those expected from the imposed test solutions. This could be the combined effect of changes in the rates of active iontransport and of restricted diffusion in the narrow lateral spaces and in the underlying connective tissues. In this way the effective osmotic driving force across the epithelial cells would be much smaller than the one imposed across the whole tissue, and consequently, the L_p could be a severe underestimate of the true epithelial L_p. There are several difficulties with this argument. Firstly, as outlined above, the water permeability obtained from vesicles studies, microelectrodes, and optical methods are not severely compromised by unstirred layers effects (Persson and Spring 1982; Cotton et al. 1989; Zeuthen 1991a). Secondly, the L_p of intact epithelia is the same in metabolizing and in poisoned tissue (Fig. 4.3). If the tissues are poisoned by cyanide or by deprivation of glucose, there is no longer any spontaneous salt and fluid transport. In other words, there is no way ion transport of the intact cells could affect the osmolarity of sub-epithelial or lateral intercellular spaces. Finally, the osmolarity of the lateral intercellular spaces is equal to that of the bathing solution under many typical transport conditions, as ascertained in amphibian gallbladder by means of ion selective microelectrodes (Zeuthen 1983; Ikonomov et al. 1985).

4.3.3   The Active Component of Water Transport

The active component of fluid transport (J_o) proceeds without any external osmotic driving force. Even if the two sides of the epithelium are bathed in the same solution a large component of fluid transport remains (Fig. 4.3). It is very well documented that the transported solution in this situation is isotonic i.e. it has the same osmolarity as the two (identical) bathing solutions (for references see Zeuthen 1996). The central question is how is the flux of water coupled to the ion fluxes? The mechanism of this coupling is still not resolved although several different types of models have been proposed. Most of these models are based upon the osmolarity of the lateral intercellular spaces. The standing gradient hypothesis and related theories assumes that this compartment has an osmolarity that is different to that of the outside solutions. In order to drive water transport in the measured direction the spaces should be hyperosmolar in "forward facing" epithelia such as gall-bladder and small intestine, and hyposmolar in "backward facing" epithelia such as the choroid plexus. Despite vigorous experimental efforts no such hyperosmolarity has been found. For example, the lateral intercellular spaces in gallbladder have been probed by ion selective microelectrodes (Zeuthen 1983; Ikonomov et al. 1985) and the contents found to be precisely isotonic to the external bathing solutions under a variety of transport situations. It has been suggested that the underlying connective tissue, basement membrane and capillary wall could form a barrier to transport and thus sustain an increased osmolarity in between and in the serosal compartment of the cells. In gall bladder the resistance of these barriers may amount to up to 30% of the intact epithelium. In contrast, in the kidney proximal tubule this barrier is negligible being permeable to large substances such as horseradish peroxidase. Accordingly, these theories cannot be the basis of any general theory; for a comprehensible review of data, see Zeuthen (1992).

Uphill water transport is a result of the dual nature of the transepithelial water transport. The osmotic component of transport changes with the osmotic gradient, but this is balanced by the active component which is relatively independent of osmolarity (Zeuthen 2010). It takes large opposing osmotic gradient to reduce water transport to zero, at which point the osmotic back flux of water is precisely matched by the active transport. In order to explain isotonic transport it is usually proposed that the water permeabilities are very high. If that was the case however, it would be very difficult to explain uphill water transport; the osmotic back fluxes would be overwhelming. This dilemma has put focus on the transport properties of the underlying connective tissues and the capillaries. However, as reviewed above, these structures offer little resistance to transport.

4.3.4 Osmotic Models

The choroid plexus epithelium has a unique role in solving the problems outlined above. This epithelium is "backward facing", and the secreted solution originates from a relatively flat ventricular membrane. There can be no significant modification of the composition of the secreted solution as suggested for "forward facing" epithelia where the secreted solution emerges from a complicated combination of lateral spaces and connective tissues. The most important parameters for an osmotic model are the passive, osmotic water permeability of the cell membranes and the location and magnitude of the putative hyper- or hyposmolar compartment that drives the transport.

The water permeability (L_p) of the ventricular membrane in the *Necturus* CPE was 1.3×10^{-4} cm s^{-1} Osm^{-1} and that of the basolateral membrane twice that, 2.6×10^{-4} cm s^{-1} Osm^{-1}. The L_p of the epithelium, the serial combination of the two values, calculates as 0.9×10^{-4} cm s^{-1} Osm^{-1} or 48 µm s^{-1} (Zeuthen 1991a, 1994). The data were obtained by microelectrodes and expressed per cm^{-2} of epithelial cells as discussed above. The pattern of the apical membrane having lower water permeability than the basolateral membrane one is similar to that of most other epithelia, despite the fact that the direction of water transport is opposite. The values were one order of magnitude smaller than those obtained for amphibian gallbladder (Zeuthen 1982; Reuss and Petersen 1985; Cotton et al. 1989; Persson and

Spring 1982; Reuss 1985); but similar to those obtained for the retinal membrane in Bullfrog pigment epithelium (La Cour and Zeuthen 1993). In the Bullfrog choroid plexus epithelium, the L_p of the whole excised epithelium was estimated to 22 μm s^{-1} (Wright et al. 1977). This value referred to the ventricular or chamber cross section. Folding of the tissue however, increases the area of the cell layer itself by a factor of 11 to 22 above that of the chamber area (Zeuthen and Wright 1981). Consequently, this L_p is in fact more than an order of magnitude smaller than that determined directly from the uppermost more flat parts of the epithelium by microelectrodes in *Necturus* (Zeuthen 1991a). This folding problem was also encountered directly in the Bullfrog in the determination of the K$^+$ permeability (Zeuthen and Wright 1981). It appears that for the excised whole epithelium, the lower and more infolded part epithelium do not participate in the transepithelial transport to the same extent as the upper parts, see Fig. 1 in Nelson and Wright (1974). Similar problems pertain to the serosal or basolateral compartment which is highly folded and invaginated. Such factors must be taken into account when transport parameters of the whole excised epithelium are derived.

In order to transport water across by osmosis at rates of 4 nl cm^{-2} s^{-1} (comparable to that of other amphibian epithelia) it would take transepithelial osmotic gradients of more than 40 mOsm given the water permeabilities above. Could the ion transport of the tissue itself generate such hyperosmolarities? For a flat membrane, this is unlikely. The salt would diffuse away too fast to build up any significant hyperosmolarity at the surface (Wright et al. 1977). However, the ventricular membranes of the amphibian CPE do have invaginations and microvilli of typically 1 μm (Carpenter 1966). This increases the actual surface area, but is much too short to induce any unstirred layer or the local osmotic gradients suggested by Diamond (1964). This mechanism assumes that solutes diffusing away from the surface and along the villous structures can generate sufficient osmotic water flows. However, it can be estimated from the formulas presented by Diamond (1964) that the effects are insufficient in the amphibian CPE, primarily due to the low water permeability of the cell plasma membrane at the ventricular membrane, see also discussion in Wright et al. (1977). We conclude that a simple osmotic mechanism is unlikely to explain water transport across the amphibian CPE. The same applies for the mammalian CPE as discussed in MacAulay and Zeuthen (2010) and below.

The problems of the osmotic model can be assessed directly, independently of precise values for the L_p and CSF composition, and without help from calculations. It has been suggested that water transport in the CPE is a consequence of a small degree of hyperosmolarity of the ventricular solution (5 to 10 mOsm) in combination with a high water permeability of the epithelial membranes, for a recent review see Praetorius and Damkier (2017). If this was the case however, water transport should be abolished if the osmolarity of the ventricular solution was lowered to that of the basolateral side, i.e. reduced by 5 to 10 mOsm, as in isotonic transport. This is not what is observed in the choroid plexus or in any other epithelium; at these reduced osmolarities the transport rate equals the active component J_o (Fig. 4.3), for references see Zeuthen and Stein (1994). These difficulties have led to the suggestion that the intercellular spaces in combination with underlying connective tissues and capillary walls constitute a hyposmolar compartment in the CPE. This would require that salt was removed from the lateral spaces and underlying connective tissues at high rates by the cells, in order to match the influx of ions and other osmolytes from the capillaries. This influx might well be prohibitively large in view of the fenestrated nature of the capillaries and the loose network of collagen fibers, see Fig. 2 in Damkier et al. (2013).

In summary, the problems of the simple osmotic model of the CPE are threefold: First, the water permeability is probably not large enough. Second, it is difficult to locate any hyper/hyposmolar compartment. Finally, it is doubtful if the underlying tissues and capillaries can uphold any barrier function. In the next paragraph we test if the introduction of a new building block, cotransporter-mediated water transport, can alleviate these problems.

4.3.5   Models Combining Osmosis and Cotransporter-Mediated Water Transport

There is good evidence and general agreement that many cotransporters do cotransport water along with the other substrates, although the precise coupling mechanism is debated (Lapointe et al. 2002; Zeuthen and MacAulay 2012b; Zeuthen et al. 2002). Cotransport of water has been demonstrated for a wide variety of cotransporters of the symport type such as KCC, Na^+-K^+-$2Cl^-$ cotransporter 1 (NKCC1), Na^+/glucose cotransporter 1 (SGLT1), and monocarboxylate transporter 1 (MCT1), see Appendix. In the choroid plexus from *Necturus*, we demonstrated a tight coupling between fluxes of K^+, Cl^-, and water in a K^+/Cl^- cotransporter located in the ventricular membrane. At each turnover of the protein one K^+ ion and one Cl^- ion were cotransported together with 460 water molecules (Zeuthen 1991a, b, 1994). The tightness of the coupling to water was evident from several types of experiment. Changes in external K^+ or Cl^- concentrations led to an immediate change in the rate of water transport. Conversely, osmotically induced effluxes of water within the protein gave rise to effluxes of K^+ and Cl^-. The coupling ratio was the same (460 water molecules per pair of ions) irrespective of the type of experiment. It should be emphasized that the cotransport was electroneutral, since exactly one K^+ ion was transported for each Cl^- ion. Na^+ had no role in the process as was tested in several lines of experiments employing Na^+ free solutions or Na^+ selective microelectrodes (Zeuthen 1994). This is important: in the mammalian choroid plexus it appears to be the Na^+ dependent NKCC1 cotransporter which is responsible for the uphill transport of water. The cotransport of water and ions had all the hallmarks of being mediated by a membrane protein undergoing conformational changes: (1) Specificity: the cotransport of water required K^+ and was specific for Cl^-, anion substitutions with gluconate, SCN^-, acetate or NO_3^- inhibited transport. (2) Saturation: the transport of water and K^+ saturated for extracellular osmotic gradients at or above 200 mOsm, and at extracellular K^+ concentrations of about 50 mM. (3) Inhibition: the transport was inhibited by the loop diuretics, furosemide at concentrations of 10^{-4} M and bumetanide at 10^{-5} M. A key experiment is shown in Fig. 4.4. Here the osmotic effects of mannitol, NaCl, and KCl were compared. When the osmolarity of the ventricular solution was increased abruptly by means of adding NaCl or mannitol, the CPE-cells shrank as would be expected from conventional osmotic behaviour. Surprisingly, if the external osmolarity was increased by means of KCl the cells *swelled*; despite the adverse osmotic gradient water moved into the cell. In the presence of furosemide the cotransport of water and ions was abolished and the osmotic effects of NaCl, KCl, and mannitol were the same. Importantly, the time resolution for the measurements of ions and water were fast (about 1 s), cotransport of water was evident before any significant changes in intracellular osmolarity were observed (Zeuthen 1994). For comparison, data from the mammalian CPE are also shown in Fig. 4.5. Here NKCC1 serves a role similar to that of KCC in the amphibian tissue (Steffensen et al. 2018).

Can inclusion of cotransport-mediated water transport alleviate the problems of the osmotic model? This can be analyzed by setting up a simple model that employs osmotic water transport across the ventricular and the basolateral membranes in combination with cotransport of water by a KCC transporter in the ventricular membrane (Fig. 4.5). The values for the various pathways are based upon the microelectrode recordings discussed above (Zeuthen 1994, 2010). In *Necturus* choroid plexus the KCC is abolished in Cl^- free solutions. The remaining water permeability of the ventricular membrane is made up from lipids, aquaporin water channels, and other membrane proteins. In the model this component is lumped into an osmotic water permeability of 0.6×10^{-4} cm s^{-1} Osm^{-1}. Presumably, the water permeability of the basolateral pathway also consists of a multitude of pathways, but in the model this is lumped together into an osmotic permeability of 2.6×10^{-4} cm s^{-1} Osm^{-1}. *The transport capacity of the KCC* has been determined quantitatively per cm^2 of epithelium as a function of the surrounding K^+, Cl^-, and water concentrations by means of ion selective microelectrodes (Zeuthen 1994, 2010). Given physiological values of intracellular concentrations of K^+ and Cl^- of 100 and 40 mM and concentrations for K^+ and Cl^- in the ventricular fluid of 2 and 110 mM, the rate of water transport by the KCC transporter can be estimated to 3.6 nl cm^{-2} s^{-1}. The cell is assumed to be hyperosmolar by 10–20 mOsm relative to the surrounding solutions, which is within the range of values determined in other amphibian epithelia (Zeuthen 1983), yet too small to affect the cotransporter to any significant extent. Under isotonic conditions, i.e. the ventricular and the basolateral side of the epithelium is bathed in identical solutions, the hyperosmolarity of the cell will cause a recirculation of 1.2 nl cm^{-2} s^{-1} across the ventricular membrane. This means that the net rate of water transport becomes 2.4 nl cm^{-2} s^{-1}. The model can transport water against significant transepithelial osmotic gradients. If the osmolarity of the ventricular solution is lowered by dilution, it takes a hyposmolarity of around 100 mOsm before the osmotic backflux is matched by the active efflux maintained by the KCC, and the combined transepithelial water transport becomes zero (Fig. 4.5). If the osmotic gradient is implemented by adding an inert osmolyte (mannitol) to the basolateral side, the epithelium can transport against 60 mOsm (not shown).

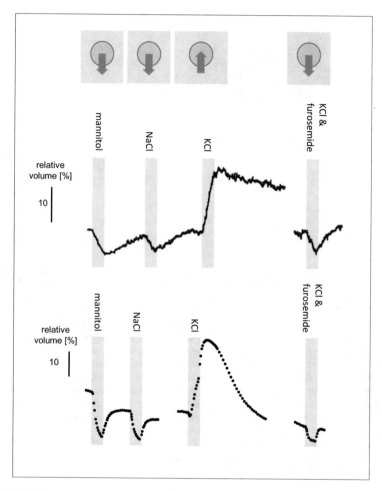

Fig. 4.4 Uphill transport of water in the luminal membrane of the choroid plexus in amphibia (upper traces) and mammals (lower traces). Addition of 100 mM of mannitol or 50 mM of NaCl to the ventricular solution (vertical blue bars) caused the epithelial cells of the choroid plexus to shrink (see symbols above) as would be expected from the outward osmotic gradient. In contrast, addition of 50 mM of KCl caused the cells *to swell* despite the fact that the osmolarity of the bath is higher than the intracellular osmolarity. In other words, water moves into the cells uphill, against the direction of the osmotic gradient. In presence of the inhibitor furosemide however, the cells shrank in response to the KCl. For the amphibian (*Necturus maculosus*), we concluded that the influx of KCl induces an influx of water via the furosemide sensitive cotransporter KCC. In this protein the water flux can proceed uphill against the direction of the osmotic gradient since it is coupled the downhill influx of salt. Importantly, osmotic effects can be excluded since the intracellular concentrations of K^+ and Cl^- did not change significantly during the tests. Data from the amphibian were obtained by means of ion and volume sensitive microelectrodes (Zeuthen 1994). Data from the mammal (mouse) were obtained from volume changes of cells loaded with fluorescence dyes and it was concluded that the uphill water transport was energized by a NKCC1 cotransporter (Steffensen et al. 2018). The duration of the changes in osmolarity for the amphibian were 20 s and for the mammal 50 s

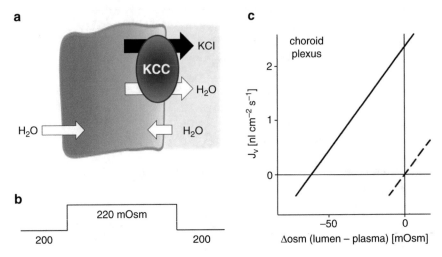

Fig. 4.5 Inclusion of cotransport-mediated water transport in an osmotic model of amphibian choroid plexus. In the model shown in (**a**), fluid transport is the combined result of active (uphill) water transport and osmotic transport. Water is exported uphill from the cell in cotransport with KCl in a K^+/Cl^- cotransporter localized in the ventricular membrane. Osmotic transport takes place across both the basolateral and the ventricular membrane. There is an osmotic influx across the basolateral membrane driven by the cellular hyperosmolarity of about 20 mOsm (**b**). In addition, the hyperosmolarity give rise to an osmotic reflux across the ventricular membrane of about 30%. The model is based on experimental data from the choroid plexus of *Necturus maculosus* and the lengths of the arrows are quantitatively correct. In this model the ability of the cell to transport water uphill is calculated in (**c**), which illustrate the rate of transport as a function of the degree of hyposmolarity of the ventricular solution. When the osmolarity is reduced by about 110 mOsm, net transport is nulled. At this point, the osmotic back-flux is exactly matched by the cotransport of water. In the absence of cotransported water (i.e. in a poisoned tissue) the rate of fluid transport simply reflects osmotic transport via the serial combination of the passive osmotic permeabilities of the two membranes (broken line). Data from Zeuthen (1991a, b, 1994), see also Zeuthen (1996). The model reflects the data obtained in living epithelia rather well (Fig. 4.3)

The model aligns with the general properties of the various epithelia summarized in Fig. 4.3. The epithelial cells have an osmotic pathway via the apical and basolateral membranes in series, which enables the epithelium to take advantage of favorable transepithelial gradients. This constitutes the water permeability (L_p) of the epithelium. A large fraction of the active component of water transport (J_o) is due to cotransport-mediated water transport. In the amphibian choroid plexus this function is maintained by the KCC located in the ventricular membrane. As outlined below, this role is probably maintained by a NKCC1 In the mammalian CPE (Steffensen et al. 2018). For intestine, kidney proximal tubule, and gall bladder, the cotransport-mediated water transport is found in the basolateral membranes. Cotransport of water would ensure fluid transport even in the face of adverse transepithelial osmotic gradients. Interestingly, the K^+ leaking out of the cell is not just wasted, but used to energize the transport of water. It should be emphasized that the model is entirely cellular; there is no need to invoke any barrier function of underlying connective tissues and capillary walls.

4.4   From Amphibians to Mammals

The results obtained in amphibians constitute a blueprint for investigations of mammalian CPE function. In the following we compare amphibian data with those of mammals.

For vertebrates with relatively large brains such as amphibians and mammals, the structure and function of the cerebrospinal fluid system are similar. The fluid is produced into a system of ventricles by several choroid plexuses. The ultrastructure of the epithelial cell display a ventricular membrane that is relatively flat and a basolateral membrane folded into a system of lateral intercellular spaces (Carpenter 1966; Scott et al. 1974; Damkier et al. 2013). The cerebrospinal fluid system is generally smaller in amphibians than in mammals, which is probably why there are fewer data for CSF production rates in amphibians than in mammals. In fact, the rate of production in amphibians has only been measured in Bullfrog (Wright et al. 1977) where a rate of 0.07 ml/min/gr tissue was found. This is significantly lower than that determined in mammals of around 0.3 ml/min/gr (Cserr 1971). The water permeabilities as well as the transport rates for ions are also smaller in amphibians compared to mammals by a factor of about ten. These differences comply with what has been found for other types of epithelia, for references see Zeuthen (1992, 1996).

The types of membrane proteins located to the two sides of the epithelial cell of amphibians and mammals have similarities, but they are not identical: similar functions may not be undertaken by the same kinds of proteins. Experiments on amphibians showed that the major link between metabolism and ion transport resides in a Na^+/K^+-ATPase located exclusively at the ventricular membrane (Wright 1972; Zeuthen and Wright 1981). Na^+ is pumped out of the cell by this ATPase, while K^+ is pumped in, only to leak out again via channels and cotransporters. A similar pump-leak system has been identified in mammals (Damkier et al. 2013). Differences between amphibians and mammals become apparent however, when other types of membrane proteins are investigated. We have suggested that fluid transport by the choroid plexus requires cotransporter-mediated water transport, both in amphibians (Zeuthen 1991b) and mammals (Steffensen et al. 2018). But the type of cotransporters involved appears to be different; in the salamander water transport was energized by cotransport in a K^+/Cl^- cotransporter (KCC) located in the ventricular membrane, in mammals this coupling to water is accomplished a Na^+-K^+-$2Cl^-$ cotransporter (isoform NKCC1), see Fig. 4.6. The ability of uphill water transport by the KCC and the NKCC1 was demonstrated by experiments like those of Fig. 4.4. Water transport is induced by the abrupt increase of external amount of KCl. Both transporters induce uphill transport of water, but in the NKCC1 this could be shown to depend on the presence of Na^+. Cotransport of water by the NKCC1 has

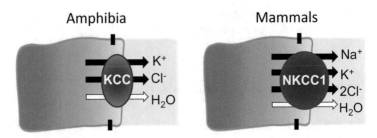

Fig. 4.6 The role of cotransporters in epithelial fluid transport. Cotransport-mediated water transport is a key element in cerebrospinal fluid production. The KCC is operating in the ventricular membrane of amphibians (Zeuthen 1991a, b, 1994), the NKCC1 in mammals (Steffensen et al. 2018)

Table 4.1 Osmolarity of cerebrospinal fluid of mammals (csf) relative to plasma

Species	csf-plasma [mOsm]	References
Rabbit	305–299 (Δ 6)	Davson and Segal (1996)
Rabbit	302–298 (Δ 4)	Pollay and Curl (1967)
Dog	305–300 (Δ 5)	Davson and Segal (1996)
Cat	315–309 (Δ 6)	Bito et al. cited in Davson and Segal (1996)
Cat	321–323 (Δ −2)	Hochwald et al. (1974), Bradbury (1969), Pollay and Curl (1967)
Rat	315–308 (Δ 7)	DePasquale et al. (1989)
Human[a]	289–289 (Δ 0)	Hendry (1962), Davson and Purvis (1954)

[a]Lumbar csf

previously been demonstrated in tissue cultures of mammalian epithelia predominantly from the eye (Hamann et al. 2005, 2010), but also by heterologous expression in Xenopus oocytes (Zeuthen and MacAulay 2012a). Interestingly the NKCC2 isoform did not cotransport water. This transporter is located to the thick ascending limb of Henle, where water transport by this protein would be problematic. The inability of water transport by the NKCC2 serves as an important control for the concept of cotransporter-mediated water transport.

As discussed in a previous section, it is difficult to explain fluid transport in the amphibian choroid plexus by models based entirely on osmosis. The same applies to mammals, where the osmotic driving forces and the water permeability of the epithelium appear to be too low. There is some uncertainty as to the exact value of the osmolarity of the cerebrospinal fluid of mammals relative to plasma (Table 4.1). For rabbit, dog, and rat the CSF is 2 to 3% hyperosmolar relative to plasma, while the data from cat are conflicting, both hyper- and hyposmolar values have been reported. Data from humans suggest that CSF is isotonic with plasma. It should be noted however, that in these studies CSF were obtained from lumbar samples. Clearly, a more conclusive and extended study of the precise value of the osmolarity of the CSF is required. Irrespective of the exact value of the osmolarity of CSF in steady state, it can be calculated that it would take much larger gradients of about 250 mOsm to pull water through osmotically at the required rates given the water permeability of mammalian epithelial cell membranes (Zeuthen 1992, 1996).

The problems of a simple osmotic model for mammals are illustrated in experiments where the CSF production proceed against oppositely directed osmotic gradients of 50 to 150 mOsm (Heisey et al. 1962; Hochwald et al. 1974; Welch et al. 1966), see Fig. 4.3. In a simple osmotic model this gradient would lead to impossibly large back-fluxes. It has been suggested that such observations are complicated by unstirred layers of high osmolarity building up in between and behind the cells. But this would require that the wall of the fenestrated capillaries and the loose underlying connective tissues constituted a significant diffusion barrier to ions, which was able to sustain gradients of up to 100 mOsm between blood plasma and the epithelial cells. This seems unlikely.

Finally, the passive water channel AQP1 has been assumed to be responsible for a major fraction of the passive water permeability of the ventricular membrane in mammalian choroid plexus. If this protein is knocked out however, the production rate of CSF reduced by only 20%. This effect should be evaluated in the light of an 80% decrease in the central venous blood pressure in these animals (Oshio et al. 2005). The lack of dependence of an intact osmotic pathway is corroborated by the fact that the human phenotype Coulton-null, which has no aquaporin of the isoform AQP1, suffers very few ill effects (Agre et al. 1995; Preston et al. 1994). The small effect of absence of AQP1 lends support to our suggestion that the exit of water from the epithelial cell is cotransporter-mediated to a significant extent (Fig. 4.5).

4.5 Conclusions

1. The amphibian CPE is a good place to begin investigating the mechanisms of cerebrospinal ion transport and fluid production. The tissues are easy to handle and long lasting, which has allowed key transport parameters to be determined.
2. The properties of fluid transport in the choroid plexus epithelium are similar to those of other epithelia, except that the direction of fluid transport relative to the anatomy is opposite. This allows the molecular mechanisms of the final fluid formation at the ventricular membrane to be studied at high precision.
3. The Na^+-K^+ ATPase is located exclusively in the ventricular membrane. $3Na^+$ are pumped into the ventricular solution for each $2K^+$ pumped into the cell. Most of the K^+ leaks back into the ventricular solution.
4. The K^+ leak is shared between electroneutral cotransport with Cl^- in a K^+/Cl^- cotransporter and electrogenic channel mediated transport. Various estimates suggest that 50 to 90% of the leak is mediated by cotransporters.
5. The osmotic water permeabilities have been determined in amphibians with high precision. In mammals they can be inferred quite accurately from vesicle studies. The values are accurate within a factor of two, not orders of magnitude as sometimes suggested. No significant unstirred layers are observed.

6. Osmosis is inadequate to explain fluid transport. The water permeabilities as well as the osmotic gradients are too small. In order to explain uphill transport of water, a simple osmotic model would have to rely upon a hypothetical barrier function of the connective tissue.
7. In cotransporter-mediated water transport water is moved along with the ions. The ion fluxes can energize uphill water fluxes. The phenomenon is generally agreed upon, but the underlying mechanism is not fully understood. A hyperosmolar compartment within the cotransporter is considered important.
8. Cotransporter-mediated water transport has been demonstrated in the ventricular membrane of the CPE. The cotransporters involved are a KCC in amphibia and NKCC1 in mammals.
9. Models of the CPE with cotransporter-mediated water transport located at the ventricular membrane give a good quantitative description of the transport in the living tissue.
10. Cotransporter-mediated water transport is not limited to epithelial transport in vertebrate epithelia. It has also been implicated in the movement of water in yeast (Zhang et al. 2011), plants (Wegner 2013), cancer cells (Springer et al. 2014), and neurons (Glykys et al. 2014).

Acknowledgements S. Christoffersen and W. Zeuthen are thanked for graphical assistance, E. K. Ørnbo and N. MacAulay for critical reading.

Appendix: Molecular Mechanism of Water Cotransport

It is generally accepted that cotransport proteins of the symport type play a key role for the coupling between ion and water fluxes, although the exact molecular mechanism is not understood. Water follows the other substrates closely in the transport protein: the molar ratio between substrate and water transport is a constant irrespective of whether the transport is driven by concentration, electrical, or osmotic gradients. Cotransport of water has been described for a number of different cotransporters by a variety of techniques, for reviews see Zeuthen (2010), Zeuthen and MacAulay (2012b). The present review deals mainly with the K^+/Cl^- cotransporter KCC, but the phenomenon has been established in several other cotransporters, such as the $Na^+/K^+/2Cl^-$ (isoform NKCC1), the SGLT1, and the $H^+/lactate$ transporter (Zeuthen et al. 1996). Interestingly, the isoform NKCC2 did not cotransport water (Hamann et al. 2005; Zeuthen and MacAulay 2012a). Many of the cotransporters have been studied heterologeously expressed in Xenopus oocytes which enables the effects of site directed mutations to be investigated (Zeuthen et al. 2016).

To understand water cotransport, the structure of each conformational states of the same protein must be known. Furthermore, the transitions between these states must be described kinetically. At present, a complete set of data for one protein is not available and a putative model can only be set up piecing together information from

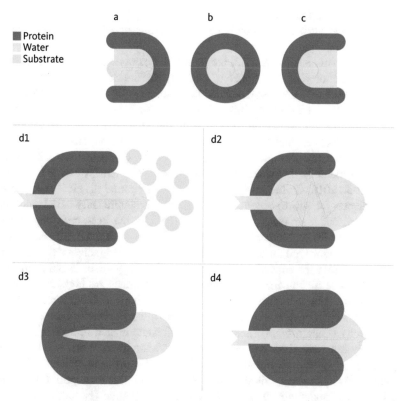

Fig. 4.7 Molecular mechanisms of water cotransport. During transport the cotransport protein changes conformation. In the outward facing confirmation (**a**) the binding site(s) is available from the outer solution via an aqueous cavity. This is followed by an occluded conformation (**b**) in which the substrate and some water molecules are internalized. In the inward facing conformation (**c**), the substrate and water gain access to the internal solution. It is generally held that the mechanism of water cotransport is associated with the exit step and various mechanisms have been suggested: (d1) Substrate concentration increases significantly in the external solution abutting the protein (unstirred layers) and water is transported by conventional osmosis through the protein. (d2) The presence of substrate in the exit funnel gives rise to large intramolecular hyperosmolarities, water transport results from osmosis. (d3) When the substrate has left the protein, the exit cavity collapses and water in the cavity is squeezed out. (d4) Water in the protein is pushed through by the substrate. Figure adapted from Zeuthen and MacAulay (2012b)

different cotransporters. In broad terms, transport is maintained by the transition between two conformational states, an outwards facing and an inward facing, the so called alternating access model (Mitchell 1957). In the first state the substrate gains access to the binding site from the outer solution via an aqueous cavity (Fig. 4.7a). This is followed by the occlusion of the substrates and some water molecules (b). Finally, in the inward open state (c) the substrate and the occluded water gain access to the inner solution via an aqueous cavity. After the substrate has left the protein, the empty transporter may attain a closed conformation from which it returns to the inward-open conformation (not shown). It is generally held that the coupling mechanism is closely associated with the exit step, four models have been suggested, d1–d4.

The first two models (d1 and d2) are essentially osmotic and require an aqueous pathway in the protein. An aqueous pathway in the sodium coupled glucose cotransporter SGLT1 has been described both functionally (Loo et al. 1996, 1999) and structurally (Faham et al. 2008). This aqueous pathway, as well as that in members of other relevant cotransporters from the SSS superfamily (solute-sodium superfamily), has been analyzed in molecular dynamics simulations (Li et al. 2013), which demonstrates opening and closing of the pathway. Furthermore, it has been shown experimentally that water and substrates share a pathway inside the SGLT1 (Zeuthen et al. 2016). In the so-called *unstirred layer model* (d1) it is assumed that during transport the substrate concentration increases significantly in the solution at the exit side, and that water transport is driven by the accompanying osmotic gradient. The major problem with this model is however, that the diffusion rates in the external solutions are far too high to sustain significant gradients; the substrates diffuse away from the transporter before any osmotic gradient is build up. Even inside cells the diffusion for sugars and smaller substrates are high, about half of those in free solution (Zeuthen and MacAulay 2002). This problem is circumvented in the *hyperosmolar-cavity model* (d2). Here the hyperosmolarity builds up inside the exit cavity of the protein. As long as the substrate is held in here in a thermo-dynamically free state, there will be an osmotic driving force. Water will finally exit the protein driven by an intramolecular hydrostatic pressure. Given reasonable parameters for the sugar cotransporter SGLT1 and the monoport GLUT, this model predicts quantitatively several experimental results (Naftalin 2008). The water transport will depend on how long the substrate is present inside the cavity before entering the outer solution. In this context, it is interesting that the state in which the substrate leaves the cotransporter has been shown to be the rate limiting step, for a review on the SGLT see Wright et al. (2011). This might explain why hyperosmolarities in the exit solutions accelerates the rate of water cotransport (Zeuthen 1994). If a large inert osmolyte, say mannitol, was permanently present in the exit solution, the water remaining in the exit cavity after the substrate has left would be under osmotic stress. This would speed up its closure and hence the return of the protein to its outward open conformation ready for a new transport cycle. In fact, the experimental evidence for water cotransport could be explained by a combination of osmo-sensitive kinetics and a *hyperosmolar-cavity model*.

The last two models do not employ osmosis as such. In the *Occlusion model* (d3) it is assumed that a given amount of water is occluded together with the substrate inside the protein. This water is transferred with the substrate to the exit cavity and squeezed out as the cavity collapses. However, in the crystal structures determined so far, the occluded state are found to contain far less than the 90 water molecules required by this model. It should be noted however, that the transport experiments were performed on proteins from amphibians or mammals, while the structures were determined in bacterial proteins. It remains to be seen how much water can be held in cotransporters from vertebrates. In the *Brownian piston model* (d4) the water is pushed through the protein (Zeuthen 1996). The idea has been tested by molecular dynamics simulations lasting 200 ns (Choe et al. 2010). It could be argued that this is too short to give a realistic picture of the transport process which may last around 10 ms.

This analysis shows that it is possible to set up a working model for cotransport of water based on well-established experimental and structural findings. At present the *hyperosmolar-cavity model* (d2) holds most promise. It describes more experimental findings and conflicts with fewer facts than the other models. A number of questions for future investigation present themselves: For example, is there a pathway for water in the protein throughout the whole transport cycle? Can thermodynamics describe the substrate-created hyperosmolarity in the exit funnel; how high is the osmolarity and how many water molecules are transported osmotically through the protein in this period? How much water is occluded in mammalian proteins? Will a difference in substrate affinity between the entry and the exit side (Eskandari et al. 2005) lead to a preferred direction of water transport?

Transporters from the cation chloride cotransporters (CCC) differ from those of the SSS family. For both families a water flux is induced when the transporter is exposed to a transmembrane osmotic gradient. But for the CCC-cotransporters the water fluxes are associated with fluxes of ions and have high activation energies (Hamann et al. 2005). This contrasts with the SSS proteins which have a channel-like aqueous pathway with low activation energy. Functionally at least, this pathway works in parallel with the Na^+/glucose/water cotransport system. It would appear that the CCC transporters are more tightly coupled to water than the SSS transporters.

References

Agre P, Smith BL, Preston GM (1995) ABH and Colton blood group antigens on aquaporin-1, the human red cell water channel protein. Transfus Clin Biol 2(4):303–308

Bradbury MW (1969) Dependence of cerebrospinal fluid potassium, calcium and magnesium on cerebrospinal fluid sodium. J Physiol 201(1):32P–33P

Brown PD, Loo DD, Wright EM (1988) Ca2+-activated K+ channels in the apical membrane of Necturus choroid plexus. J Membr Biol 105(3):207–219

Carpenter SJ (1966) An electron microscopic study of the choroid plexuses of *Necturus maculosus*. J Comp Neurol 127(3):413–433

Choe S, Rosenberg JM, Abramson J, Wright EM, Grabe M (2010) Water permeation through the sodium-dependent galactose cotransporter vSGLT. Biophys J 99(7):L56–L58

Christensen O, Zeuthen T (1987) Maxi K+ channels in leaky epithelia are regulated by intracellular Ca2+, pH and membrane potential. Pflugers Arch 408(3):249–259

Cotton CU, Weinstein AM, Reuss L (1989) Osmotic water permeability of Necturus gallbladder epithelium. J Gen Physiol 93(4):649–679

Cserr HF (1971) Physiology of the choroid plexus. Physiol Rev 51(2):273–311

Damkier HH, Brown PD, Praetorius J (2013) Cerebrospinal fluid secretion by the choroid plexus. Physiol Rev 93(4):1847–1892

Davson H, Purvis C (1954) Cryoscopic apparatus suitable for studies on aqueous humour and cerebro-spinal fluid. J Physiol 124(2):12–3P

Davson H, Segal MB (1996) Physiology of the CSF and blood-brain barriers. CRC, Boca Raton, FL

DePasquale MICH, Patlak CS, Cserr HF (1989) Brain ion and volume regulation during acute hypernatremia in Brattleboro rats. Am J Physiol Renal Physiol 256(6):F1059–F1066

Diamond JM (1962) The mechanism of water transport by the gall-bladder. J Physiol 161 (3):503–527

Diamond JM (1964) The mechanism of isotonic water transport. J Gen Physiol 48(1):15–42

Diamond JM (1979) Osmotic water flow in leaky epithelia. J Membr Biol 51(3–4):195–216

Eskandari S, Wright EM, Loo DDF (2005) Kinetics of the reverse mode of the Na+/glucose cotransporter. J Membr Biol 204(1):23–32

Faham S, Watanabe A, Besserer GM, Cascio D, Specht A, Hirayama BA, Wright EM, Abramson J (2008) The crystal structure of a sodium galactose transporter reveals mechanistic insights into Na+/sugar symport. Science 321(5890):810–814

Glykys J, Dzhala V, Egawa K, Balena T, Saponjian Y, Kuchibhotla KV, Bacskai BJ, Kahle KT, Zeuthen T, Staley KJ (2014) Local impermeant anions establish the neuronal chloride concentration. Science 343(6171):670–675

Green R, Giebisch G, Unwin R, Weinstein AM (1991) Coupled water transport by rat proximal tubule. Am J Physiol Renal Physiol 261(6):F1046–F1054

Hamann S, Herrera-Perez JJ, Bundgaard M, Alvarez-Leefmans FJ, Zeuthen T (2005) Water permeability of Na+-K+-2Cl- cotransporters in mammalian epithelial cells. J Physiol 568 (1):123–135

Hamann S, Herrera-Perez JJ, Zeuthen T, Alvarez-Leefmans FJ (2010) Cotransport of water by the Na+-K+-2Cl- cotransporter NKCC1 in mammalian epithelial cells. J Physiol 588 (21):4089–4101

Heisey SR, Held D, Pappenheimer JR (1962) Bulk flow and diffusion in the cerebrospinal fluid system of the goat. Am J Physiol 203(5):775–781

Hendry EB (1962) The osmotic pressure and chemical composition of human body fluids. Clin Chem 8(3):246–265

Hochwald GM, Wald A, DiMattio J, Malhan C (1974) The effects of serum osmolarity on cerebrospinal fluid volume flow. Life Sci 15(7):1309–1316

Ikonomov O, Simon M, Frömter E (1985) Electrophysiological studies on lateral intercellular spaces of Necturus gallbladder epithelium. Pflugers Arch 403(3):301–307

La Cour M, Zeuthen T (1993) Osmotic properties of the frog retinal pigment epithelium. Exp Eye Res 56(5):521–530

Lapointe J-Y, Gagnon M, Poirier S, Bissonnette P (2002) The presence of local osmotic gradients can account for the water flux driven by the Na+-glucose cotransporter. J Physiol 542(1):61–62

Li J, Shaikh SA, Enkavi G, Wen PC, Huang Z, Tajkhorshid E (2013) Transient formation of water-conducting states in membrane transporters. Proc Natl Acad Sci U S A 110(19):7696–7701

Loo DD, Brown PD, Wright EM (1988) Ca 2+-activated K+ currents in Necturus choroid plexus. J Membr Biol 105(3):221–231

Loo DD, Zeuthen T, Chandy G, Wright EM (1996) Cotransport of water by the Na+/glucose cotransporter. Proc Natl Acad Sci U S A 93(23):13367–13370

Loo DD, Hirayama BA, Meinild A-K, Chandy G, Zeuthen T, Wright EM (1999) Passive water and ion transport by cotransporters. J Physiol 518(1):195–202

MacAulay N, Zeuthen T (2010) Water transport between CNS compartments: contributions of aquaporins and cotransporters. Neuroscience 168(1873-7544 (Electronic)):941–956

Mitchell P (1957) A general theory of membrane transport from studies of bacteria. Nature 180 (4577):134–136

Naftalin RJ (2008) Osmotic water transport with glucose in GLUT2 and SGLT. Biophys J 94 (10):3912–3923

Nelson DJ, Wright EM (1974) The distribution, activity, and function of the cilia in the frog brain. J Physiol 243(1):63–78

Oshio K, Watanabe H, Song Y, Verkman AS, Manley GT (2005) Reduced cerebrospinal fluid production and intracranial pressure in mice lacking choroid plexus water channel Aquaporin-1. FASEB J 19(1):76–78

Parsons DS, Wingate DL (1961) The effect of osmotic gradients on fluid transfer across rat intestine in vitro. Biochim Biophys Acta 46(1):170–183

Persson BE, Spring KR (1982) Gallbladder epithelial cell hydraulic water permeability and volume regulation. J Gen Physiol 79(3):481–505

Pollay M, Curl F (1967) Secretion of cerebrospinal fluid by the ventricular ependyma of the rabbit. Am J Physiol 213(4):1031–1038

Praetorius J, Damkier HH (2017) Transport across the choroid plexus epithelium. Am J Physiol Cell Physiol 312(6):C673–C686

Preston GM, Smith BL, Zeidel ML, Moulds JJ, Agre P (1994) Mutations in aquaporin-1 in phenotypically normal humans without functional CHIP water channels. Science 265 (5178):1585–1587

Reuss L (1985) Changes in cell volume measured with an electrophysiologic technique. Proc Natl Acad Sci U S A 82(17):6014–6018

Reuss L, Petersen KU (1985) Cyclic AMP inhibits Na+/H+ exchange at the apical membrane of Necturus gallbladder epithelium. J Gen Physiol 85(3):409–429

Scott DE, Van Dyke DH, Paull WK, Kozlowski GP (1974) Ultrastructural analysis of the human cerebral ventricular system. Cell Tissue Res 150(3):389–397

Spector R, Keep RF, Snodgrass SR, Smith QR, Johanson CE (2015) A balanced view of choroid plexus structure and function: focus on adult humans. Exp Neurol 267:78–86

Springer CS, Li X, Tudorica LA, Oh KY, Roy N, Chui SY-C, Naik AM, Holtorf ML, Afzal A, Rooney WD, Huang W (2014) Intra-tumor mapping of intra-cellular water lifetime: metabolic images of breast cancer? NMR Biomed 27:760–773

Steffensen AB, Oernbo EK, Stoica A, Gerkau NJ, Barbuskaite D, Tritsaris K, Rose CR, MacAulay N (2018) Cotransporter-mediated water transport underlying cerebrospinal fluid formation. Nat Commun 9(1):2167

Wegner LH (2013) Root pressure and beyond: energetically uphill water transport into xylem vessels? J Exp Bot 65(2):381–393

Welch KEAS, Sadler KEIT, Gold GERA (1966) Volume flow across choroidal ependyma of the rabbit. Am J Physiol 210(2):232–236

Wright EM (1972) Mechanisms of ion transport across the choroid plexus. J Physiol 226 (2):545–571

Wright EM, Wiedner G, Rumrich G (1977) Fluid secretion by the frog choroid plexus. Exp Eye Res 25:149–155

Wright EM, Loo DD, Hirayama BA (2011) Biology of human sodium glucose transporters. Physiol Rev 91(2):733–794

Zeuthen T (1982) Relations between intracellular ion activities and extracellular osmolarity in Necturus gallbladder epithelium. J Membr Biol 66(1):109–121

Zeuthen T (1983) Ion activities in the lateral intercellular spaces of gallbladder epithelium transporting at low external osmolarities. J Membr Biol 76(2):113–122

Zeuthen T (1991a) Secondary active transport of water across ventricular cell membrane of choroid plexus epithelium of Necturus maculosus. J Physiol 444(1):153–173

Zeuthen T (1991b) Water permeability of ventricular cell membrane in choroid plexus epithelium from Necturus maculosus. J Physiol 444(1):133–151

Zeuthen T (1992) From contractile vacuole to leaky epithelia. Coupling between salt and water fluxes in biological membranes. Biochim Biophys Acta Rev Biomembr 1113(2):229–258

Zeuthen T (1994) Cotransport of K^+, cl^- and H_2O by membrane proteins from choroid plexus epithelium of Necturus maculosus. J Physiol 478(2):203–219

Zeuthen T (1996) Molecular mechanisms of water transport. RG Landes Company, Texas, p 170

Zeuthen T (2010) Water-transporting proteins. J Membr Biol 234(2):57–73

Zeuthen T, MacAulay N (2002) Cotransporters as molecular water pumps. Int Rev Cytol:259–284

Zeuthen T, MacAulay N (2012a) Cotransport of water by Na^+-K^+-$2Cl^-$ cotransporters expressed in Xenopus oocytes: NKCC1 versus NKCC2. J Physiol 590(5):1139–1154

Zeuthen T, MacAulay N (2012b) Transport of water against its concentration gradient: fact or fiction? WIREs Membr Transp Signal 1(4):373–381

Zeuthen T, Stein WD (1994) Cotransport of salt and water in membrane proteins: membrane proteins as osmotic engines. J Membr Biol 137(3):179–195

Zeuthen T, Wright EM (1978) An electrogenic Na+/K+ pump in the choroid plexus. Biochim Biophys Acta Biomembr 511(3):517–522

Zeuthen T, Wright EM (1981) Epithelial potassium transport: tracer and electrophysiological studies in choroid plexus. J Membr Biol 60(2):105–128

Zeuthen T, Hamann S, La Cour M (1996) Cotransport of H+, lactate and H2O by membrane proteins in retinal pigment epithelium of bullfrog. J Physiol 497(1):3–17

Zeuthen T, Zeuthen E, Klaerke DA (2002) Mobility of ions, sugar, and water in the cytoplasm of Xenopus oocytes expressing Na+-coupled sugar transporters (SGLT1). J Physiol 542(1):71–87

Zeuthen T, Gorraitz E, Her K, Wright EM, Loo DD (2016) Structural and functional significance of water permeation through cotransporters. Proc Natl Acad Sci U S A 113(44):E6887–E6894

Zhang Y, Poirier-Quinot M, Springer CS, Balschi JA (2011) Active trans-plasma membrane water cycling in yeast is revealed by NMR. Biophys J 101(11):2833–2842

Chapter 5
The Significance of the Choroid Plexus for Cerebral Iron Homeostasis

Lisa Juul Routhe, Maj Schneider Thomsen, and Torben Moos

Abstract The homeostasis of the brain interstitial fluid is sustained by two brain barriers, i.e. the blood-brain barrier (BBB) formed by brain capillary endothelial cells and the blood-cerebrospinal fluid (CSF) barrier, which both selectively allow transport of nutrients and metals into the brain. Iron is essential for normal brain function but detrimental in excess, hence justifying a tightly regulated cerebral iron homeostasis based on the transport of iron from the periphery across the brain barriers. Regulation of iron uptake involves proteins like transferrin receptor, divalent metal transporter 1 (DMT1), ferritin and ferroportin. It is commonly accepted that iron uptake and transport across the blood-CSF barrier are mediated by the transferrin receptor although by two principally different mechanisms. Transferrin complexed with iron can engage with the transferrin receptor and traverse the choroid plexus epithelial cells by receptor-mediated transcytosis. Moreover, Fe^{3+} can detach from transferrin within slightly acidic endosomes, escape into the cytosol by DMT1-mediated pumping after Fe^{3+} has been reduced to Fe^{2+}. In the cytosol, iron is either stored in ferritin or exported into the CSF mediated by ferroportin. Within the CSF, iron binds to transferrin primarily synthesized or secreted by the choroid plexus. Non-transferrin-bound iron (NTBI) is present in the CSF, favoring the argument that the iron-binding capacity of transferrin is exceeded in CSF. We hypothesize that transferrin derived from the CSF play a crucial role for binding of iron transported through the BBB, and for further transport of iron within the brain interstitium and subsequent delivery to transferrin receptor-expressing neurons.

Abbreviations

BBB   Blood-brain barrier
Blood-CSF barrier   Blood-cerebrospinal fluid barrier

L. J. Routhe · M. S. Thomsen · T. Moos (✉)
Laboratory for Neurobiology, Biomedicine, Institute of Health Science and Technology,
Aalborg University, Aalborg Ø, Denmark
e-mail: tmoos@hst.aau.dk

© The American Physiological Society 2020
J. Praetorius et al. (eds.), *Role of the Choroid Plexus in Health and Disease*,
Physiology in Health and Disease, https://doi.org/10.1007/978-1-0716-0536-3_5

C/EBP	CCAAT/enhancer-binding protein
CNS	Central nervous system
CSF	Cerebrospinal fluid
DcytB	Duodenal cytochrome B
DMT1	Divalent metal transporter
Fe^{2+}	Ferrous iron
Fe^{3+}	Ferric iron
GPI	Glycosylphosphatidylinositol
IgG	Immunoglobulin G
IRE	Iron-responsive element
IRP	Iron-regulatory protein
NTBI	Non-transferrin-bound iron
Steap	Six-transmembrane epithelial antigen of prostate

5.1 Introduction

The blood-brain barrier (BBB) formed by brain capillary endothelial cells and the blood-cerebrospinal fluid (CSF) barrier (blood-CSF barrier) formed by choroid plexus epithelial cells denote the morphological and functional bases for communication between the circulation and the brain parenchyma. In contrast to brain capillary endothelial cells, the choroid plexus transport large molecules of the blood plasma from the circulation into the ventricular system by means of non-specific vesicular transport (Moos and Møllgaard 1993b). This means that the concentration of large molecules like albumin and immunoglobulins are present in CSF with a CSF/plasma ratio of 0.1–1%, which is 10 to 100 times higher than in the brain interstitial fluid (Pardridge 2016).

To ensure transport of vital components across the brain barriers, selective transporters for nutrients like amino acids, glucose, vitamins, and essential metals, including iron exist (Lichota et al. 2010). Despite, the very low concentrations of iron found in the brain, iron plays a critical role in the development and normal function of the brain. The concentrations vary dramatically among different brain regions ranging from approximately 50 μg iron/g wet weight in the cerebral cortex to 130 μg iron/g wet weight in the caudate nucleus, and 200 μg iron/g wet weight in the substantia nigra (Fernández et al. 2017). Based on these data (Fernández et al. 2017), a rough estimate on total iron in the brain with an average concentration of 70 μg iron/g wet weight and a brain weight of 1400 g indicates that the total amount of iron in the brain is approximately 100 mg of the 3–4 g iron in the human body.

The uptake and transport of iron into the brain are developmentally regulated being significantly higher early in life when the brain is growing. Later in life, uptake and excretion of iron cease, which explains the extremely low turnover of iron occurring in the brain compared to many other organs. Nonetheless, the brain continues to acquire iron via transport across the BBB and the blood-CSF barrier throughout life (Moos and Morgan 2000, 2001). There is good agreement that iron is taken up at the brain barriers by receptor-mediated uptake of iron complexed with transferrin. The choroid plexus epithelial cells, contrary to the brain capillary endothelial cells of the BBB, enables both receptor-mediated uptake and non-specific transcellular transport of transferrin (Moos and Morgan 2000).

This chapter elucidates the significance of the choroid plexus for iron transport and homeostasis in the brain. Initially, we describe the chemistry of iron. Due to the importance of balanced iron homeostasis, its regulation at a systemic and cellular level will also be elucidated. Transport of iron into the brain across the BBB and the blood-CSF barrier is described with special emphasis placed on the expression of iron-transporting proteins in the choroid plexus. This is followed by coverage on the significance of transferrin transported through the blood-CSF barrier, and a suggestion that extracellular transferrin is playing an important role within the brain interstitial fluid to ensure sufficient iron homeostasis.

5.2   Iron in Biological Systems

Iron is an essential metal for the mammalian body. The necessity of iron for normal body function is reflected in its participation as a co-factor in basic cellular processes ranging from oxygen transport, mitochondrial respiration, and DNA synthesis. Hence, enzymes such as cytochrome oxidase (mitochondrial respiration), tyrosine hydroxylase (neurotransmitter formation), and ribonucleotide reductase (cell division) are important proteins that depend on iron binding (Zecca et al. 2004; Zhang 2014).

Iron exists in various oxidation states ranging from $^{-2}$ to $^{+6}$. In aqueous solutions, it occurs on its reduced form as ferrous iron (Fe^{2+}) or on its oxidized form as ferric iron (Fe^{3+}). The solubility of Fe^{2+} is much higher than that of Fe^{3+}, as Fe^{3+} is hardly soluble in water at neutral pH. Fe^{2+}, however, is rapidly oxidized to Fe^{3+} in aerobic conditions, meaning that Fe^{3+} can be considered the sole form of ionic iron (Morgan 1996). To compensate for the low solubility of Fe^{3+}, proteins with major affinity for Fe^{3+} has evolved with capability for neutralization, extracellular transport, uptake, efflux and intracellular binding.

5.2.1   Iron Transport

Binding and transport of Fe^{3+} in the extracellular space is mediated by transferrin that consists of two identical parts, each with capacity for binding of a single Fe^{3+} atom, meaning that each molecule can bind two iron atoms. Plasma transferrin has a molecular weight of approximately 80 g/mol and a concentration of 25–50 μM. It is mainly synthesized in the liver, but other cell types are also able to synthesize transferrin, including the choroid plexus epithelial cells (Bloch et al. 1985, 1987).

Iron circulating in plasma is bound to transferrin with a saturation of approximately 15–50% in human adults (Luck and Mason 2012; Anderson and Vulpe 2009). The binding of transferrin to two Fe^{3+} ions is referred to as holo-transferrin, the binding of a single iron atom as mono-transferrin, while transferrin not bound to iron is referred to as apo-transferrin (Moos and Morgan 2000). The affinity of holo-transferrin for the transferrin receptor at physiological pH (7.4) is much higher than those of mono- or apo-transferrin (Luck and Mason 2012; Eckenroth et al. 2011). With decreasing pH, the affinity of transferrin for iron lowers, which subsequently leads to the release of iron from transferrin and detachment of apo-transferrin from the transferrin receptor (Luck and Mason 2013).

5.2.2 Iron Uptake Mechanisms

Cells import iron from the extracellular space through different import mechanisms, i.e. transferrin receptor-mediated endocytosis of holo-transferrin, uptake of non-transferrin-bound iron (NTBI) by divalent metal transporter 1 (DMT1) expressed in the cellular membrane, or non-specific endocytosis (Roberts et al. 1993).

The transmembrane transporter DMT1 is a member of the natural resistance-associated macrophage protein family expressed in various cell types. It is well functioning at low pH and acts as a transporter for divalent metal ions, especially Fe^{2+} (Burkhart et al. 2016; Simpson et al. 2014; Gunshin et al. 1997). DMT1 either functions by taking up Fe^{2+} at the luminal membrane of enterocytes or by pumping Fe^{2+} across the endosomal membrane in cells that have taken up Fe^{3+} via receptor-mediated uptake of transferrin in clathrin-coated vesicles as observed in the BBB and choroid plexus (Gunshin et al. 1997, 2005; Fleming et al. 1998; Rouault et al. 2009; Burkhart et al. 2016; Skjørringe et al. 2015). Within the acidic compartment of endosomes, Fe^{3+} dissociates from transferrin and is reduced to Fe^{2+} by a ferrireductase. Subsequently, Fe^{2+} is pumped out of the endosome via DMT1 and enters the cytosol while the resulting apo-transferrin is recycled to the blood (Burkhart et al. 2016). Fe^{2+} in the cytosol is then either rapidly oxidized and stored in ferritin or transported out of the cell (Moos and Morgan 2000). The efflux of Fe^{2+} is facilitated by ferroportin under tight interaction with membrane-bound copper-containing ferroxidases, such as hephaestin and ceruloplasmin (McKie et al. 2000; Donovan et al. 2000; Abboud et al. 2000; Musci 2014). They both oxidize Fe^{2+} to Fe^{3+} to prevent deleterious reactions in the cellular membrane, as Fe^{2+} rapidly participates in oxidation-reduction reactions (see Sect. 5.3).

5.2.3 Storage of Iron

Ferritin and hemosiderin are important for intracellular binding and storage of residual Fe^{3+}. Ferritin consists of 24 polypeptide subunits that form a hollow shell with the capacity of storing 4500 Fe^{3+} molecules in a non-toxic form (Harrison and

Arosio 1996). Ferritin is comprised of two isoforms, i.e. the heavy chain (H) and the light chain (L). The H chain is a 21 kDa protein expressed mostly by the heart and brain, while the L chain is a 19 kDa protein expressed in tissues that store iron, such as the liver and spleen. Together any copies of these isoforms form a functional ferritin protein with a molecular weight around 480 kDa (Harrison and Arosio 1996). The two ferritin isoforms have different functions. Thus, the H chain contains a center with ferroxidase activity that catalyzes the oxidation of Fe^{2+}, while the L chain offers sites for iron nucleation and mineralization, thereby assisting the activity of the H chain (Levi et al. 1994). Interestingly, manipulating ferritin L in mammalian cells has no effect on cellular iron homeostasis (Cozzi et al. 2004). This has also been confirmed in clinical studies, where ferritin L insufficiency or excess had no effect on body iron status (Cremonesi et al. 2004; Martin et al. 1998). In contrast, lack of ferritin H is lethal at early stages of embryonic development in mice (Ferreira et al. 2000). Mobilization of iron from ferritin occurs by lysosomal degradation of ferritin or by a process called autophagy mediated by the protein nuclear receptor coactivator 4 (Mancias et al. 2014). Ferritin enriched in H subunits occurs extracellularly, and within the brain interstitial fluid expression of Tim-2, a receptor for ferritin H, suggests directed uptake of iron-containing ferritin H in oligodendrocytes (Todorich et al. 2008).

Hemosiderin, which is also involved in storage of iron, binds even more iron per molecule than ferritin and is believed to arise from ferritin that has been partly degraded within the lysosomes (Hoy and Jacobs 1981; Miyazaki et al. 2002). Hemosiderin forms membrane-enclosed granules, siderosomes, that are detectable using the Prussian Blue histochemical stain (Iancu et al. 1995; Collingwood and Dobson 2006). These siderosomes increases in number in organs like the liver and spleen during conditions with circulatory iron-overload (Morgan 1996). The siderosomes can release their iron content following a rapid lowering of the iron-content after a blood loss. It is believed that iron stored in ferritin and hemosiderin in the brain is without the capability of forming free radical unless destabilized by oxidative modifications (Quintana et al. 2006; Carocci et al. 2018).

5.3   Iron in Redox Reactions with Free Radical Formation

Iron overload can be detrimental and result in oxidative stress, which is one of the conspirators in the pathology of many neurodegenerative diseases (Ward et al. 2014). Being a transition metal, iron can act as a reversible reducer and oxygenator shifting between Fe^{2+} and Fe^{3+}, while donating or accepting an electron, which makes iron an essential component of oxygen-binding molecules and cytochromes (Andrews 1999). The transition from Fe^{2+} to Fe^{3+} under the generation of hydroxyl radicals ($\bullet OH$) is also known as the Fenton reaction (Duck and Connor 2016; Wardman and Candeias 1996):

$$Fe^{2+} + H_2O_2 \rightarrow Fe^{3+} + \bullet OH + OH^-$$

Fe^{3+} can be reduced to Fe^{2+} under the consumption of superoxide ($\bullet O_2^-$) and the formation of oxygen:

$$Fe^{3+} + \bullet O_2^- \rightarrow Fe^{2+} + O_2$$

The net reaction of the two reactions, the Haber-Weiss reaction, states that the free radical of superoxide is converted to hydroxyl radicals ($\bullet OH$):

$$\bullet O_2^- + H_2O_2 \rightarrow \bullet OH + OH^- + O_2$$

In contrast to superoxide, hydroxyl radicals ($\bullet OH$) are highly reactive and demands the concurrent presence of anti-oxidative molecules to prevent cellular stress and damage to DNA, lipids, and proteins (Anderson et al. 2009; Chevion 1988; Stohs and Bagchi 1995). The transitions of iron take place in the mitochondria and scavenging of free radicals during the formation of oxygen is kept within limits of anti-oxidants like ascorbic acid, catalase, glutathione reductase, and vitamins D and E. Excess iron is neutralized by storage in ferritin, which prevents further propagation of free radical formation (Focht et al. 1997).

5.4   The Regulation of Systemic and Cellular Iron Homeostasis

As the body evidently suffers in conditions with iron deprivation or excess, it is of importance to maintain a sufficient nutritional balance of iron.

5.4.1   Systemic Regulation

Iron homeostasis is regulated in a strict combination of gastrointestinal uptake, mobilization from hepatic storage and recycling of dying red blood cells phagocytosed by macrophages (Ganz 2013). Excretion of iron, on the other hand, is fairly unregulated and does not contribute to the regulation of total body iron.

Absorption of iron mainly takes place in the duodenum and is facilitated by intestinal enterocytes.

To enter the circulation, iron must pass both the apical and basolateral membranes of the enterocytes.

This absorption is primarily regulated by the liver-derived hormone hepcidin (Anderson et al. 2009) secreted from hepatocytes in response to systemic iron-overload and by stimulation of pro-inflammatory cytokines (Ganz and Nemeth

2012; Nemeth et al. 2004). Hepcidin present in the 10–100 nM range interacts directly with ferroportin at the basolateral membrane causing its proteolysis in lysosomes (Nemeth et al. 2004). Increasing levels of systemic hepcidin significantly lower the expression of ferroportin in the intestinal cells and hence lower the gastrointestinal iron absorption. Conversely, in iron-deficiency without inflammation, the expression of hepcidin decreases leading to sustained expression of ferroportin, which facilitates gastrointestinal uptake and transport of iron into the circulation (Frazer et al. 2002).

5.4.2 Molecular Regulation of Cellular Iron Homeostasis

Cellular iron homeostasis is maintained by regulation of influx via transferrin receptor-mediated uptake, intracellular storage in ferritin, and efflux via ferroportin. This regulation is accomplished by regulatory mechanisms that orchestrate post-transcriptional changes leading to often dramatic changes in cellular protein levels without altering the mRNA expression. The post-transcriptional regulation includes stabilization or degradation of transferrin receptor (*TFR*) mRNA, *DMT1* mRNA, ferritin (*FTH* and *FTL*) mRNA, and ferroportin (*FPN1*) mRNA and is primarily controlled by the iron-responsive element (IRE)/iron-regulatory protein (IRP) system (Pantopoulos 2004). IRPs that consist of two forms, IRP1 and IRP2, act as either a translational enhancer or inhibitor in response to changes in intracellular iron levels.

In cellular iron deficiency, IRP1 loses contact to labile iron and assumes a [3Fe-4S] configuration that binds with high affinity to IREs stem-loop structures located in the $3'$ or $5'$ untranslated regions (UTR) of the mRNA transcripts of iron-handling proteins like transferrin receptor and DMT1 ($3'$) and ferritin and ferroportin ($5'$) (Aziz and Munro 1987; Müllner and Kühn 1988; Pantopoulos 2004). The binding of IRPs to IREs in the $3'$ UTR stabilizes mRNA of transferrin receptor mRNA and DMT1 to promote uptake of available iron. Conversely, binding of IRPs to IREs in the $5'$ UTR inhibits translation of mRNA encoding ferritin and ferroportin, and thus indirectly facilitates a demise of ferritin and ferroportin to ensure higher availability of iron within the cells' labile iron pool. In cellular iron overload, the binding of IRPs to IREs is blocked by iron resulting in reduced mRNA translation and decreased levels of TfR and DMT1, but increased ferritin and ferroportin as IRP dissociates from its binding site, which results in improved association between ferritin and ferroportin mRNA and ribosomes with a resulting increased protein synthesis (Kühn and Hentze 1992).

5.5   Iron Transport into the Brain Occurs
       at the Blood-Brain Barrier and the Blood-CSF Barrier

Iron is transported into the brain via the BBB and the blood-CSF barrier (Rouault et al. 2009; Wang et al. 2008a). Unique for the brain, the transferrin receptor is expressed on the luminal side of the brain capillary endothelial cells (Jefferies et al. 1984) meaning the brain performs a preferential extraction of iron from the circulation compared to other organs of the body. In the choroid plexus, the quite leaky fenestrated capillary endothelial cells are devoid of transferrin receptors, but the choroid plexus epithelial cells express transferrin receptors indicating their capacity to engage with holo-transferrin of the plasma (Crowe and Morgan 1992; Moos 1996; Giometto et al. 1990).

5.5.1   Transport of Iron Through the Blood-Brain Barrier

The main knowledge about iron transport into the brain originates from studies on its transport across the BBB. It is commonly accepted that the most important mechanism for iron uptake by brain capillary endothelial cells is mediated by the transferrin receptor on their luminal side that internalizes holo-transferrin in clathrin-coated endosomes (Fig. 5.1) (Jefferies et al. 1984). A dispute prevails on the following transport of holo-transferrin into the brain and this has led to two different hypotheses. We recently reviewed the rationales and evidence of these hypotheses (Johnsen and Moos 2016; Skjørringe et al. 2015). One hypothesis claims that holo-transferrin is transported through the BBB via receptor-mediated transcytosis. Endosomes containing holo-transferrin are postulated to traverse the brain endothelial cells and release holo-transferrin at the abluminal side of the endothelial cells into the brain interstitium (Fishman et al. 1987; Descamps et al. 1996; Leitner and Connor 2012; Moos et al. 2007). The first indications of transferrin transcytosis across the BBB were published by Fishman et al., who investigated the distribution of ^{125}I-labeled transferrin in the brain following intravenous injection (Fishman et al. 1987). Although most of the labeled transferrin was associated with the vasculature, indications of transferrin transcytosis were ascribed to the presence of intact transferrin within the brain parenchyma. Using the transferrin receptor-specific antibody OX26, other studies supported these findings by their presumable demonstration of OX26 transport through the BBB although no morphological evidence was published to suggest transendothelial transport (Pardridge et al. 1991; Friden et al. 1991; Lee et al. 2000).

Later isotope studies using double labeling with ^{59}Fe and ^{125}I-labeled transferrin clearly indicated that iron reaching the brain surpasses that of transferrin, leading to the hypothesis that iron is detached from transferrin within the acidic compound of the endosomes on the route towards the lysosomes (Crowe and Morgan 1992; Taylor et al. 1991; Morris et al. 1992). Supporting this notion, the lower appearance of

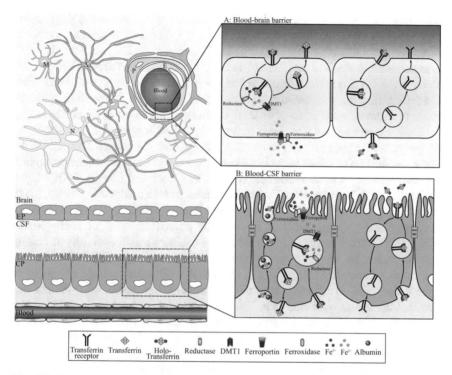

Fig. 5.1 Illustration of iron transport at the blood-brain barrier (BBB) (**a**) and the blood-cerebrospinal fluid (blood-CSF) barrier (**b**). In the overview, the BBB and the blood-CSF barrier are depicted. The BBB in the brain parenchyma is composed of endothelial cells (E) supported by pericytes (P) and astrocytes (A). Microglia (M) and neurons (N) are also presented in the brain compartment. The blood-CSF barrier consists of choroid plexus epithelial cells (CP) in contact with plasma filtered through the fenestrated capillaries. The choroid plexus transports nutrients into the ventricular space filled with CSF. The brain parenchyma and the ventricular space are separated by ependymal cells (EP). (**a**) Iron uptake by the BBB is hypothesized to occur via a receptor-mediated endocytosis (**a**, left) or transcytosis (**a**, right). In the endocytotic pathway, iron is transported through the BBB via transferrin-receptor mediated uptake (**a**, left). Fe^{3+} detaches from transferrin in the acidic endosomal compartment, is reduced to Fe^{2+} by a ferrireductase, and is pumped through the endosomal membrane via divalent metal transporter 1 (DMT1). Fe^{2+} is subsequently exported through the abluminal membrane of the endothelial cells by ferroportin. Fe^{2+} is immediately re-oxidized to Fe^{3+} by the ferroxidase ceruloplasmin that is present in both the cellular membrane and the brain interstitium. The re-oxidation is then followed by binding of Fe^{3+} to transferrin. In the transcytotic pathway, holo-transferrin is also taken up by transferrin-receptor and transported across the endothelial cell to the abluminal side, where holo-transferrin is released into the brain interstitium. Afterwards, the transferrin-receptor is recycled to the luminal side of the endothelial cell (**a**, right). (**b**) Iron is transported through the blood-CSF barrier by three different mechanisms, i.e. non-specific vesicular transport (**b**, left), receptor-mediated endocytosis (**b**, middle) and receptor-mediated transcytosis (**b**, right). Large molecules such as albumin, immunoglobulin and transferrin are transported through the choroid plexus by non-specific vesicular transport (**b**, left). However, the main contribution for iron uptake is mediated by the transferrin-receptor by a mechanism like that occurring in the BBB (**a**, left), i.e. Fe^{3+} is detached from transferrin in the endosome, reduced to Fe^{2+} by ferrireductase and transported into the cytosol by DMT1 (**b**, middle). Cytosolic Fe^{2+} is then exported into the ventricular CSF by ferroportin and immediately oxidized to Fe^{3+} by ferroxidases such as hephaestin and ceruloplasmin, which are both expressed by choroid plexus epithelial cells. Fe^{3+} in the CSF is mainly bound by transferrin, but Fe^{3+} is also present as

transferrin compared to iron could not be explained by transferrin degradation (Strahan et al. 1992). Additionally, horseradish peroxidase-conjugated to transferrin examined at the ultrastructural level after intravascular administration not only failed to undergo transcytosis at the BBB (Crowe and Morgan 1992), the complex was also absent from the abluminal side of the brain endothelium (Roberts et al. 1993), thus suggesting that the transferrin receptor-containing vesicles fail to fuse with the abluminal membrane (Siupka et al. 2017). Similarly, later studies in rats exploiting antibodies targeting the transferrin receptor failed to support the hypothesis on transcytosis of transferrin receptor containing vesicles (Moos and Morgan 2001). The accumulation of OX26 in the brain parenchyma was simply too small to support any conclusions on transcytosis across the BBB (Moos and Morgan 2001). Upon endocytosis, Fe^{3+} is reduced to Fe^{2+} by molecules with ferrireductase activity, likely Steap or DcytB, and pumped into the cytosol through DMT1 (Ohgami et al. 2006; McCarthy and Kosman 2012). A recent study proved evidence for expression of Steap, DMT1 and ferroportin in purified brain capillary endothelial cells indicating that the molecules needed to justify this hypothesis on iron transferrin are indeed present (Burkhart et al. 2016). Fe^{2+} reaches the abluminal surface of the brain endothelium and is exported from the cell through ferroportin in a process that involves immediate oxidation to Fe^{3+} by ceruloplasmin or hephaestin produced by endothelial cells, pericytes and astrocytes (McCarthy and Kosman 2013; Burkhart et al. 2016). In the brain extracellular space, Fe^{3+} is bound to apo-transferrin of the extracellular compartment. The Fe^{3+} may also bind to ATP or citrate that are abundant in the extracellular space, although their affinity for Fe^{3+} is lower than that of apo-transferrin (Leitner and Connor 2012). Clearly, the hypothesis on iron transport through the brain capillary endothelial cells mediated by detachment of iron from transferrin within the acidic environment of the endosome and recycling of apo-transferrin to the blood gets the best support [cf. Skjørringe et al. (2015) and Johnsen and Moos (2016)].

A newly introduced idea suggests that the endothelial cells of the brain sense the iron level within the brain by communicating with the brain interstitium (Simpson et al. 2014; Chiou et al. 2019). This implies that transferrin with its varying saturation is taken up from the brain interstitium, and the resulting contribution to the intracellular concentration of iron within the endothelium, in turn, leads to changes in transferrin receptor. The latter could thereby affect the uptake of holo-transferrin from the plasma (Chiou et al. 2019). Discouraging this hypothesis, there has never been any convincing demonstration of transferrin receptors on the abluminal side of the endothelium, thus it might be the amount of NTBI, which is sensed by the endothelium after uptake from the interstitium and not transferrin.

Fig. 5.1 (continued) non-transferrin bound iron. Holo-transferrin can be transported across the choroid plexus endothelium by receptor-mediated transport (**b**, right). Holo-transferrin engages with the transferrin receptor thus forming an endosome that traverses the choroid plexus and releases holo-transferrin into the CSF, while recycling the transferrin receptor-containing vesicle to the abluminal surface

Recent studies focusing on the transport of low-affinity antibodies against the transferrin receptor also indicate that targeted antibodies can obtain receptor-mediated uptake and transport through the brain endothelium when detached from their receptor due to the slightly acidic environment in the endosomal compartment (Yu et al. 2014; Freskgård and Urich 2017). In the process of receptor-mediated endocytosis, the endosome could also capture and contain large molecules of the blood plasma, including albumin. The transport through the brain endothelium and release into the brain is unclarified but could theoretically represent the fusion of vesicles containing antibodies and albumin with the abluminal membrane. Thereby providing an explanation for the diminutive appearances of albumin in the brain after intravenous injection (Crowe and Morgan 1992; Taylor et al. 1991; Morris et al. 1992).

5.5.2 Iron and Iron-Proteins of the Choroid Plexus

To understand how iron is transported across the blood-CSF barrier a description of both localization of iron and iron-proteins of the choroid plexus is needed. The subcellular distribution of iron in the choroid plexus epithelium is not so well-described compared to other tissues like the liver and spleen where iron is detected in lysosomes, siderosomes and residual bodies (Blissett et al. 2017; Miyata et al. 2009; Fernandes et al. 2018; Kurz et al. 2011). The choroid plexus epithelium contains histochemically detectable iron in both the developing and adult human and rodent brain (Fig. 5.2) (Moos 1995; Fernandes et al. 2018; Connor et al. 2011; Moos and Møllgaard 1993a). This is contrasted in brain endothelial cells in the rat in which histological iron is only seen on embryonic day 14 (E14) until early postnatally, and not in the adult (Moos 1995).

The choroid plexus epithelial cells contain transferrin that is morphologically confined to both a dot-like appearance indicating a subcellular localization in the endosomal-lysosomal system (Moos, unpublished) but also a more uniform distribution that clearly differs from albumin and likely reflects its synthesis (Zakin 1992) (Fig. 5.3). Supporting that transferrin is taken up by receptor-mediated endocytosis, the choroid plexus epithelial cells profoundly express transferrin receptors, which give the epithelial cells the capacity for transcytosis of holo-transferrin (Giometto et al. 1990; Moos 1996).

Apart from transcytosis in choroid plexus epithelial cells, the expression of the transferrin receptor also enables the epithelial cells to release iron from holo-transferrin after the receptor-mediated uptake. The epithelial cells contain DMT1, which display a dot-like distribution within the cytosol thereby suggesting that clathrin-coated, transferrin receptor-containing vesicles can fuse with acidic endosomes and mediate pumping of iron into the cytosol (Fig. 5.2) (Moos and Morgan 2004). This is further underscored by the expression of ferrireductase like DcytB in the epithelial cells (Rouault et al. 2009).

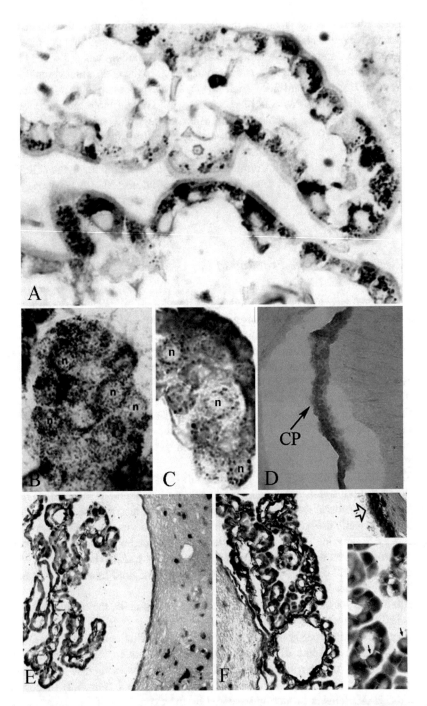

Fig. 5.2 Iron and iron-proteins in the choroid plexus. (**a**) Iron (black) localizes to the cytosol without nuclear labeling (Moos and Møllgaard 1993a). (**b**) Transferrin receptor (Moos 1996). (**c**) DMT1 (Skjørringe et al. 2015). Both proteins are present in the cytosol with a dot-like appearance corresponding to vesicles. *n* unlabeled nuclei. (**d**) Ferroportin in the choroid plexus (CP) shown at

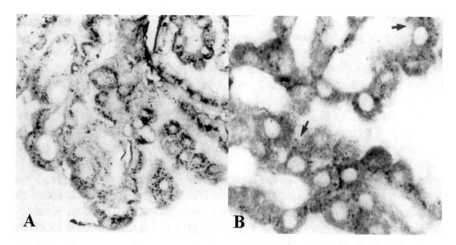

Fig. 5.3 Choroid plexus epithelial cells of the third ventricle in the adult rat stained for albumin (**a**) and transferrin (**b**) (Moos, unpublished). Albumin reveals a dot-like appearance corresponding to the endosomal-lysosomal system reflecting the morphological appearance of transport through the blood-CSF barrier by non-specific endocytosis. In comparison, transferrin reveals an immunoreaction product seen as uniform staining with dot-like appearance in some cells (arrows). The different immunoreaction product of transferrin likely reflects the simultaneous synthesis and vesicular transport. Magnification ×500 (**a**), ×800 (**b**)

The presence of ferroportin and ferroxidase in the apical membrane indicates that the choroid plexus epithelial cells have the full capacity for transport of cytosolic iron into the CSF (Wang et al. 2008b; Rouault et al. 2009; Boserup et al. 2011). Ferroxidase activity is thought to be mediated by hephaestin or ceruloplasmin, both found in the choroid plexus (Rouault et al. 2009). Ceruloplasmin exists in two isoforms, a soluble and a glycosylphosphatidylinositol (GPI) membrane-anchored form (Patel and David 1997; Burkhart et al. 2016). Within the central nervous system (CNS), the GPI-anchored isoform is expressed in astrocytes and the soluble form is expressed in astrocytes and the choroid plexus epithelium (Klomp et al. 1996; Patel and David 1997). Ceruloplasmin plays an essential role in iron homeostasis and in the absence of ceruloplasmin, ferroportin is degraded causing intracellular iron accumulation (Musci 2014; De Domenico et al. 2007). A recent study suggests that the choroid plexus epithelial cells are particularly prone to this iron accumulation (Zanardi et al. 2017). The authors demonstrate profound iron sequestration in the choroid plexus of 10-month old ceruloplasmin knock out mice, which

Fig. 5.2 (continued) low-power magnification (Boserup et al. 2011). (**e–f**) Ferritin H and L mRNA in the choroid plexus demonstrated by non-radioactive in situ hybridization, respectively. The ferritin mRNA is limited to the cytoplasm of the choroid plexus epithelial cells, leaving the nucleus unstained (small arrows). Labeling is also seen in the ependyma (arrowhead) (Hansen et al. 1999). Magnification: **a–c** × 900, **d–f** × 40 (inset × 1800). Permission to reproduce material is obtained from licence holders

was abolished by ceruloplasmin administration (Zanardi et al. 2017). Hence, ceruloplasmin may play an essential role in iron homeostasis and secretion of iron into the CSF.

The choroid plexus epithelial cells also profoundly express ferritin protein and mRNA (Hansen et al. 1999; Rouault et al. 2009) suggesting iron is also kept in these cells for storage. A recent study conferred the subcellular distribution of the choroid plexus epithelium to iron-containing ferritin or hemosiderin granules of the lysosomal system (Fernandes et al. 2018; Case 1959; Connor et al. 2011). Collectively, these observations validate that choroid plexus epithelial cells accumulate iron that is incorporated in ferritin for storage or mobilized by the degradation of ferritin in lysosomes. Ferritin H, in particular, seems to play an important role in the functioning of the choroid plexus epithelium (Schweizer et al. 2014). For instance, mice with tissue-specific deletion of ferritin H in the choroid plexus of the lateral ventricles developed hydrocephalus (Schweizer et al. 2014). This led the authors to speculate that ferritin H expression, and hence maintained capacity for iron storage, could be essential for normal CSF production and ventricle formation.

5.5.3   Mechanisms for Iron and Transferrin Transport at the Choroid Plexus

Previously, the choroid plexus was considered less important in maintaining iron homeostasis in the brain, as the surface area of the choroid plexus is lower than that of the BBB (Keep and Jones 1990; Pardridge 2016), hence allowing less iron to enter the brain via the blood-CSF barrier (Keep and Jones 1990; Moos et al. 2007).

Similar to the capillaries forming the BBB, the choroid plexus express transferrin receptors (Rouault et al. 2009; Giometto et al. 1990). In the following, three mechanisms for transferrin transport through the blood-CSF are described, (1) non-specific vesicular transport, (2) receptor-mediated transcytosis, and (3) endocytosis of holo-transferrin followed by a detachment of iron from transferrin, recycling of apo-transferrin to the abluminal surface for recycling in blood plasma (Fig. 5.1).

1. The non-specific vesicular transport and transport of large molecules of the blood plasma like albumin, immunoglobulins, and transferrin through the choroid plexus are quite dramatic (Crowe and Morgan 1992). The concentrations of large molecules of the plasma is thus much higher in CSF than in the brain interstitium, indicating that transport of large molecules across the blood-CSF barrier is significantly higher than at the BBB (Pardridge 2016; Spector et al. 2015). Morphologically, these large molecules like albumin and transferrin exhibit a dot-like appearance that indicates vesicular trafficking in the choroid plexus (Fig. 5.3).

2. Opposite to that occurring at the BBB, uptake of holo-transferrin by the choroid plexus also leads to transferrin-receptor mediated transcytosis and transport into

the CSF. Supporting this notion, peripheral administration of ^{125}I-labeled holo-transferrin and ^{131}I-labeled albumin led to a higher accumulation of transferrin in the CSF compared to that of albumin (Crowe and Morgan 1992). Similarly, co-injection of antibodies against the transferrin receptor and non-immune isotypic antibodies leads to a substantially higher appearance of the former in the CSF (Moos and Morgan 2001).

3. Results from the study of Crowe and Morgan (1992) also indicate that iron is transported through the blood-CSF barrier by receptor-mediated endocytosis of transferrin with detachment of iron within endosomes of the choroid plexus epithelial cells and further transport of iron into the CSF. This suggestion originated from a higher transport of radioactive iron in the CSF compared to uptake of radiolabeled transferrin minutes after peripheral injection of [[125]I-^{59}Fe] holo-transferrin (Crowe and Morgan 1992). The observations could be attributed to receptor-mediated uptake of the complex followed by the detachment of iron from transferrin, which is clearly supported by the choroid plexus' expression of many iron-handling proteins described above. Thus, within the endosomes, Fe^{3+} is released from transferrin and transported into the cytosol by DMT1 after reduction to Fe^{2+} by a ferrireductase, such as DcytB. Subsequently, Fe^{2+} is exported to the CSF via ferroportin and oxidized by ceruloplasmin or hephaestin (Gunshin et al. 1997; Rouault et al. 2009; Klomp et al. 1996). Further acknowledging this statement, NTBI is present in CSF suggesting that iron also enters the ventricles detached from transferrin (Moos and Morgan 1998a). Since NTBI is present, the synthesis and secretion of transferrin into the CSF by the choroid plexus epithelium is probably diminutive and unable to capture all iron released from the choroid plexus.

5.6   Transport of Iron Across the Blood-CSF Barrier During Development and Iron-Deficiency

The CSF concentration of plasma proteins, including transferrin, is higher in the developing brain and decreases with increasing postnatal age. This phenomenon ascribes to lower synthesis and a lower turnover of the CSF, which retains its content of solutes within the brain and leads to a more widespread diffusion of iron and transferrin from the ventricles and deeper into the brain parenchyma (Moos and Morgan 1998b). Delivery of holo-transferrin to the brain simultaneously compensates a high need for iron in very early development where brain endothelial cells retain iron and ferritin seemingly without transporting much iron from the capillaries and further into the brain (Moos 1995). Dramatic changes follow thereafter, and the transport of iron into the brain across the BBB and blood-CSF barriers peaks in the rodent already few weeks after birth where the growth of the brain maximizes (Morgan and Moos 2002). Thereafter the uptake decreases to levels clearly lower than in the developing brain.

In iron deficiency, the transport of holo-transferrin across the blood-CSF barrier is elevated (Crowe and Morgan 1992) attributable to a higher expression of transferrin receptors by choroid plexus epithelial cells, which leads to raised basolateral transport of holo-transferrin into the ventricular system (Moos and Morgan 2001; Connor et al. 2011; Moos et al. 1998). To underline the specificity of the transport in iron-deficiency, there are no changes in the transfer of albumin across the blood-CSF barrier (Crowe and Morgan 1992).

5.7   Iron and Iron-Binding in CSF

The concentration of transferrin in CSF is estimated to 0.1–0.28 µM in human and approximately 0.52 µM in rats, thus reaching values approximately 1:100 of the plasma concentration (Bradbury 1997; Moos and Morgan 1998a). Total iron in CSF of the rat has been measured to 1.1 µM, which is comparable with human CSF values ranging from 0.29 to 1.12 µM (Wan et al. 2006; Moos and Morgan 1998a; Bradbury 1997; Knutson 2019). This suggests that despite low iron concentrations, the saturation of transferrin is significantly higher in CSF than in plasma, and the values in fact also indicate that iron might occur in CSF as NTBI. Supporting the latter, we demonstrated that following intravenous injection of holo-transferrin a fraction of iron occurred in the CSF and brain extracellular fluid as NTBI, suggesting that iron enters the brain without complete capture by extracellular transferrin of the extracellular space (Moos and Morgan 1998a). Other candidates for iron binding in the CSF could be ATP, citrate and larger proteins such as ferritin and albumin (Bradbury 1997; Moos and Morgan 1998a; Moos et al. 2007). Although the CSF is enriched with ascorbic acid, extracellular iron in the brain is thought to appear as Fe^{3+} (Bradbury 1997). Ascorbic acid is not synthesized in the body, and its transport into the brain occurs at the blood-CSF barrier mediated by the sodium-dependent vitamin C transporter-2 that is highly expressed by the choroid plexus epithelium (Tsukaguchi et al. 1999; Sotiriou et al. 2002). A possible function of ascorbic acid within the brain related to iron could be to prevent damage to cellular membranes generated by an excess of NTBI (Moos et al. 2007). The anti-oxidative activity of ascorbic acid theoretically maintains iron on its ferrous state, and it should not be overlooked that the presence of Fe^{2+} has been reported within the brain extracellular fluid (Sävman et al. 2005), but this observation has not gained support elsewhere.

5.8   Sources of Extracellular Transferrin in the Brain and Its Implications for Cellular Iron Delivery

The principal sources for extracellular transferrin in the brain are a mixture of secretion from hepatocytes, oligodendrocytes, and the choroid plexus epithelial cells (Tsutsumi et al. 1989). The passage of liver-derived transferrin into the brain is restricted due to the BBB but occurs via transport through the choroid plexus as described above. However, other sources of transferrin in the brain interstitium should also be considered. It is hypothesized that oligodendrocytes contribute significantly to transferrin synthesis in the brain (Fishman et al. 1987; Bloch et al. 1985). Theoretically, the transferrin from oligodendrocytes could make a significant contribution to the transferrin levels in the CSF. However, the evidence for the secretion of transferrin from oligodendrocytes is vague. Iron deficiency increases secretion of transferrin in the liver but does not affect transferrin synthesis in the brain (Idzerda et al. 1986; Moos et al. 2001). The increased synthesis of transferrin protein in the liver in iron deficiency relies on increased transferrin mRNA expression due to the interaction between transferrin mRNA promoter and cytosolic protein, CCAAT/enhancer-binding protein (C/EBP) (Zakin 1992). Only the transferrin promoter of liver cells interacts with this C/EBP protein, whereas the promoter of oligodendrocytes is without affinity for the C/EBP protein. This suggests that transferrin synthesis within oligodendrocytes is irresponsive to iron deficiency and that their transferrin synthesis is of importance for intracellular functions but without significance for handling iron in the brain interstitial space (Theisen et al. 1993; Zakin et al. 2002). The choroid plexus also synthesize transferrin (Bloch et al. 1987) but unfortunately, the responsiveness of transferrin synthesis by the choroid plexus in iron deficiency has not been examined in rodent studies. The human choroid plexus, however, has higher transferrin expression in iron-deficiency in patients suffering from restless leg syndrome, suggesting that the choroid plexus senses iron-deficiency and replies by increasing transferrin secretion into the CSF (Connor et al. 2011).

Following single dosing of holo-transferrin into the lateral ventricle, iron is subsequently detectable in the lateral ventricles and transported along an expected route into the third ventricle, through the cerebral aqueduct terminating in the fourth ventricle, and from there out into the subarachnoid space (Moos and Morgan 1998b). Iron deposits alongside the surfaces of the ventricular system and subarachnoid space and does not protrude deeply into the brain parenchyma, e.g. the red nucleus of the mesencephalon rich in neuronal transferrin receptors remains unlabeled (Fig. 5.4). A limitation of the single injection paradigm compared with repeated dosing is that repeated dosing may lead to more profound labeling of the brain parenchyma (Moos et al. 2007; Leitner and Connor 2012). This could occur as holo-transferrin like other proteins of the CSF is expected to recirculate to some extent from the subarachnoid space alongside large vessels to reach the brain interstitium (Thomsen et al. 2017), where holo-transferrin can engage with neurons expressing transferrin receptor. The neurons mainly express transferrin receptors on their dendrites and soma, and from there iron redistributes significantly by intracellular transport (Dwork et al. 1990; Moos and Morgan 1998b; Lasiecka et al. 2010).

At steady state levels, major plasma proteins like albumin, immunoglobulin G (IgG) and holo-transferrin are all detectable in neurons in spite of the existence of the BBB (Moos and Høyer 1996; Moos and Morgan 2001). The plasma proteins appear in neurons corresponding to the distribution of the endosomal-lysosomal system. It is likely that the presence of such plasma proteins relies on their transport through the

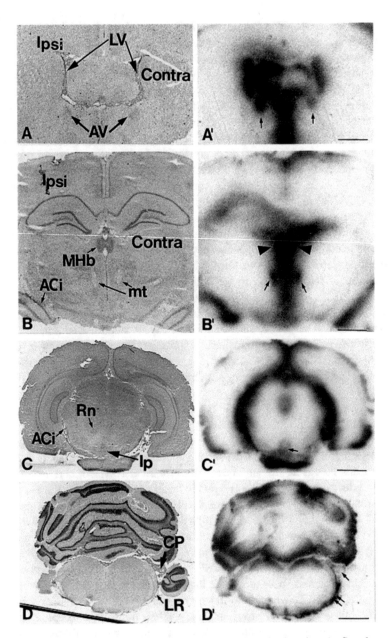

Fig. 5.4 Representative images of adult rat brain Cresyl violet stained sections (**a–d**) and autoradiograms after injection of [⁵⁹Fe]-Tf (**a′–d′**). The telencephalic (**a–a′**), di- and telencephalic (**b–b′**), mesencephalic (**c–c′**), and cerebellar and pontine (**d–d′**) regions were analyzed 24 h after the injection of ⁵⁹Fe-Tf. The radioactivity is higher in the ipsilateral side (Ipsi) compared to the contralateral side (Contra) in the forebrain (**a–b′**). Radioactivity is observed in the anterior ventral thalamic nucleus (AV in **a**, arrows in **a′**), the mammillothalamic tract (mt in **b**, arrows in **b′**), the medial habenular nucleus (MHb in **b**, arrowheads in **b′**) and the interpeduncular nucleus (Ip in **c**, arrow in **c′**). It reaches a maximum in the lateral ventricles (LV in **a**, **a′**), the third ventricle (**b–b′**), and at the brain surface (**c–d′**). In the latter, radioactivity is seen in cisterns such as the ambient cistern

choroid plexus followed by circulation in the ventricular system with recirculation deeper into the brain (Moos and Høyer 1996). Extracellular transferrin could play an important role to ferry iron of brain interstitium into neurons, meaning that transferrin of the choroid plexus could be instrumental for transport and delivery of iron to transferrin receptor-containing neurons.

5.9 Conclusions

Transport of iron across the BBB favor that iron is detached from transferrin inside brain endothelial cells before being transported into the brain. Despite a scarce transendothelial vesicular traffic across the brain endothelial cells, a small amount of holo-transferrin might transfer the BBB non-specifically. Similarly, the choroid plexus epithelial cells allow holo-transferrin to pass the blood-CSF barrier by non-specific vesicular transport. They are also able to take up holo-transferrin by transferrin receptor-mediated endocytosis with subsequent detachment of the iron within the endosome and reduction of Fe^{3+} to Fe^{2+} followed by export to the CSF. The involvement of this pathway is supported by the fact that the choroid plexus epithelial cells express proteins necessary for this pathway i.e. DcytB, DMT1, hephaestin, ceruloplasmin, and ferroportin. Iron storage is detected within the choroid plexus epithelium in both the developing and adult brain. This is supported by profound mRNA expression of ferritin, further indicating that iron is released from the endosome and incorporated in ferritin. Lastly, data also indicate that holo-transferrin can be transported across the choroid plexus epithelial cells by transferrin receptor-mediated transcytosis.

The CSF contains a lower concentration of transferrin compared to that of iron, hence supporting that the iron-binding capacity of transferrin is exceeded in CSF leading to the presence of NTBI in CSF. The presence of holo-transferrin extracellularly in the brain most likely originates from hepatic-derived transferrin that passes the blood-CSF barrier and to some extent the BBB through the described pathways.

In addition to the capability of the choroid plexus epithelial cells to transport iron across the blood-CSF barrier, the cells are also able to synthesize and secrete transferrin into the CSF. Transferrin from CSF primarily distribute to the brain parenchyma adjacent to the ventricular surface, however, we hypothesize that choroid plexus-derived transferrin might also contribute significantly to the amount of transferrin in the brain interstitial fluid by distribution via the subarachnoid space along major vessels. Within the brain parenchyma, transferrin might bind Fe^{3+}

Fig. 5.4 (continued) (ACi) and lateral recess (LR, double arrow in **d′**). The deeper regions of the brain, such as the red nucleus (Rn) and the choroid plexus (CP, single arrow in **d′**), are almost unlabeled. Bars: **a** = 1.1 mm, **b** = 1.7 mm, **c** = 2.5 mm, **d** = 2.2 mm (Moos and Morgan 1998b). Permission to reproduce material is obtained from licence holder

transported through the BBB, and subsequently deliver holo-transferrin to transferrin receptor-expressing neurons.

It has been shown that the uptake and transport of holo-transferrin at the blood-CSF barrier is up-regulated during development and in iron-deficiency. However, the circulation and internalization of holo-transferrin at the choroid plexus are not thoroughly studied in conditions with brain pathology. Important questions to answer will be if the choroid plexus senses iron-challenges of the brain with changes in transferrin synthesis and other iron-scavenger proteins, and to address how neurons adapt to changes in extracellularly iron levels in the brain interstitium.

Acknowledgements The most recent results obtained and described by the authors were generated by generous grants from the Lundbeck Foundation (Grant no: R191-2015-1360) and *Fonden til Lægevidenskabens Fremme* (Grant no: 14-191).

References

Abboud S et al (2000) A novel mammalian iron-regulated protein involved in intracellular iron metabolism. J Biol Chem 275(26):19906–19912

Anderson GJ, Vulpe CD (2009) Mammalian iron transport. Cell Mol Life Sci 66(20):3241–3261

Anderson GJ, Frazer DM, McLaren GD (2009) Iron absorption and metabolism. Curr Opin Gastroenterol 25(2):129–135

Andrews NC (1999) Disorders of iron metabolism. N Engl J Med 341(26):1986–1995

Aziz N, Munro HN (1987) Iron regulates ferritin mRNA translation through a segment of its 5′ untranslated region. Proc Natl Acad Sci U S A 84(23):8478–8482

Blissett AR et al (2017) Magnetic mapping of iron in rodent spleen. Nanomedicine: NBM 13:977–986

Bloch B et al (1985) Transferrin gene expression visualized in oligodendrocytes of the rat brain by using in situ hybridization and immunohistochemistry. Proc Natl Acad Sci U S A 82 (19):6706–6710

Bloch B et al (1987) Transferrin gene expression in choroid plexus of the adult rat brain. Brain Res Bull 18(4):573–576

Boserup MW et al (2011) Heterogenous distribution of ferroportin-containing neurons in mouse brain. Biometals 24(2):357–375

Bradbury MWB (1997) Transport of iron in the blood-brain-cerebrospinal fluid system. J Neurochem 69(2):443–454

Burkhart A et al (2016) Expression of iron-related proteins at the neurovascular unit supports reduction and reoxidation of iron for transport through the blood-brain barrier. Mol Neurobiol 53(10):7237–7253

Carocci A et al (2018) Oxidative stress and neurodegeneration: the involvement of iron. BioMetals 31(5):715–735

Case NM (1959) Hemosiderin granules in the choroid plexus. J Biophys Biochem Cytol 6:527–530

Chevion M (1988) A site-specific mechanism for free radical induced biological damage: the essential role of redox-active transition metals. Free Radic Biol Med 5(1):27–37

Chiou B et al (2019) Endothelial cells are critical regulators of iron transport in a model of the human blood–brain barrier. J Cereb Blood Flow Metab 39(11):2117–2131

Collingwood J, Dobson J (2006) Mapping and characterization of iron compounds in Alzheimer's tissue. J Alzheimers Dis 10(2–3):215–222

Connor JR et al (2011) Profile of altered brain iron acquisition in restless legs syndrome. Brain 134 (Pt 4):959–968

Cozzi A et al (2004) Analysis of the biologic functions of H- and L-ferritins in HeLa cells by transfection with siRNAs and cDNAs: evidence for a proliferative role of L-ferritin. Blood 103 (6):2377–2383

Cremonesi L et al (2004) Case report: a subject with a mutation in the ATG start codon of L-ferritin has no haematological or neurological symptoms. J Med Genet 41(6):e81

Crowe A, Morgan EH (1992) Iron and transferrrin uptake by brain and cerebrospinal fluid in the rat. Brain Res 592(1–2):8–16

De Domenico I et al (2007) Ferroxidase activity is required for the stability of cell surface ferroportin in cells expressing GPI-ceruloplasmin. EMBO J 26(12):2823–2831

Descamps L et al (1996) Receptor-mediated transcytosis of transferrin through blood-brain barrier endothelial cells. Am J Physiol 270(4):1149–1158

Donovan A et al (2000) Positional cloning of zebrafish ferroportin1 identifies a conserved vertebrate iron exporter. Nature 403(6771):776–781

Duck KA, Connor JR (2016) Iron uptake and transport across physiological barriers. Biometals 29 (4):573–591

Dwork AJ et al (1990) An autoradiographic study of the uptake and distribution of iron by the brain of the young rat. Brain Res 518(1–2):31–39

Eckenroth BE et al (2011) How the binding of human transferrin primes the transferrin receptor potentiating iron release at endosomal pH. Proc Natl Acad Sci U S A 108(32):13089–13094

Fernandes C et al (2018) Iron localization in the guinea pig choroid plexus: a light and transmission electron microscopy study. J Trace Elem Med Biol 49(May):128–133

Fernández B et al (2017) Biomonitorization of iron accumulation in the substantia nigra from Lewy body disease patients. Toxicol Rep 4:188–193

Ferreira C et al (2000) Early embryonic lethality of H ferritin gene deletion in mice. J Biol Chem 275(5):3021–3024

Fishman JB et al (1987) Receptor-mediated transcytosis of transferrin across the blood-brain barrier. J Neurosci Res 18(2):299–304

Fleming MD et al (1998) Nramp2 is mutated in the anemic Belgrade (b) rat: evidence of a role for Nramp2 in endosomal iron transport. Genetics 95(3):1148–1153

Focht SJ et al (1997) Regional distribution of iron, transferrin, ferritin, and oxidatively-modified proteins in young and aged Fischer 344 rat brains. Neuroscience 79(1):255–261

Frazer DM et al (2002) Hepcidin expression inversely correlates with the expression of duodenal iron transporters and iron absorption in rats. Gastroenterology 123(3):835–844

Freskgård P-O, Urich E (2017) Antibody therapies in CNS diseases. Neuropharmacology 120:38–55

Friden PM et al (1991) Anti-transferrin receptor antibody and antibody-drug conjugates cross the blood-brain barrier. Proc Natl Acad Sci U S A 88(11):4771–4775

Ganz T (2013) Systemic iron homeostasis. Physiol Rev 93(4):1721–1741

Ganz T, Nemeth E (2012) Hepcidin and iron homeostasis. Biochim Biophys Acta Mol Cell Res 1823(9):1434–1443

Giometto B et al (1990) Transferrin receptors in rat central nervous system. An immunocytochemical study. J Neurol Sci 98(1):81–90

Gunshin H et al (1997) Cloning and characterization of a mammalian proton-coupled metal-ion transporter. Nature 388(6641):482–488

Gunshin H et al (2005) Slc11a2 is required for intestinal iron absorption and erythropoiesis but dispensable in placenta and liver. J Clin Invest 115(5):1258–1266

Hansen TM et al (1999) Expression of ferritin protein and subunit mRNAs in normal and iron deficient rat brain. Brain Res Mol Brain Res 65(2):186–197

Harrison PM, Arosio P (1996) The ferritins: molecular properties, iron storage function and cellular regulation. Biochim Biophys Acta Bioenerg 1275(3):161–203

Hoy TG, Jacobs A (1981) Ferritin polymers and the formation of haemosiderin. Br J Haematol 49 (4):593–602

Iancu TC et al (1995) The hypotransferrinaemic mouse: ultrastructural and laser microprobe analysis observations. J Pathol 177(1):83–94

Idzerda RL et al (1986) Rat transferrin gene expression: tissue-specific regulation by iron deficiency. Proc Natl Acad Sci U S A 83(11):3723–3727

Jefferies WA et al (1984) Transferrin receptor on endothelium of brain capillaries. Nature 312 (5990):162–163

Johnsen KB, Moos T (2016) Revisiting nanoparticle technology for blood-brain barrier transport: unfolding at the endothelial gate improves the fate of transferrin receptor-targeted liposomes. J Control Release 222:32–46

Keep RF, Jones HC (1990) A morphometric study on the development of the lateral ventricle choroid plexus, choroid plexus capillaries and ventricular ependyma in the rat. Dev Brain Res 56 (1):47–53

Klomp LWJ et al (1996) Ceruloplasmin gene expression in the murine central nervous system. J Clin Invest 98(1):207–215

Knutson MD (2019) Non-transferrin-bound iron transporters. Free Radic Biol Med 133:101–111

Kühn LC, Hentze MW (1992) Coordination of cellular iron metabolism by post-transcriptional gene regulation. J Inorg Biochem 47(3–4):183–195

Kurz T, Eaton JW, Brunk UT (2011) The role of lysosomes in iron metabolism and recycling. Int J Biochem Cell Biol 43(12):1686–1697

Lasiecka ZM et al (2010) Neuronal early endosomes require EHD1 for L1/NgCAM trafficking. J Neurosci 30(49):16485–16497

Lee HJ et al (2000) Targeting rat anti-mouse transferrin receptor monoclonal antibodies through blood-brain barrier in mouse. J Pharmacol Exp Ther 292(3):1048–1052

Leitner DF, Connor JR (2012) Functional roles of transferrin in the brain. Biochim Biophys Acta Gen Subj 1820(3):393–402

Levi S et al (1994) The role of the L-chain in ferritin iron incorporation. J Mol Biol 238(5):649–654

Lichota J et al (2010) Macromolecular drug transport into the brain using targeted therapy. J Neurochem 113(1):1–13

Luck AN, Mason AB (2012) Transferrin-mediated cellular iron delivery. Curr Top Membr 69:3–35

Luck AN, Mason AB (2013) Structure and dynamics of drug carriers and their interaction with cellular receptors: focus on serum transferrin. Adv Drug Deliv Rev 65(8):1012–1019

Mancias JD et al (2014) Quantitative proteomics identifies NCOA4 as the cargo receptor mediating ferritinophagy. Nature 509(7498):105–109

Martin ME et al (1998) A point mutation in the bulge of the iron-responsive element of the L ferritin gene in two families with the hereditary hyperferritinemia-cataract syndrome. Blood 91 (1):319–323

McCarthy RC, Kosman DJ (2012) Mechanistic analysis of iron accumulation by endothelial cells of the BBB. BioMetals 25(4):665–675

McCarthy RC, Kosman DJ (2013) Ferroportin and exocytoplasmic ferroxidase activity are required for brain microvascular endothelial cell iron efflux. J Biol Chem 288(24):17932–17,940

McKie AT et al (2000) A novel duodenal iron-regulated transporter, IREG1, implicated in the basolateral transfer of iron to the circulation. Mol Cell 5(2):299–309

Miyata K et al (2009) Spontaneous iron accumulation in hepatocytes of a 7-week-old female rat. J Toxicol Pathol 22(3):199–203

Miyazaki E et al (2002) Denatured H-ferritin subunit is a major constituent of haemosiderin in the liver of patients with iron overload. Gut 50(3):413–419

Moos T (1995) Developmental profile of non-heme iron distribution in the rat brain during ontogenesis. Brain Res Dev Brain Res 87(2):203–213

Moos T (1996) Immunohistochemical localization of intraneuronal transferrin receptor immunoreactivity in the adult mouse central nervous system. J Comp Neurol 375(4):675–692

Moos T, Høyer PE (1996) Detection of plasma proteins in CNS neurons conspicuous influence of tissue-processing parameters and the utilization of serum for blocking nonspecific reactions. J Histochem Cytochem 44(6):591–603

Moos T, Møllgaard K (1993a) A sensitive post-DAB enhancement technique for demonstration of iron in the central nervous system. Histochemistry 99(6):471–475

Moos T, Møllgaard K (1993b) Cerebrovascular permeability to azo dyes and plasma proteins in rodents of different ages. Neuropathol Appl Neurobiol 19(2):120–127

Moos T, Morgan EH (1998a) Evidence for low molecular weight, non-transferrin-bound iron in rat brain and cerebrospinal fluid. J Neurosci Res 54(4):486–494

Moos T, Morgan EH (1998b) Kinetics and distribution of [59Fe-125I]transferrin injected into the ventricular system of the rat. Brain Res 790(1–2):115–128

Moos T, Morgan EH (2000) Transferrin and transferrin receptor function in brain barrier systems. Cell Mol Neurobiol 20(1):77–95

Moos T, Morgan EH (2001) Restricted transport of anti-transferrin receptor antibody (OX26) through the blood-brain barrier in the rat. J Neurochem 79(1):119–129

Moos T, Morgan EH (2004) The significance of the mutated divalent metal transporter (DMT1) on iron transport into the Belgrade rat brain. J Neurochem 88(1):233–245

Moos T, Oates PS, Morgan EH (1998) Expression of the neuronal transferrin receptor is age dependent and susceptible to iron deficiency. J Comp Neurol 398(3):420–430

Moos T, Oates PS, Morgan EH (2001) Expression of transferrin mRNA in rat oligodendrocytes is iron-independent and changes with increasing age. Nutr Neurosci 4(1):15–23

Moos T et al (2007) Iron trafficking inside the brain. J Neurochem 103(5):1730–1740

Morgan EH (1996) Iron metabolism and transport. In: Zakim D, Boyer TD (eds) Hepatology: a textbook of liver disease, vol 1. Saunders, Philadelphia, pp 526–554

Morgan EH, Moos T (2002) Mechanism and developmental changes in iron transport across the blood-brain barrier. Dev Neurosci 24(2–3):106–113

Morris CM et al (1992) Uptake and distribution of iron and transferrin in the adult rat brain. J Neurochem 59(1):300–306

Müllner EW, Kühn LC (1988) A stem-loop in the 3′ untranslated region mediates iron-dependent regulation of transferrin receptor mRNA stability in the cytoplasm. Cell 53(5):815–825

Musci MC (2014) Ceruloplasmin-ferroportin system of iron traffic in vertebrates. World J Biol Chem 5(2):204–215

Nemeth E et al (2004) IL-6 mediates hypoferremia of inflammation by inducing the synthesis of the iron regulatory hormone hepcidin. J Clin Investig 113(9):1271–1276

Ohgami RS et al (2006) The Steap proteins are metalloreductases. Blood 108(4):1388–1394

Pantopoulos K (2004) Iron metabolism and the IRE/IRP regulatory system: an update. Ann N Y Acad Sci 1012:1–13

Pardridge WM (2016) CSF, blood-brain barrier, and brain drug delivery. Expert Opin Drug Deliv 13(7):963–975

Pardridge WM, Buciak JL, Friden PM (1991) Selective transport of an anti-transferrin receptor antibody through the blood-brain barrier in vivo. J Pharmacol Exp Ther 259(1):66–70

Patel BN, David S (1997) A novel glycosylphosphatidylinositol-anchored form of ceruloplasmin is expressed by mammalian astrocytes. J Biol Chem 272(32):20185–20190

Quintana C et al (2006) Study of the localization of iron, ferritin, and hemosiderin in Alzheimer's disease hippocampus by analytical microscopy at the subcellular level. J Struct Biol 153 (1):42–54

Roberts RL, Fine RE, Sandra A (1993) Receptor-mediated endocytosis of transferrin at the blood-brain barrier. J Cell Sci 104(Pt 2):521–532

Rouault TA, Zhang DL, Jeong SY (2009) Brain iron homeostasis, the choroid plexus, and localization of iron transport proteins. Metab Brain Dis 24(4):673–684

Sävman K et al (2005) Non-protein-bound iron in brain interstitium of newborn pigs after hypoxia. Dev Neurosci 27(2–4):176–184

Schweizer C, Fraering PC, Kühn LC (2014) Ferritin H gene deletion in the choroid plexus and forebrain results in hydrocephalus. Neurochem Int 71(1):17–21

Simpson IA et al (2014) A novel model for brain iron uptake: introducing the concept of regulation. J Cereb Blood Flow Metab 35(1):48–57

Siupka P et al (2017) Bidirectional apical-basal traffic of the cation-independent mannose-6-phosphate receptor in brain endothelial cells. J Cereb Blood Flow Metab 37(7):2598–2613

Skjørringe T et al (2015) Divalent metal transporter 1 (DMT1) in the brain: implications for a role in iron transport at the blood-brain barrier, and neuronal and glial pathology. Front Mol Neurosci 8:1–13

Sotiriou S et al (2002) Ascorbic-acid transporter Slc23a1 is essential for vitamin C transport into the brain and for perinatal survival. Nat Med 8(5):514–517

Spector R et al (2015) A balanced view of choroid plexus structure and function: focus on adult humans. Exp Neurol 267:78–86

Stohs SJ, Bagchi D (1995) Oxidative mechanisms in the toxicity of metal ions. Free Radic Biol Med 18(2):321–336

Strahan ME, Crowe A, Morgan EH (1992) Iron uptake in relation to transferrin degradation in brain and other tissues of rats. Am J Physiol 263(4 Pt 2):R924–R929

Taylor EM, Crowe A, Morgan EH (1991) Transferrin and iron uptake by the brain: effects of altered iron status. J Neurochem 57(5):1584–1592

Theisen M et al (1993) A C/EBP-binding site in the transferrin promoter is essential for expression in the liver but not the brain of transgenic mice. Mol Cell Biol 13(12):7666–7676

Thomsen MS, Routhe LJ, Moos T (2017) The vascular basement membrane in the healthy and pathological brain. J Cereb Blood Flow Metab 37(10):3300–3317

Todorich B et al (2008) Tim-2 is the receptor for H-ferritin on oligodendrocytes. J Neurochem 107 (6):1495–1505

Tsukaguchi H et al (1999) A family of mammalian Na+-dependent L-ascorbic acid transporters. Nature 399(6731):70–75

Tsutsumi M, Skinner MK, Sanders-Bush E (1989) Transferrin gene expression and synthesis by cultured choroid plexus epithelial cells. Regulation by serotonin and cyclic adenosine 3′,5-′-monophosphate. J Biol Chem 264(16):9626–9631

Wan S et al (2006) Deferoxamine reduces CSF free iron levels following intracerebral hemorrhage. Acta Neurochir Suppl 96:199–202

Wang X, Li GJ, Wei Z (2008a) Efflux of iron from the cerebrospinal fluid to the blood at the blood-CSF barrier: effect of manganese exposure. Computer 144(5):724–732

Wang X, Miller DS, Zheng W (2008b) Intracellular localization and subsequent redistribution of metal transporters in a rat choroid plexus model following exposure to manganese or iron. Toxicol Appl Pharmacol 230(2):167–174

Ward RJ et al (2014) The role of iron in brain ageing and neurodegenerative disorders. Lancet Neurol 13(10):1045–1060

Wardman P, Candeias LP (1996) Fenton chemistry: an introduction. Radiat Res 145(5):523

Yu YJ et al (2014) Therapeutic bispecific antibodies cross the blood-brain barrier in nonhuman primates. Sci Transl Med 6(261):261ra154

Zakin MM (1992) Regulation of transferrin gene expression. FASEB J 6(14):3253–3258

Zakin MM, Baron B, Guillou F (2002) Regulation of the tissue-specific expression of transferrin gene. Dev Neurosci 24(2–3):222–226

Zanardi A et al (2017) Ceruloplasmin replacement therapy ameliorates neurological symptoms in a preclinical model of aceruloplasminemia. EMBO Mol Med 10(1):91–106

Zecca L et al (2004) Iron, brain ageing and neurodegenerative disorders. Nat Rev Neurosci 5 (11):863–873

Zhang C (2014) Essential functions of iron-requiring proteins in DNA replication, repair and cell cycle control. Protein Cell 5(10):750–760

Chapter 6
Acid/Base Transporters in CSF Secretion and pH Regulation

Dagne Barbuskaite, Helle Damkier, and Jeppe Praetorius

Abstract Early functional studies on the choroid plexus established that the cerebrospinal fluid (CSF) secretion rate was sensitive to inhibitors of acid/base transporters and carbonic anhydrase. In fact, the CO_2/HCO_3^- buffer system is required for effective CSF secretion. Unlike blood, the CSF contains very little protein and yet it responds efficiently to changes in pH utilizing primarily the CO_2/HCO_3^- buffer system. Besides a central role in CSF secretion, recent studies indicate a central role for the choroid plexus epithelium in CSF pH regulation. The luminal plasma membrane expresses a variety of acid and base transport proteins of which one, the NBCe2, was directly shown to regulate CSF pH. This chapter describes the current knowledge on the molecular mechanisms in choroid plexus acid/base transport and the understanding of their possible involvement in both CSF secretion and pH regulation.

6.1 Acid/Base Regulation in Brain Extracellular Fluid Compartments

The brain is composed of two extracellular fluid compartments: the brain extracellular fluid (BECF) surrounding cells in the parenchyma as well as the cerebrospinal fluid (CSF) that fills the ventricles and covers the brain and spinal cord (Hladky and Barrand 2016).

The fluid compartments are separated from the blood by two distinct barriers. The blood-brain barrier (BBB) separates the BECF from the blood, and the blood-cerebrospinal fluid barrier (BCSFB) separates blood from cerebrospinal fluid

D. Barbuskaite
Department of Neuroscience, Faculty of Health and Medical Sciences, University of Copenhagen, Copenhagen, Denmark
e-mail: dagne@sund.ku.dk

H. Damkier · J. Praetorius (✉)
Department of Biomedicine, Faculty of Health, Aarhus University, Aarhus, Denmark
e-mail: jp@biomed.au.dk

© The American Physiological Society 2020
J. Praetorius et al. (eds.), *Role of the Choroid Plexus in Health and Disease*,
Physiology in Health and Disease, https://doi.org/10.1007/978-1-0716-0536-3_6

(CSF). These barriers have distinct barrier properties that hinder diffusion of many substances. The BECF and the CSF in the ventricles are only separated by a single layer of ependymal cells which does not constitute a barrier per se. The CSF in the subarachnoid space and the space surrounding the penetrating vessels from the surface of the brain, the Virchow-Robin space, is separated from the BECF by the pia-glial membrane. This membrane consists of a thin layer of connective tissue covering the surface of the brain parenchyma and the glia limitans which is a layer of astrocytic end-feet. This membrane is permeable to diffusion of substances between the CSF and BECF. Both the BBB and the BCSFB are permeable to CO_2 that enters the brain freely and if not buffered will lead to a decrease in brain pH. In order to buffer the pH, bases such as HCO_3^- need to be transported from the blood across the barriers. This requires the presence of transporters in the membranes of the barrier.

Neuronal excitability is greatly affected by pH changes in the brain parenchyma (Baron et al. 1985). Firing of action potentials builds up the acid content in the interstitium. These acids are usually buffered and cleared and do not lead to functional damage (Kazemi et al. 1967). In some circumstances, acid production can even protect brain function. The high neuronal activity during seizures builds up an acidic environment. This activates acid sensing ion channels which leads to inhibition of the neuronal activity, and thus to halting the seizure (Ziemann et al. 2008). In febrile seizures, the fever causes the child to hyperventilate and thereby respiratory alkalosis develops in the immature brain (Schuchmann et al. 2006). The seizure is broken by the central pH decrease that follows the increased neuronal activity.

Proteins play a central role in buffering the effects of sudden changes in pH of the blood plasma and other fluid spaces. In contrast, CSF contains only small amounts of protein but brain pH is nevertheless maintained within a quite narrow range. Fluctuations in plasma pH and pCO_2 are followed over time by similar changes in CSF pH (Lee et al. 1969). While CO_2 crosses all brain barriers rapidly, H^+ and HCO_3^- mainly cross the CPE via membrane transporters. One would expect this to result in sizable fluctuations in brain pH during respiratory acidosis because of the lack of protein buffers. However, in respiratory acidosis the decrease in blood pH surpasses that of CSF pH due to the presence of efficient buffers such as the open CO_2/HCO_3^- buffer system. In support for such buffering, Hasan and collaborators found that inhalation of 5% CO_2 for 4 h lead to a threefold higher increase in CSF HCO_3^- compared to plasma HCO_3^- in dogs (Hasan and Kazemi 1976).

CSF pH is known to directly influence BECF pH in the areas close to the ventricle system (Okada et al. 1993). Also, the CSF in the subarachnoid space enters the brain parenchyma via the blood vessels. The composition of CSF thereby greatly affects the composition of the BECF and thereby the pH regulation of the brain parenchyma.

6.2   Choroid Plexus Acid-Base Transport

The production of CSF amounts to approximately 500 ml CSF daily in humans, of which the major proportion is secreted by the epithelial cells of the choroid plexus (Cserr 1971). The CSF formation by the choroid plexus depends primarily on the

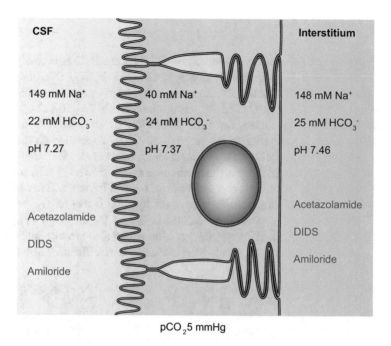

CSF   **Interstitium**

149 mM Na⁺   40 mM Na⁺   148 mM Na⁺

22 mM HCO₃⁻   24 mM HCO₃⁻   25 mM HCO₃⁻

pH 7.27   pH 7.37   pH 7.46

                                      Acetazolamide

Acetazolamide

                                      DIDS

DIDS

                                      Amiloride

Amiloride

pCO₂5 mmHg

Fig. 6.1 Schematic diagram of normal physiological values and inhibitor actions relevant to choroid plexus epithelial acid-base transport. Acetazolamide and DIDS inhibits CSF secretion both when applied from the CSF side and the interstitial side. Amiloride inhibits CSF secretion when applied from the blood/interstitial side as well as from the luminal side

transepithelial movement of Na^+, Cl^-, HCO_3^- and H_2O (Pollay and Curl 1967; Welch 1963; Davson and Segal 1970; Wright 1972). Apart from the central involvement of typical transport proteins such as e.g. the Na^+, K^+-ATPase (Welch 1963; Davson and Segal 1970; Wright 1972) in CSF production by the choroid plexus, the CO_2/HCO_3^- buffering system and acid-base transporters seem to be of pivotal importance for the process of secretion. Firstly, the carbonic anhydrase inhibitor acetazolamide inhibits CSF secretion significantly (Welch 1963; Tschirgi et al. 1954; Kister 1956; Davson and Luck 1957; Ames et al. 1965; Segal and Burgess 1974; Pollay and Davson 1963; Vogh et al. 1987; Murphy and Johanson 1989a) (Fig. 6.1). Secondly, CSF secretion is sensitive to manipulations of basolateral pH and HCO_3^- concentration (Saito and Wright 1983), and basolateral application of the stilbene-type HCO_3^- transport inhibitor DIDS reduces the secretion rate by choroid plexus epithelial cells (Deng and Johanson 1989). Finally, the NHE inhibitor amiloride inhibits Na^+ import into the choroid plexus from the blood side as well as the CSF secretion (Davson and Segal 1970; Murphy and Johanson 1989a, b; Johanson and Murphy 1990), suggesting a dependence of CSF secretion on basolateral Na^+/H^+ exchange. Nevertheless, injection of amiloride into the ventricles had a similar effect on CSF secretion (Segal and Burgess 1974). In the following, we summarize the current knowledge on choroid plexus carbonic anhydrases and acid-base transport proteins.

6.2.1 Carbonic Anhydrases

In many biological systems, the hydration of CO_2 to form H_2CO_3 is catalyzed by carbonic anhydrases (Boron and Boulpaep 2012), and H_2CO_3 then dissociates spontaneously to HCO_3^- and H^+. Carbonic anhydrases are among the fastest enzymes studied and are blocked by acetazolamide (Maren 1962). Carbonic anhydrase activity is important for CSF production by the choroid plexus, as acetazolamide reduces secretion in the range of 50–100% depending on the experimental setup (Welch 1963; Davson and Segal 1970; Ames et al. 1965). In the early experiments, carbonic anhydrase activity was mainly considered in relation to the cytosol where HCO_3^- would be produced within the cell by intracellular carbonic anhydrase such as CAII (Fig. 6.2). This led to the hypothesis that HCO_3^- secretion by the CPE depended on intracellular formation from CO_2 and H_2O (Vogh and Maren 1975; Maren 1972). However, this reaction would also produce intracellular H^+, and the choroid plexus cell would therefore require an efficient basolateral H^+

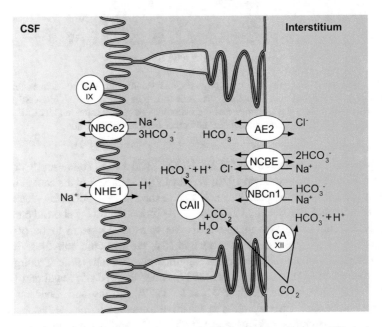

Fig. 6.2 Schematic diagram of choroid plexus epithelial carbonic anhydrases (CAs). Acetazolamide and high concentrations inhibit CAs efficiently, but it is impossible to tell which of the three forms are most relevant for CSF formation and pH regulation. Interestingly, enzyme histochemistry is only detected in the plasma membrane domains (Masuzawa et al. 1981) i.e. corresponding to CAIX and CAXII. The traditional hypothesis predicts intracellular conversion of CO_2 and H_2O to H^+ and HCO_3^- to sustain bicarbonate secretion. The alternative hypothesis involves CAXII in extracellular activity to sustain HCO_3^- import into the epithelial cells and perhaps CAIX for sustaining luminal HCO_3^- extrusion. Thus, like in the duodenal villus cells, there is a notion that HCO_3^- secretion is sustained by transcellular HCO_3^- movement rather than intracellular HCO_3^- generation from CO_2

extrusion mechanism such as a Na^+/H^+ exchanger (NHE) in order to sustain CSF secretion. There is, indeed, an apparent lack of such transport mechanism in the basolateral membrane of choroid plexus epithelial cells, i.e. there is no basolateral membrane expression of NHE's and only minimal acid extrusion capacity in the absence of Na^+ or CO_2/HCO_3^- [see below Bouzinova et al. (2005), Damkier et al. (2009), Christensen et al. (2017)]. The only known basolateral "acid-extruders" are actually two HCO_3^- import proteins (Ncbe and NBCn1, see below).

Thus, the model of intracellular formation of HCO_3^- for secretion and basolateral H^+ extrusion is difficult to sustain despite Maren's pioneering work. One indication for a transcellular route for secreted HCO_3^- came from the clear enzyme histochemical demonstration that choroid plexus epithelial carbonic anhydrase activity was almost entirely restricted to the microvilli and the basal infoldings (Masuzawa et al. 1981). This is consistent with the detection of plasma membrane carbonic anhydrases CAIX and CAXII in the choroid plexus (Christensen et al. 2017; Kallio et al. 2006), where CAXII is localized to the basolateral membrane and CAIX may be enriched in the luminal domain (Fig. 6.2). Both CAIX and CAXII are single-pass transmembrane proteins with extracellular catalytic domains (Tureci et al. 1998; Opavsky et al. 1996). We are not aware of any other normal epithelium expressing both of these CA forms. The close proximity between these enzymes and the plasma membrane HCO_3^- transporters may increase transport rates by supplying sufficient substrate (HCO_3^- or even CO_3^{2-}) for the transport (McMurtrie et al. 2004). The putative direct physical and functional associations of external membrane-bound and/or cytosolic carbonic anhydrases with HCO_3^- transporters such as AE2 and NBCe1 is compelling although still debated (McMurtrie et al. 2004; Boron 2010). Nevertheless, it remains possible that membrane bound CAs with extracellular enzyme activity are the actual targets of acetazolamide in the choroid plexus, and that a reduced HCO_3^- transport rate results in the observed decline in CSF secretion. The primary target for acetazolamide in the choroid plexus needs to be investigated in order to establish to what extent secreted HCO_3^- arise from cytosolic production and from transcellular carrier mediated movement.

6.2.2 Base Transporters

As mentioned above, the presence of the CO_2/HCO_3^- buffering system is necessary for efficient CSF secretion thereby indirectly implicating the participation of HCO_3^- transporters in the process of CSF secretion (Haselbach et al. 2001). Indeed, DIDS reduces HCO_3^- and Na^+ transport rates to the luminal side of the epithelium both after basolateral and luminal application (Deng and Johanson 1989; Nattie and Adams 1988; Mayer and Sanders-Bush 1993). Bicarbonate transporters arise from two gene families: the solute carrier (*Slc*) 4 and 26 families (Romero et al. 2013; Cordat and Reithmeier 2014). The *Slc4* derived polypeptides include the classical Cl^-/HCO_3^- exchangers of which Ae2 and, the electrogenic sodium bicarbonate exchanger NBCe2, the electroneutral NBCn1, and Ncbe/NBCn2 expressed in the choroid plexus (Damkier et al. 2013). Apart from epithelial NBCn1, these transporters are DIDS sensitive and candidate targets mediating the inhibitory effect of stilbene derivatives on CSF secretion.

6.2.2.1   The Anion Exchanger Ae2, Slc4a2

Ae2 mediates electroneutral exchange of extracellular Cl^- for intracellular HCO_3^-, and thus extrudes base from the cell (Alper 2009),is more active at alkaline pH, and is inhibited by DIDS (4). The import of Cl^- into the choroid plexus epithelial cell from the interstitium is necessary for continuous CSF secretion, and most likely involves classical anion exchangers as judged from the sensitivity towards basolateral DIDS application (Deng and Johanson 1989; Frankel and Kazemi 1983) and the requirement for CO_2/HCO_3^- (Hughes et al. 2010). Ae2 is expressed at the basolateral membrane domain of choroid plexus epithelial cells with the highest signal at the basal labyrinth (Lindsey et al. 1990; Alper et al. 1994; Praetorius and Nielsen 2006) (Fig. 6.3). The localization and direction of transport suggests that Ae2 is involved in transcellular Cl^- transport and is in fact the only known basolateral Cl^- entry mechanism for sustaining CSF secretion. Ae2 may also be indirectly involved in CSF pH regulation, as it could facilitate continued H^+ excretion across the luminal membrane as e.g. in CSF pH regulation during alkalosis.

6.2.2.2   The Electroneutral Na^+ and HCO_3^- Transporter Ncbe, NBCn2, Slc4a10

Polypeptides derived from *slc4a10* have all been characterized as DIDS sensitive electroneutral Na^+ driven and HCO_3^- importers (Giffard et al. 2003; Wang et al. 2000; Damkier et al. 2010; Parker et al. 2008). First, the HCO_3^- transport was shown to depend on intracellular Cl^- in mammalian expression systems transfected with *slc4a10* (Giffard et al. 2003; Wang et al. 2000). In one study, the stoichiometry for Na^+ to acid-base equivalents was estimated to 1:2 and a net Cl^- efflux was DIDS-sensitive (Damkier et al. 2010). Thus, the transporter was characterized as a Na^+-dependent Cl^-/HCO_3^- exchanger and named Ncbe. However, convincing results from a study using the human *SLC4A10* gene product expressed in *Xenopus laevis* oocytes did not support the dependence of transport on intracellular Cl^- nor was a net Cl^- extrusion detected by Parker et al. (2008). The protein was therefore renamed NBCn2 as an electroneutral $Na^+:HCO_3^-$ cotransporter. We prefer to apply the transporter names Ncbe in rodents and NBCn2 in humans as long as this controversy is not resolved.

In the choroid plexus, the Ncbe/NBCn2 transporter is localized on the basolateral membrane (Praetorius and Nielsen 2006; Praetorius et al. 2004a), and is therefore proposed to supply the CPE with both Na^+ and HCO_3^- from the blood side of the epithelium (Fig. 6.4). Consistent with this idea, the Na^+ and CO_2/HCO_3^-dependent base uptake (or acid extrusion) in the choroid plexus epithelium is partly DIDS

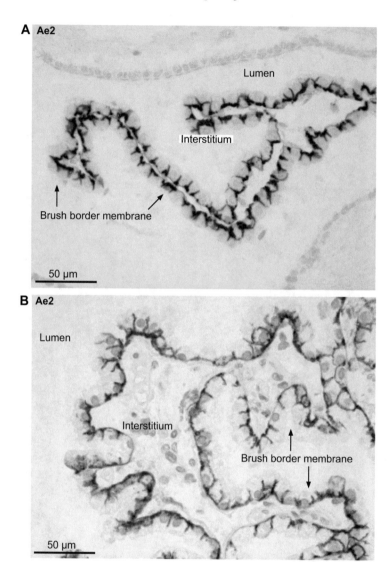

Fig. 6.3 Bright-field micrograph of anti-Ae2 immuno-peroxidase reactivity in the choroid plexus. (**a**) The brown reaction product is confined to the basolateral membrane domains including the basal labyrinth in mouse tissue. (**b**) A similar staining pattern was observed in the human choroid plexus. Nuclei are counterstained with hematoxylin (blue). For methods, see previous publications (Praetorius and Nielsen 2006; Praetorius et al. 2004a)

sensitive, probably reflecting Ncbe activity (Bouzinova et al. 2005). Direct involvement of Ncbe in the pH_i regulation of choroid plexus epithelial cells was demonstrated in studies of Ncbe knockout mice. The choroid plexus from Ncbe knockout mice showed a severely diminished Na^+-dependent acid extrusion compared to normal littermates (Jacobs et al. 2008), and up to 70% of the DIDS-sensitive pH_i

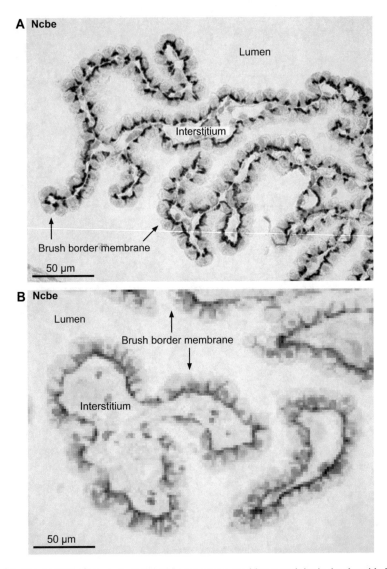

Fig. 6.4 Bright-field micrograph of anti-Ncbe immuno-peroxidase reactivity in the choroid plexus. (**a**) The brown reaction product is confined to the basolateral membrane domains including the basal labyrinth in mouse tissue. (**b**) A similar staining pattern was observed in the human choroid plexus. Nuclei are counterstained with hematoxylin (blue). For methods, see previous publications (Praetorius and Nielsen 2006; Praetorius et al. 2004a)

recovery from acid load depends on this transporter (Damkier et al. 2009). Besides regulating pH_i of the choroid plexus epithelial cells, Ncbe mediated HCO_3^- loading may well facilitate the luminal bicarbonate extrusion. In this way, Ncbe could sustain luminal HCO_3^- secretion and indirectly participate in CSF pH regulation.

As one of a very few Na^+ transporters in the basolateral membrane, the Ncbe/NBCn2 probably constitutes the bottleneck for Na^+ in CSF secretion, although this has not yet been demonstrated directly. Nevertheless, the genetic disruption of Ncbe resulted in a reduction in brain ventricle size to almost the same degree as the Na^+: HCO_3^- cotransport (Jacobs et al. 2008). This may suggest that Ncbe knockout also greatly diminished the rate of CSF secretion. One should take caution suggesting a central role for Ncbe in CSF secretion, although the studies in Ncbe knockout mice are compelling in that regard. Large changes in abundance of other transport proteins of interest may also explain the decreased ventricular volume and the feasibly diminished CSF secretion in Ncbe knockout mice. Ncbe knockout critically affected the expression levels of the Na^+, K^+-ATPase, AQP1, Ae2, and e.g. membrane protein anchoring proteins (Damkier et al. 2009; Christensen et al. 2013; Damkier and Praetorius 2012). The luminal Na^+/H^+ exchanger NHE1 of the normal choroid plexus was demonstrated in the basolateral membrane domain of the epithelial cells from Ncbe knockout mice (Damkier et al. 2009). Thus, studies directly demonstrating the involvement of Ncbe in CSF secretion are highly warranted.

6.2.2.3 The Electrogenic Na^+:HCO_3^- Cotransporter NBCe2, Slc4a5

The electrogenic cotransporter NBCe2 transports 1 Na^+ with 2 or 3 HCO_3^- in or out of cells across the plasma membrane depending on tissue-specific factors (Sassani et al. 2002; Virkki et al. 2002). Typical electrochemical gradients for Na^+ and HCO_3^- favor outward transport in the 1:3 transport mode and inward transport in the 1:2 transport mode. Electrogenic HCO_3^- transport in the luminal membrane of choroid plexus epithelial cells was established in bull frog (Saito and Wright 1984), and the Na^+ dependence of transport indicated an outwardly directed Na^+:HCO_3^- cotransport mechanism was shown in mouse and rat (Banizs et al. 2007; Johanson et al. 1992). These observations could be explained by the expression of NBCe2 in the luminal plasma membrane domain of rat (Bouzinova et al. 2005) and mouse (Christensen et al. 2018) choroid plexus (Fig. 6.5). NBCe2 mRNA was detected in human choroid plexus as well (Praetorius and Nielsen 2006; Damkier et al. 2007), but the protein was not immunolocalized to the epithelium, probably reflecting amino acid sequence differences compared to the rodent immunogenic epitope applied thus far. NBCe2 exports Na^+ and HCO_3^- across the luminal membrane of mouse CPE in an apparent 1:3 stoichiometry (Millar and Brown 2008). Thus, NBCe2 is believed to take part in the Na^+ secretion of the choroid plexus epithelium as well as the regulation of intracellular pH and even extracellular pH.

Inhalation of 11% CO_2 causes a significant increase in the Na^+ content in rodent CSF (Nattie 1980). This effect may represent indirect evidence for NBCe2 function in the choroid plexus. More recently, the physiological significance of NBCe2 has been studied more directly in three different NBCe2 knockout mouse strains. The first study reported that NBCe2 deficiency resulted in decreased ventricular volume and intracranial pressure, changes in CSF electrolyte composition, and resistance to seizure induction (Kao et al. 2011). Taken together, the study suggested NBCe2 to

D. Barbuskaite et al.

Fig. 6.5 Confocal micrograph of anti-NBCe2 immuno-fluorescence reactivity in the choroid plexus. (**a**) The green immunoreaction is confined to the luminal membrane domain in mouse tissue. (**b**) A similar staining pattern was observed in the rat choroid plexus. Nuclei are counterstained with Topro3 (red). For methods, see previous publications (Bouzinova et al. 2005; Christensen et al. 2018)

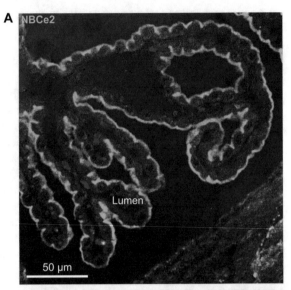

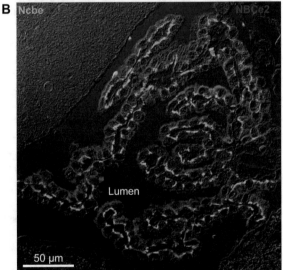

be involved in both the regulation of CSF [HCO_3^-] and in CSF secretion. However, the interpretation of the findings was compromised by the changes in expression profiles of other transporter proteins in choroid plexus. For example, the changes in expression and localization of Ncbe and Na,K-ATPase units in this gene trap NBCe2 knockout would most likely contribute considerably to the observed phenotype. The second study focused on the renal consequences of NBCe2 knockout in the kidney. It was, however, noted that the ventricle volume did not seem to be greatly affected by the gene disruption (Groger et al. 2012) and therefore this study seems to contradict a central role for NBCe2 in CSF secretion. In the latest study of NBCe2 ko mice, caution was again taken to validate the model by describing the expression profiles of various transporters in the choroid plexus epithelium (Christensen et al. 2018). This model did not reveal any changes in abundance or subcellular distribution of other transport proteins or gross changes in ventricle dimensions in the knockouts. However, NBCe2 was directly involved in base extrusion after intracellular alkalization ex vivo, and critically implicated in CSF pH recovery during hypercapnia-induced acidosis in vivo (Christensen et al. 2018). Thus, this study provided direct mechanistic evidence of the implication of the choroid plexus epithelium in CSF pH regulation, and established NBCe2 as the central mediators of this physiological process.

6.2.2.4 The Electroneutral Na^+:HCO_3^- Cotransporter NBCn1, Slc4a7

NBCn1 is an electroneutral sodium bicarbonate cotransporter that mediates Na^+ and HCO_3^- uptake into cells (Romero et al. 2013). The epithelial NBCn1 form is relatively DIDS insensitive, and in the rat choroid plexus the Na-dependent HCO_3^- influx was partly ascribed to this transporter (Bouzinova et al. 2005). Experiments in Ncbe and NBCn1 knockout mice indicate that NBCn1 may contribute less to pH recovery from acidosis in this species (Jacobs et al. 2008; Damkier, unpublished observation). The polarized plasma membrane distribution of NBCn1 in the choroid plexus varies among species and even among stains (Fig. 6.6). NBCn1 was localized to basolateral membrane domain in rats and humans (Praetorius and Nielsen 2006; Praetorius et al. 2004a), but can also be detected in the luminal membrane domain within the same human choroid plexus (Praetorius and Nielsen 2006). In the mouse, the luminal or basolateral NBCn1 expression seems to depend on the strain (Damkier et al. 2009; Praetorius et al. 2004a; Kao et al. 2011).

It is unlikely that NBCn1 takes part significantly in the vectorial HCO_3^- transport across the choroid plexus epithelium during correction of CSF pH, as it is not always expressed in the same membrane domain. As a HCO_3^- loader, NBCn1 could be involved in regulating pH within the choroid plexus epithelium, which may require separate acid-base transporters than those involved in the secretory process. It is unknown whether choroid plexus NBCn1 abundance and function is regulated in acid-base disturbances as elsewhere (Praetorius et al. 2004b). In neurons, NBCn1 is upregulated during metabolic acidosis (Park et al. 2010), thus this transporter could possibly have a similar role in the choroid plexus. Further studies are warranted to elucidate the roles of NBCn1 in this tissue.

6.2.2.5 The *slc4a11* Gene Product

The transport function of *SLC4A11* gene products is much debated, and consensus has not yet been reached regarding a unifying name for the transporter. The controversy probably results from the application of varying experimental

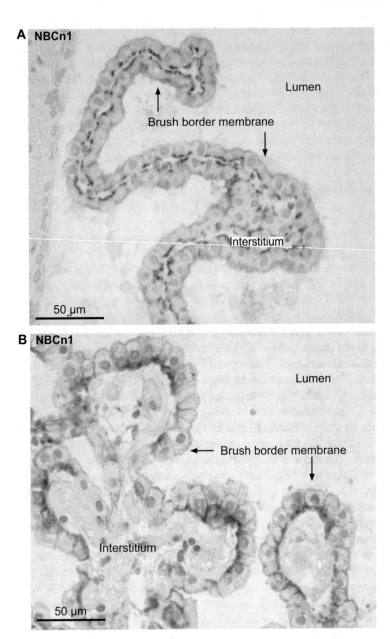

Fig. 6.6 Bright-field micrograph of anti-NBCn1 immuno-peroxidase reactivity in the choroid plexus. (**a**) The brown reaction product is confined to the basolateral membrane domains including the basal labyrinth in mouse tissue. (**b**) A similar staining pattern was observed in the human choroid plexus. Nuclei are counterstained with hematoxylin (blue). For methods, see previous publications (Praetorius and Nielsen 2006; Praetorius et al. 2004a)

techniques and species differences among genes and expression systems in these studies. In bovine corneal endothelial cells, the Na^+-coupled OH^- transport depended on the expression of *SLC4A11* (Jalimarada et al. 2013). The human *SLC4A11* forms seem to transport H^+ in both Na^+-dependent and -independent modes. In the presence of ammonia, *SLC4A11* generated inward currents that were comparable in magnitude (Kao et al. 2016). In another study, however, human *SLC4A11* seems to mediate electroneutral NH_3 transport in a Na^+ independent manner (Loganathan et al. 2016). By contrast, mouse *slc4a11* mediates a OH^- (or H^+) selective conductance without cotransport characteristics (Myers et al. 2016). Both the rat and the human choroid plexus epithelia express *SLC4A11* derived proteins in the luminal membrane domains (58), but *slc4a11* was not detected in the mouse choroid plexus (Damkier et al. 2018). Thus, the significance of NaBC1 in relation to choroid plexus function is unknown.

6.2.2.6   Anion Conductances

As mentioned above, HCO_3^- is secreted across the luminal membrane by electrogenic processes and can be stimulated by cAMP. This transport could be mediated by NBCe2 alone or in concert with HCO_3^- conductive anion channels. Indeed, protein kinase A stimulates inward-rectifying anion conductances in mammalian choroid plexus epithelial cells (Kibble et al. 1996, 1997). In the amphibian choroid plexus, a HCO_3^- conductive and cAMP-regulated ion channel was proposed as the major HCO_3^- efflux mechanism of the luminal membrane (Saito and Wright 1983, 1984). It seems that the inward-rectifying anion channels in the mammalian choroid plexus possess such high HCO_3^- permeability (Kibble et al. 1996). The molecular identities of the anion channels have not been fully established yet, but CFTR and Clc-2 have been ruled out as mediators of the inward-rectifying anion channels (Kibble et al. 1996, 1997; Speake et al. 2002). Recent mass spectrometry analysis of mouse choroid plexus epithelial cells indicated the expression of voltage-dependent anion channels (Vdac1–3) (Damkier et al. 2018).

6.2.2.7   *Slc26*-Derived HCO_3^- Transporters

It is possible that membrane proteins belonging to the *slc26* gene family could serve as HCO_3^- transporters in the choroid plexus epithelium. The mRNA for *slc26a2* (Kant et al. 2018), *slc26a4* (Saunders et al. 2015), *slc26a7*, *slc26a10*, and *slc26a11* (own unpublished observations) have all been detected at the mRNA level in the choroid plexus, but the corresponding proteins have not been detected in isolated epithelial cells by mass spectrometry (Damkier et al. 2018). Thus, it remains to be established to what extent these transporters contribute to base transport in the choroid plexus epithelium.

6.2.3   Acid Transporters

6.2.3.1   The Na$^+$/H$^+$ Exchangers, NHE1, slc9a1 and NHE6, slc9a6

The Na$^+$/H$^+$ exchanger NHE1 is almost ubiquitously expressed and mediates the electroneutral exchange of Na$^+$ and H$^+$, driven by the inward Na$^+$ gradient. It is inhibited by amiloride derivatives, such as EIPA (233). Na$^+$/H$^+$ exchange has been suggested as basolateral entry pathway for Na$^+$ and therefore centrally involved in CSF secretion, as amiloride applied from the blood side of the CPE inhibits Na$^+$ accumulation and CSF secretion (Davson and Segal 1970; Murphy and Johanson 1989a, b). This type of transporter would at the same time provide a basolateral H$^+$ extrusion mechanism to support luminal HCO$_3^-$ secretion. NHE1 mRNA was demonstrated by RT-PCR analysis (Kalaria et al. 1998) and EIPA-sensitive Na/H exchange was detected in isolated epithelial cells (Bouzinova et al. 2005; Damkier et al. 2009). Thus, for some time NHE1 was thought as the primary Na$^+$ loader in the choroid plexus epithelium.

Some evidence seems to contradict a central role of basolateral NHE activity in choroid plexus function. Firstly, amiloride applied form the luminal side was as efficient as basolateral amiloride in inhibiting CSF secretion (Segal and Burgess 1974). Secondly, amiloride failed to inhibit the Na$^+$-dependent pH recovery in the choroid plexus cultures in the presence of the CO$_2$/HCO$_3^-$ buffer system (Mayer and Sanders-Bush 1993). Finally, NHE1 was immunolocalized to the luminal membrane domain of the choroid plexus epithelium in both mice and humans (Damkier et al. 2009; Praetorius and Nielsen 2006; Kao et al. 2011) (Fig. 6.7). NHE1 expression was required for detection of any Na$^+$-dependent pH$_i$ recovery from acid load in choroid plexus, as judged from experiments on NHE1 knockout mice (Damkier et al. 2009). At first glance, this is indicative of NHE1 being the only NHE in the choroid plexus. However, the abundance of NHE1 is low in the choroid plexus as judged from the immunoreactivity in the referred studies and from mass spectrometry (Damkier et al. 2018). Another Na$^+$/H$^+$ exchanger detected in the CPE is NHE6. Although this transporter is usually found in the endosomes, it has been immunolocalized partly in the luminal membrane domain in mouse CPE (Damkier et al. 2018) (Fig. 6.8a). No mRNA for NHE2–5 and NHE7–9 have been detected in the CPE in the same study. NHEs are usually considered as regulators of intracellular pH$_i$, but since NHE1 and NHE6 transport H$^+$ out of the cell into the CSF, they could partake in CSF pH regulation. Future studies of the respective roles of NHE1 and NHE6 in intracellular pH and/or CSF pH regulation are warranted.

6.2.3.2   H$^+$-ATPases

Transmembrane ATPases catalyze the "uphill" movement of H$^+$ across membranes utilizing the energy generated by the hydrolysis of ATP into ADP and free phosphate. The vacuolar H$^+$-ATPase (V-ATPase) lowers the pH of vesicles in the endo-

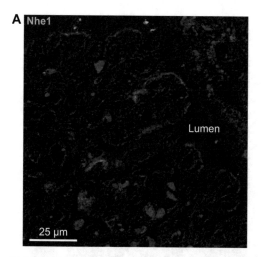

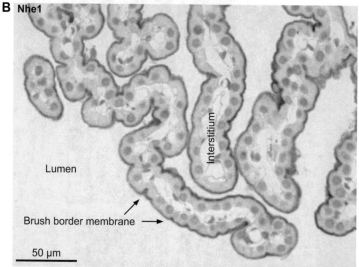

Fig. 6.7 Immunolocalization of Nhe1 in the choroid plexus epithelium. (**a**) Confocal micrograph of luminal membrane domain anti-Nhe1 immuno-fluorescence reactivity (Orlowski AB, green) in the mouse choroid plexus. Nuclei are counterstained with Topro3 (red). (**b**) Bright-field micrograph of anti-NBCn1 immuno-peroxidase reactivity in the rat choroid plexus produced a similar staining pattern with a separate antibody (Nöel AB, brown). Nuclei are counterstained with hematoxylin (blue). For methods, see previous publications (Damkier et al. 2009)

lysosomal system, but can also be plasma membrane acid extruders in epithelia. The B2 isoform of the V-ATPase is usually restricted to the lysosomes and endosomes (Puopolo et al. 1992), yet a small population of the polypeptide was immunolocalized to the luminal membrane domain in the mouse choroid plexus epithelium (Christensen et al. 2017) (Fig. 6.8b). However, the contribution of

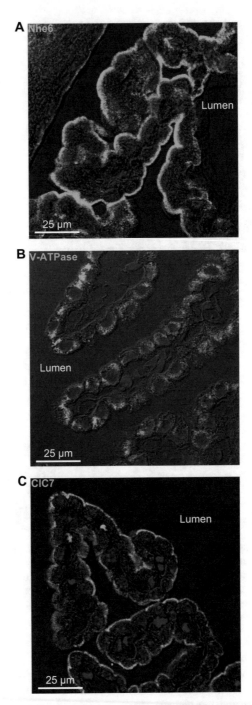

Fig. 6.8 Immunolocalization of additional acid transport proteins in the mouse choroid plexus. Varying degrees of intracellular and luminal membrane domain immunoreactivity (green) is shown for (**a**) Nhe6, (**b**) V-ATPase, and (**c**) ClC-7. Nuclei are counterstained with Topro3 (red). For methods, see Christensen et al. (Christensen et al. 2017; Damkier et al. 2018)

luminal V-ATPase to acid-extrusion in the choroid plexus of mice seems to be insignificant compared to NHE activity (Christensen et al. 2017). H^+/K^+-ATPases comprise a separate group of cellular acid extruders, that might contribute to choroid plexus acid-base transport. In rabbits, application of omeprazole, an inhibitor of gastric H^+/K^+-ATPase from the luminal side reduced CSF secretion rate (Lindvall-Axelsson et al. 1992). Thus far, only the non-gastric isoform of H^+/K^+-ATPase has been detected in the rabbit and mouse choroid plexus (Damkier et al. 2018; Pestov et al. 1998). In the unstimulated mouse choroid plexus the Na^+-independent acid extrusion is negligible (Christensen et al. 2017), leaving little room for H^+/K^+-ATPase to have a significant role in this tissue.

6.2.3.3   Acid Transporters Related to Cl^- Channel

Several polypeptides belonging to the ClC group of Cl^- channels have been characterized as Cl^-/H^+ exchangers. Among these, ClC3, ClC4, ClC5 and ClC7 were detected by mass spectroscopy in isolated epithelial cells from the mouse choroid plexus (Damkier et al. 2018). A fraction of the ClC7 immunoreactivity was detected in the luminal CPE membrane domain in the same study, where the expected vesicular ClC7 was also described (Fig. 6.8c). Cl^-/H^+ exchange activity was demonstrated in isolated mouse CPE clusters, as removal and reintroduction of extracellular Cl^- in the absence of Na^+ and HCO_3^- caused the relevant changes in pH_i. Whether ClC7 is the main Cl^-/H^+ exchanger, or other ClC isoforms also contribute to pH regulation in the CPE in vivo requires further investigations. It is peculiar that, like osteoclasts, choroid plexus epithelial cells seems to express a fraction of endo-lysosomal acid transport proteins in the luminal membrane (Fig. 6.9).

6.3   Choroid Plexus Acid-Base Transporters and Cerebrospinal Fluid Secretion

Some evidence supports the central involvement of choroid plexus acid-base transporters in the secretion of CSF. Firstly, the concentrations of Na^+ and HCO_3^- in CSF are higher than expected from an ultrafiltrate (Hughes et al. 2010). Secondly, the CSF secretion and the choroid plexus transport rates are highly affected by acetazolamide, amiloride, and DIDS (Welch 1963; Davson and Segal 1970; Tschirgi et al. 1954; Davson and Luck 1957; Ames et al. 1965; Murphy and Johanson 1989a; Johanson and Murphy 1990; Johanson et al. 1992; McCarthy and Reed 1974; Vogh and Godman 1985). Thirdly, the ventricle volume is affected by genetic disruption of Ncbe and in one report of NBCe2 (Jacobs et al. 2008; Kao et al. 2011). AE2 and Ncbe are most likely mediating the major fraction of Na^+, Cl^- and HCO_3^- import to the epithelial cells from the blood side that is necessary for secretion across the

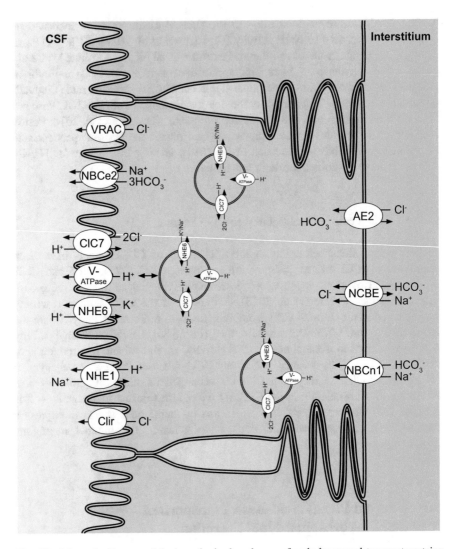

Fig. 6.9 Schematic diagram of the hypothesized exchange of endo-lysosomal transport proteins with the luminal plasma membrane, as a model for the observation of partial luminal expression of Nhe6, the V-ATPase, and ClC-7

luminal membrane. These proteins are expressed in high abundance in the basal labyrinth of the choroid plexus across species and very few alternative absorptive pathways exist in the basolateral membrane. Thus, there are indications for a dependence of fluid secretion by the CPE on acid-base transporters.

6.4   Choroid Plexus and Cerebrospinal Fluid pH Regulation

The acute CSF pH greatly depends on the breathing rate, yet over time CSF pH is maintained within narrow limits (Lee et al. 1969). Upon changes in respiration frequency or tidal volume, pCO_2 in the blood changes accordingly. Since CO_2 can easily cross the blood-brain and blood-CSF barriers, increased pCO_2 in the blood results in increased pCO_2 in the CSF and brain ECF, and thus low pH (Johnson et al. 1983). Changes in metabolic acid-base balance are not reflected in CSF pH, as HCO_3^- and H^+ have to be actively transported over these barriers. As mentioned previously, contrary to the blood plasma, CSF contains very little protein and phosphate buffers, thus should be subjected to a great level of pH variation during respiratory acidosis. However, several studies have demonstrated that increased CSF pCO_2 is compensated by an increase in CSF HCO_3^-, independently of blood acid-base balance. The robust expression of a HCO_3^- export protein NBCe2 in the luminal plasma membrane of the CPE spurred the idea that this tissue could serve a CSF pH regulatory function (Bouzinova et al. 2005) (Fig. 6.5).

NBCe2 transports Na^+ and HCO_3^- out of the CPE cells, most likely with a 1:3 stoichiometry (Millar and Brown 2008), and with its membrane localization it constitutes an ideal candidate for the HCO_3^- secretion into the CSF. As discussed above, two sources of HCO_3^- for the secretion has been proposed. The classic view is an origin from the hydration of intracellular CO_2. The high levels of carbonic anhydrase II in the CPE would efficiently convert CO_2 to HCO_3^- to supply the luminal transporter with HCO_3^- (Johanson et al. 1992). The problem with this explanation is the apparent lack of NHE proteins in the basolateral membrane. The alternative view is transcellular passage of HCO_3^-, where e.g. Ncbe mediates the import of HCO_3^- across the basolateral membrane (Praetorius et al. 2004c).

CSF pH seems to respond more efficiently to respiratory acidosis than to alkalosis, indicating that the acid extrusion by the CPE is less effective than the alkaline export (Kazemi et al. 1967). Nevertheless, the CPE cells express several proteins that could be involved in acid extrusion to the CSF. In the luminal membrane, Na^+/H^+ exchangers NHE1 (and NHE6) export H^+ in exchange for Na^+ (Damkier et al. 2009) and thereby candidate as mechanisms to lower CSF pH. NHE1 is involved in the regulation of intracellular pH in the CPE cells (Damkier et al. 2009), but its putative role in CSF pH regulation has not been investigated. Recently, we demonstrated that a fraction of the vacuolar H^+-ATPase (V-ATPase) is expressed in the luminal membrane of CPE cells (Christensen et al. 2017). Although this pump is involved in extracellular acidification elsewhere, our study rejects the V-ATPase as a significant contributor to CSF pH regulation (Christensen et al. 2017). The $2Cl^-/H^+$ antiporter ClC-7 was also immunolocalized to the luminal membrane of CPE cells in the same study. However, the highly electrogenic nature of the transport by this protein makes ClC7 highly unlikely as an efficient acid extruder. It is unknown whether AE2 plays a role on CSF pH regulation, but the basolateral expression and the export HCO_3^- in exchange for Cl^- (Alper et al. 1994) could be required sustain luminal H^+ secretion.

References

Alper SL (2009) Molecular physiology and genetics of Na+-independent SLC4 anion exchangers. J Exp Biol 212(Pt 11):1672–1683

Alper SL et al (1994) The fodrin-ankyrin cytoskeleton of choroid plexus preferentially colocalizes with apical Na+K(+)-ATPase rather than with basolateral anion exchanger AE2. J Clin Invest 93 (4):1430–1438

Ames A III, Higashi K, Nesbett FB (1965) Effects of Pco2 acetazolamide and ouabain on volume and composition of choroid-plexus fluid. J Physiol 181(3):516–524

Banizs B et al (2007) Altered pH(i) regulation and Na(+)/HCO3(-) transporter activity in choroid plexus of cilia-defective Tg737(orpk) mutant mouse. Am J Physiol Cell Physiol 292(4):C1409–C1416

Baron R et al (1985) Cell-mediated extracellular acidification and bone resorption: evidence for a low pH in resorbing lacunae and localization of a 100-kD lysosomal membrane protein at the osteoclast ruffled border. J Cell Biol 101(6):2210–2222

Boron WF (2010) Evaluating the role of carbonic anhydrases in the transport of HCO3--related species. Biochim Biophys Acta 1804(2):410–421

Boron WF, Boulpaep EL (2012) Medical physiology. Saunders, Philadelphia, PA

Bouzinova EV et al (2005) Na+-dependent HCO_3^- uptake into the rat choroid plexus epithelium is partially DIDS sensitive. Am J Physiol Cell Physiol 289(6):C1448–C1456

Christensen IB et al (2013) Polarization of membrane associated proteins in the choroid plexus epithelium from normal and slc4a10 knockout mice. Front Physiol 4:344

Christensen HL et al (2017) The V-ATPase is expressed in the choroid plexus and mediates cAMP-induced intracellular pH alterations. Physiol Rep 5(1):e13072

Christensen HL et al (2018) The choroid plexus sodium-bicarbonate cotransporter NBCe2 regulates mouse cerebrospinal fluid pH. J Physiol 596(19):4709–4728

Cordat E, Reithmeier RA (2014) Structure, function, and trafficking of SLC4 and SLC26 anion transporters. Curr Top Membr 73:1–67

Cserr HF (1971) Physiology of the choroid plexus. Physiol Rev 51(2):273–311

Damkier HH, Praetorius J (2012) Genetic ablation of Slc4a10 alters the expression pattern of transporters involved in solute movement in the mouse choroid plexus. Am J Physiol Cell Physiol 302(10):C1452–C1459

Damkier HH, Nielsen S, Praetorius J (2007) Molecular expression of SLC4-derived Na+-dependent anion transporters in selected human tissues. Am J Physiol Regul Integr Comp Physiol 293(5): R2136–R2146

Damkier HH et al (2009) Nhe1 is a luminal Na+/H+ exchanger in mouse choroid plexus and is targeted to the basolateral membrane in Ncbe/Nbcn2-null mice. Am J Physiol Cell Physiol 296 (6):C1291–C1300

Damkier HH, Aalkjaer C, Praetorius J (2010) Na+-dependent HCO_3^- import by the slc4a10 gene product involves Cl⁻ export. J Biol Chem 285(35):26998–27007

Damkier HH, Brown PD, Praetorius J (2013) Cerebrospinal fluid secretion by the choroid plexus. Physiol Rev 93(4):1847–1892

Damkier HH et al (2018) The murine choroid plexus epithelium expresses the 2Cl(-)/H(+) exchanger ClC-7 and Na(+)/H(+) exchanger NHE6 in the luminal membrane domain. Am J Physiol Cell Physiol 314(4):C439–C448

Davson H, Luck CP (1957) The effect of acetazoleamide on the chemical composition of the aqueous humour and cerebrospinal fluid of some mammalian species and on the rate of turnover of 24Na in these fluids. J Physiol 137(2):279–293

Davson H, Segal MB (1970) The effects of some inhibitors and accelerators of sodium transport on the turnover of 22Na in the cerebrospinal fluid and the brain. J Physiol 209(1):131–153

Deng QS, Johanson CE (1989) Stilbenes inhibit exchange of chloride between blood, choroid plexus and the cerebrospinal fluid. Brain Res 510:183–187

Frankel H, Kazemi H (1983) Regulation of CSF composition--blocking chloride-bicarbonate exchange. J Appl Physiol 55(1 Pt 1):177–182

Giffard RG et al (2003) Two variants of the rat brain sodium-driven chloride bicarbonate exchanger (NCBE): developmental expression and addition of a PDZ motif. Eur J Neurosci 18 (11):2935–2945

Groger N et al (2012) Targeted mutation of SLC4A5 induces arterial hypertension and renal metabolic acidosis. Hum Mol Genet 21(5):1025–1036

Hasan FM, Kazemi H (1976) Dual contribution theory of regulation of CSF HCO3 in respiratory acidosis. J Appl Physiol 40(4):559–567

Haselbach M et al (2001) Porcine choroid plexus epithelial cells in culture: regulation of barrier properties and transport processes. Microsc Res Tech 52(1):137–152

Hladky SB, Barrand MA (2016) Fluid and ion transfer across the blood-brain and blood-cerebrospinal fluid barriers; a comparative account of mechanisms and roles. Fluids Barriers CNS 13(1):19

Hughes AL, Pakhomova A, Brown PD (2010) Regulatory volume increase in epithelial cells isolated from the mouse fourth ventricle choroid plexus involves Na^+-H^+ exchange but not Na^+-K^+-$2Cl^-$ cotransport. Brain Res 1323:1–10

Jacobs S et al (2008) Mice with targeted Slc4a10 gene disruption have small brain ventricles and show reduced neuronal excitability. Proc Natl Acad Sci U S A 105(1):311–316

Jalimarada SS et al (2013) Ion transport function of SLC4A11 in corneal endothelium. Invest Ophthalmol Vis Sci 54(6):4330–4340

Johanson CE, Murphy VA (1990) Acetazolamide and insulin alter choroid plexus epithelial cell [Na +], pH, and volume. Am J Phys 258(6 Pt 2):F1538–F1546

Johanson CE, Parandoosh Z, Dyas ML (1992) Maturational differences in acetazolamide-altered pH and HCO3 of choroid plexus, cerebrospinal fluid, and brain. Am J Phys 262(5 Pt 2):R909–R914

Johnson DC, Hoop B, Kazemi H (1983) Movement of CO2 and HCO-3 from blood to brain in dogs. J Appl Physiol Respir Environ Exerc Physiol 54(4):989–996

Kalaria RN et al (1998) Identification and expression of the Na^+/H^+ exchanger in mammalian cerebrovascular and choroidal tissues: characterisation by amiloride-sensitive [^{3}H]MIA binding and RT-PCR analysis. Brain Res Mol Brain Res 58:178–187

Kallio H et al (2006) Expression of carbonic anhydrases IX and XII during mouse embryonic development. BMC Dev Biol 6:22

Kant S et al (2018) Choroid plexus genes for CSF production and brain homeostasis are altered in Alzheimer's disease. Fluids Barriers CNS 15(1):34

Kao L et al (2011) Severe neurologic impairment in mice with targeted disruption of the electrogenic sodium bicarbonate cotransporter NBCe2 (Slc4a5 gene). J Biol Chem 286 (37):32563–32574

Kao L et al (2016) Multifunctional ion transport properties of human SLC4A11: comparison of the SLC4A11-B and SLC4A11-C variants. Am J Physiol Cell Physiol 311(5):C820–C830

Kazemi H, Shannon DC, Carvallo-Gil E (1967) Brain CO2 buffering capacity in respiratory acidosis and alkalosis. J Appl Physiol 22(2):241–246

Kibble JD, Tresize AO, Brown PD (1996) Properties of the cAMP-activated Cl^- conductance in choroid plexus epithelial cells isolated from the rat. J Physiol 496:69–80

Kibble JD et al (1997) Whole-cell Cl^- conductances in mouse choroid plexus epithelial cells do not require CFTR expression. Am J Phys 272:C1899–C1907

Kister SJ (1956) Carbonic anhydrase inhibition. VI. The effect of acetazolamide on cerebrospinal fluid flow. J Pharmacol Exp Ther 117(4):402–405

Lee JE et al (1969) Buffering capacity of cerebrospinal fluid in acute respiratory acidosis in dogs. Am J Phys 217(4):1035–1038

Lindsey AE et al (1990) Functional expression and subcellular localization of an anion exchanger cloned from choroid plexus. Proc Natl Acad Sci U S A 87(14):5278–5282

Lindvall-Axelsson M et al (1992) Inhibition of cerebrospinal fluid formation by omeprazole. Exp Neurol 115(3):394–399

Loganathan SK et al (2016) Functional assessment of SLC4A11, an integral membrane protein mutated in corneal dystrophies. Am J Physiol Cell Physiol 311(5):C735–C748

Maren TH (1962) The binding of inhibitors to carbonic anhydrase in vivo: drugs as markers for enzyme. Biochem Pharmacol 9:39–48

Maren TH (1972) Bicarbonate formation in cerebrospinal fluid: role in sodium transport and pH regulation. Am J Phys 222(4):885–899

Masuzawa T et al (1981) Ultrastructural localization of carbonic anhydrase activity in the rat choroid plexus epithelial cell. Histochemistry 73(2):201–209

Mayer SE, Sanders-Bush E (1993) Sodium-dependent antiporters in choroid plexus epithelial cultures from rabbit. J Neurochem 60:1308–1316

McCarthy KD, Reed DJ (1974) The effect of acetazolamide and furosemide on cerebrospinal fluid production and choroid plexus carbonic anhydrase activity. J Pharmacol Exp Ther 189 (1):194–201

McMurtrie HL et al (2004) The bicarbonate transport metabolon. J Enzyme Inhib Med Chem 19 (3):231–236

Millar ID, Brown PD (2008) NBCe2 exhibits a 3 HCO_3^-:1 Na^+ stoichiometry in mouse choroid plexus epithelial cells. Biochem Biophys Res Commun 373:550–554

Murphy VA, Johanson CE (1989a) Acidosis, acetazolamide, and amiloride: effects on 22Na transfer across the blood-brain and blood-CSF barriers. J Neurochem 52(4):1058–1063

Murphy VA, Johanson CE (1989b) Alteration of sodium transport by the choroid plexus with amiloride. Biochim Biophys Acta 979(2):187–192

Myers EJ et al (2016) Mouse Slc4a11 expressed in Xenopus oocytes is an ideally selective H+/OH-conductance pathway that is stimulated by rises in intracellular and extracellular pH. Am J Physiol Cell Physiol 311(6):C945–C959

Nattie EE (1980) Brain and cerebrospinal fluid ionic composition and ventilation in acute hyper-capnia. Respir Physiol 40(3):309–322

Nattie EE, Adams JM (1988) DIDS decreases CSF HCO3- and increases breathing in response to CO2 in awake rabbits. J Appl Physiol (1985) 64(1):397–403

Okada Y et al (1993) Depth profiles of pH and PO2 in the isolated brain stem-spinal cord of the neonatal rat. Respir Physiol 93(3):315–326

Opavsky R et al (1996) Human MN/CA9 gene, a novel member of the carbonic anhydrase family: structure and exon to protein domain relationships. Genomics 33(3):480–487

Park HJ et al (2010) Neuronal expression of sodium/bicarbonate cotransporter NBCn1 (SLC4A7) and its response to chronic metabolic acidosis. Am J Physiol Cell Physiol 298(5):C1018–C1028

Parker MD et al (2008) Characterization of human SLC4A10 as an electroneutral Na/HCO3 cotransporter (NBCn2) with Cl- self-exchange activity. J Biol Chem 283(19):12777–12788

Pestov NB et al (1998) Ouabain-sensitive H,K-ATPase: tissue-specific expression of the mamma-lian genes encoding the catalytic alpha subunit. FEBS Lett 440(3):320–324

Pollay M, Curl F (1967) Secretion of cerebrospinal fluid by the ventricular ependyma of the rabbit. Am J Phys 213(4):1031–1038

Pollay M, Davson H (1963) The passage of certain substances out of the cerebrosphinal fluid. Brain 86:137–150

Praetorius J, Nielsen S (2006) Distribution of sodium transporters and aquaporin-1 in the human choroid plexus. Am J Physiol Cell Physiol 291(1):C59–C67

Praetorius J, Nejsum LN, Nielsen S (2004a) A SCL4A10 gene product maps selectively to the basolateral plasma membrane of choroid plexus epithelial cells. Am J Physiol Cell Physiol 286 (3):C601–C610

Praetorius J et al (2004b) NBCn1 is a basolateral Na+-HCO3- cotransporter in rat kidney inner medullary collecting ducts. Am J Physiol Renal Physiol 286(5):F903–F912

Praetorius J, Nejsum LN, Nielsen S (2004c) A SLC4A10 gene product maps selectively to the basolateral membrane of choroid plexus epithelial cells. Am J Phys 286:C601–C610

Puopolo K et al (1992) Differential expression of the "B" subunit of the vacuolar H(+)-ATPase in bovine tissues. J Biol Chem 267(6):3696–3706

Romero MF et al (2013) The SLC4 family of bicarbonate (HCO(3)(-)) transporters. Mol Asp Med 34(2–3):159–182

Saito Y, Wright EM (1983) Bicarbonate transport across the frog choroid plexus and its control by cyclic nucleotides. J Physiol 336:635–648

Saito Y, Wright E (1984) Regulation of bicarbonate transport across the brush border membrane of the bull-frog choroid plexus. J Physiol 350:327–342

Sassani P et al (2002) Functional characterization of NBC4: a new electrogenic sodium-bicarbonate cotransporter. Am J Physiol Cell Physiol 282(2):C408–C416

Saunders NR et al (2015) Influx mechanisms in the embryonic and adult rat choroid plexus: a transcriptome study. Front Neurosci 9:123

Schuchmann S et al (2006) Experimental febrile seizures are precipitated by a hyperthermia-induced respiratory alkalosis. Nat Med 12(7):817–823

Segal MB, Burgess AM (1974) A combined physiological and morphological study of the secretory process in the rabbit choroid plexus. J Cell Sci 14(2):339–350

Speake T et al (2002) Inward-rectifying anion channels are expressed in the epithelial cells of choroid plexus isolated from ClC-2 'knock-out' mice. J Physiol 539:385–390

Tschirgi RD, Frost RW, Taylor JL (1954) Inhibition of cerebrospinal fluid formation by a carbonic anhydrase inhibitor, 2-acetylamino-1,3,4-thiadiazole-5-sulfonamide (diamox). Proc Soc Exp Biol Med 87(2):373–376

Tureci O et al (1998) Human carbonic anhydrase XII: cDNA cloning, expression, and chromosomal localization of a carbonic anhydrase gene that is overexpressed in some renal cell cancers. Proc Natl Acad Sci U S A 95(13):7608–7613

Virkki LV et al (2002) Functional characterization of human NBC4 as an electrogenic Na+-HCO cotransporter (NBCe2). Am J Physiol Cell Physiol 282(6):C1278–C1289

Vogh BP, Godman DR (1985) Timolol plus acetazolamide: effect on formation of cerebrospinal fluid in cats and rats. Can J Physiol Pharmacol 63(4):340–343

Vogh BP, Maren TH (1975) Sodium, chloride, and bicarbonate movement from plasma to cerebrospinal fluid in cats. Am J Phys 228(3):673–683

Vogh BP, Godman DR, Maren TH (1987) Effect of AlCl₃ and other acids on cerebrospinal fluid production: a correction. J Pharmacol Exp Ther 243(1):35–39

Wang CZ et al (2000) The Na+−driven cl-/HCO3- exchanger. Cloning, tissue distribution, and functional characterization. J Biol Chem 275(45):35486–35490

Welch K (1963) Secretion of cerebrospinal fluid by choroid plexus of the rabbit. Am J Phys 205:617–624

Wright EM (1972) Mechanisms of ion transport across the choroid plexus. J Physiol 226 (2):545–571

Ziemann AE et al (2008) Seizure termination by acidosis depends on ASIC1a. Nat Neurosci 11 (7):816–822

Chapter 7
TRPV4, a Regulatory Channel in the Production of Cerebrospinal Fluid by the Choroid Plexus

Alexandra E. Hochstetler, Makenna M. Reed, and Bonnie L. Blazer-Yost

Abstract Transient receptor potential, vanilloid 4 (TRPV4) channel is an osmo- and mechano-sensitive nonselective cation channel expressed throughout the body. TRPV4 is activated and regulated by a wide variety of endogenous and exogenous stimuli. In addition to its relatively ubiquitous distribution and regulation by diverse stimuli including pressure, pH, inflammation, cell swelling and cytokines, TRPV4 appears to be a hub protein in many signaling cascades integrating input from multiple receptors and kinases. TRPV4 is present in the apical membrane of choroid plexus epithelial cells and emerging studies have linked this channel with changes in transepithelial ion and water flux.

7.1 Introduction

Cerebrospinal fluid (CSF) is the cushioning and nutritive fluid component of the central nervous system. The osmolyte composition of the CSF is acutely controlled by the action of many channels and transporters located in the membranes of the choroid plexus epithelial cells, as was elegantly described in Chap. 1. One of the more recently described of these transporters is the transient receptor potential, vanilloid 4 (TRPV) channel. TRPV4 has been characterized as a relatively ubiquitous, polymodal, nonselective cation channel and emerging data suggest that regulation of this channel in the choroid plexus epithelial cells may contribute to CSF production and composition.

TRPV4 is a member of a large family of functionally diverse TRP channels. The founding member of the TRP family was characterized in *Drosophila* as being a calcium permeable channel constituting a phototransducing component (Cosens and Manning 1969). Subsequently, many other transporters have been added under the TRP family umbrella and, to date, the TRP channels can be divided into six sub-families: TRPA (Ankyrin), TRPC (Canonical), TRPM (Melastin), TRPML

A. E. Hochstetler · M. M. Reed · B. L. Blazer-Yost (✉)
Department of Biology, Indiana University, Purdue University, Indianapolis, IN, USA
e-mail: aehochst@iupui.edu; reedmak@iu.edu; bblazer@iupui.edu

© The American Physiological Society 2020
J. Praetorius et al. (eds.), *Role of the Choroid Plexus in Health and Disease*,
Physiology in Health and Disease, https://doi.org/10.1007/978-1-0716-0536-3_7

(Mucolipin), TRPP (Polycystin), and TRPV (Vanilloid). The TRP family is broadly classified under the voltage-gated superfamily of ion channels inclusive of voltage-gated K^+, Na^+, and Ca^{2+} channels, though most TRP channels also possess ligand-gating capabilities (Yu et al. 2005). Mammalian TRP channels are expressed throughout the body and are activated and regulated by a wide variety of endogenous and exogenous stimuli. Most TRP channels share sequence homology in their six membrane spanning domains, but there are many structural and functional differences in their C- and N-termini. TRP channels normally assemble as homo- or hetero-tetramers where each subunit includes six membrane-spanning α helices (Gees et al. 2010; Cheng et al. 2011).

7.2   TRPV4 Channelopathies

There are a number of human diseases and disorders that have been linked to mutations in TRPV4 which are broadly divided into the categories of skeletal dysplasias and arthropathy, motor-sensory neuropathy, and motor-neuropathies. In humans, these channelopathies show remarkable phenotypic diversity. Within the last decade, numerous mutations in TRPV4 have been shown to cause varying severities in five skeletal dysplasias and FDAB (Familial Digital Arthropathy-Brachydactyly). These are characterized by abnormalities of bone and cartilage growth, causing various forms of dwarfism, short trunk, and progressive osteoarthropathy (Inda et al. 2012; Nishimura et al. 2010; Lamande et al. 2011). While many different TRPV4 mutations have been described within skeletal dysplasias and arthropathy, the majority occur in the ankyrin-repeat domain.

There have also been mutations in TRPV4 linked to two different motor-neuropathies: congenital distal and scapuloperoneal spinal muscle atrophies (Dubowitz et al. 1995; Main et al. 2003). Additionally, several TRPV4 mutations have been shown to cause Charcot-Marie-Tooth disease type 2C which causes limb, diaphragm, and laryngeal muscle weakness (McEntagart 2012). Several pathophysiological mechanisms of disease have been proposed as the cause of these TRPV4 channelopathies in humans. These include, briefly: altered TRPV4-dependent neuritogenesis (Jang et al. 2012), altered TRPV4-dependent gene expression via SOX9, a chondrocyte differentiating transcription factor (Kornak and Mundlos 2003), dysregulated cellular trafficking of TRPV4 due to improper binding and ubiquitination (Cuajungco et al. 2006; Wang et al. 2007; Zerangue et al. 1999; Owsianik et al. 2003; Verma et al. 2010), and dysregulated protein-protein interactions such as those with Na-K-ATPase, Kir4.1, syntrophin, and aquaporin 4 (Lanciotti et al. 2012).

Interestingly, despite the heterogeneity and severity in human diseases, there is little evidence that a global knockout of TRPV4 in rodents results in any severe phenotype in the absence of stressors. Due to the mild phenotypes exhibited by these models, they have been widely used to study the physiological role of TRPV4 in multiple tissues.

7.3   Animal Models of TRPV4 Deletion

Liedtke and colleagues created the first *trpv4–/–* mouse by flanking Exon 12 with a neo-cassette and loxP sites and excising exon 12, rendering the TRPV4 polypeptide nonfunctional and targeted for degradation. These mice are viable and fertile up to 1 year of age, though they exhibit reduced water intake due to altered osmosensation under a stressed hyperosmotic environment (Liedtke and Friedman 2003; Liedtke 2008). Additionally, these animals exhibit a larger bladder capacity due to the impairment of TRPV4 as a stretch sensor in the bladder wall (Gavaert et al. 2007). A second *trpv4–/–* mouse was generated in the laboratory of Suzuki et al. (2003) in which Exon 4 was excised using a neo-cassette marker resulting in a lack of TRPV4 production. These mice have slightly differing, albeit mild, phenotypes. They exhibit thicker bones due to impaired osteoclast differentiation (Masuyama et al. 2008), and compromised vascular endothelial function as demonstrated by a preference for warmer temperatures and an overall inability to thermoregulate in stressed conditions (Vriens et al. 2005; Lee et al. 2005; Saliez et al. 2008; Loot et al. 2008; Sonkusare et al. 2012). They also exhibit some sensory defects such as hearing alteration (Tabuchi et al. 2005), impaired pressure sensation (Suzuki et al. 2003), and compromised pain sensing (Cortright and Szallasi 2009). Recently, Deruyver et al. (2018) successfully created a *Trpv4–/–* rat and used this model to study induced urinary bladder dysfunction. Unfortunately, no data were provided regarding the overall phenotype of this rat model.

   Though these models have been informative for TRPV4 functionality, they do not increase understanding of human diseases which are often due to alteration rather than elimination of TRPV4 functionality. There have been several gain-of-function mutations in mice and zebrafish which have demonstrated skeletal dysplasia-like phenotypes similar to the human diseases (Wang et al. 2007; Masuyama et al. 2012). However, these models still do not accurately reflect the variety of disease presentation found in human patients. Regardless, the findings that TRPV4 deletion causes mild phenotypes, which are unmasked under stressed conditions, is indicative that TRPV4 may be a redundant channel in rodents. The mild phenotypes in the knockout animals combined with the demonstration that TRPV4 antagonism has no aberrant physiological effects in rodent models has led to the suggestion that TRPV4 is a promising therapeutic target (Everaerts et al. 2010; Thorneloe et al. 2012).

7.4   TRPV4 Structure and Expression

The TRPV4 protein has been shown to be expressed in many epithelia as well as nervous, liver, heart, and vascular tissue (https://www.proteinatlas.org/ENSG00000111199-TRPV4/tissue). Emerging findings suggest tissue-specific effects, which can be modulated by the type of stimulus, the isoform present, or differences in intracellular signaling effectors. Uniprot recognizes TRPV4 isoforms 1, 2, 4, 5, and 6 as characterized by Arniges et al. (2006). These authors described the

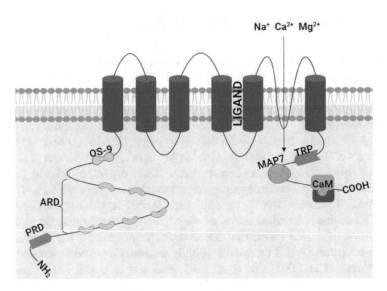

Fig. 7.1 Structure of TRPV4 and important binding domains. PRD, proline rich domain; ARD, ankyrin repeat containing domain; OS-9, binding site for OS-9; TRP, TRP box; MAP7, MAP7 binding domain; CaM, calmodulin/actin/tubulin binding domain; LIGAND, ligand binding domain

original TRPV4-A (98 kDa) as well as four splice variants TRPV4-B (Δ384–444), TRPV4-C (Δ237–284), TRPV4-D (Δ27–61), and TRPV4-E (Δ237–284 and Δ384–444) cloned from human airway epithelial cells and expressed in HeLa cells. They demonstrated that TRPV4-A and D can be either homo- or hetero-multimerized in the endoplasmic reticulum (ER), and are targeted to the membrane upon stimulation. A commonality among the B, C and E splice variants is that they all have deletions in the cytoplasmic N-terminal region which affect an ankyrin domain. Consequently, these variants do not oligomerize and are retained in the ER. The first four helices of TRPV4 (S1–S4) form the voltage-sensing or voltage-sensing like domain, and the last two segments (S5, S6) form the ion-conducting pore along with the reentrant pore loop (Liedtke et al. 2000; Deng et al. 2018). TRPV4 protein is composed of 871 residues, and normally assembles as a homotetramer (Stewart et al. 2010). However, there have been reports of TRPV4 assembling as heterotetramers such as TRPC1-TRPV4 (Ma et al. 2011); TRPP2-TRPV4 (Stewart et al. 2010); and even TRPC1-TRPP2-TRPV4 (Du et al. 2014).

TRPV4 contains many well characterized binding domains within both its amino and carboxy termini (Fig. 7.1). In the N-terminus, there is evidence that phosphorylation at Tyr110 increases the abundance of TRPV4 at the membrane (Wegierski et al. 2009). Just following Tyr110, the proline rich domain (PRD) has been shown to be an important binding site for PACSIN-3, which also increases membrane expression (Cuajungco et al. 2006; D'hoedt et al. 2007). Next to the PRD, there is an Ankyrin Repeat Domain (ARD) composed of 6 Ankyrin repeats which have been shown to be an important interaction domain for OS-9 (Everaerts et al. 2010; Inda et al. 2012). OS-9 is an endoplasmic-reticulum-resident protein involved in protein folding and ubiquitination (Wang et al. 2007). Ubiquitination can also occur in conjunction with E3 ubiquitin ligase AIP-4 (atrophin-1-interacting protein 4), which causes a reduction in membrane expression by promoting endocytosis without inducing degradation of TRPV4 (Wegierski et al. 2006). Therefore, ubiquitination is an important regulatory mechanism of TRPV4 wherein its basal activity can be directly regulated.

In the C-terminus, there is a well characterized TRP box which helps to stabilize the closed conformation and may be responsible for the interaction with the widely used agonist GSK10106790A (Teng et al. 2015). There are also binding sites preferential for calmodulin, MAP7, actin, and tubulin (Strotmann et al. 2003; Suzuki et al. 2003; Goswami et al. 2010; Garcia-Elias et al. 2008). Important phosphorylation sites for several regulatory kinases are located within these latter domains. Thus, both termini of the TRPV4 channel structure are intracellular and serve important roles in phosphorylation, activation, and vesicular trafficking of the protein.

TRPV4, like many ionotropic channels, is located in a vesicle pool just below the cell surface allowing for the regulation of expression between the membrane and cytosolic domains (Becker et al. 2008; Lei et al. 2013; Ferrandiz-Hertas et al. 2014). There are a number of proteins which are postulated to regulate this expression. Specifically, phosphorylation at Serine 824 within the calmodulin binding site by serum and glucocorticoid-induced kinase I (SGK1) has been shown to be important for the interaction with F-actin (Shin et al. 2012) and subsequently, membrane targeting. In addition to membrane expression, SGK1 phosphorylation regulates agonist sensitivity, single channel activity, and Ca^{2+} influx (Shin et al. 2012; Lee et al. 2010). Phosphorylation via the protein kinase A (PKA) pathway, also targeting Serine 824, has been shown to increase the amount of TRPV4 which is trafficked to the membrane (Mamenko et al. 2013; Fan et al. 2009). Conversely, phosphorylation via the PKC (via Serine 165, Tyrosine 175, and Serine 189) pathway doesn't appear to increase the amount of trafficked TRPV4, but it does increase the activity/sensitivity of membrane-bound TRPV4 (Mamenko et al. 2013; Fan et al. 2009) in renal cells, dorsal root ganglia (DRG) as well as when exogenously expressed in cultured cell lines. Though TRPV1 and TRPV4 co-express in the DRG, the TRPV4 modulation was shown to be independent of TRPV1 activity as modulated by capsaicin (Cao et al. 2018). Similarly, activation of the PLC pathway has been shown to up-regulate TRPV4 activity at the membrane (Garcia-Elias et al. 2013). With no lysine deficient protein kinases 1 and 4 (WNK1 and WNK4) down regulate membrane expression of TRPV4 in renal cells (Fu et al. 2006). Depletion of intracellular Ca^{2+} stores stimulates the insertion of TRPV4 into the plasma membrane thereby increasing Ca^{2+} influx. However, in contrast to most other studies, the form of the Ca^{2+}-sensitive channel is a TRPV4-TRPVC1 heteromeric channel (Ma et al. 2010, 2011).

7.5 Activation and Modulation of TRPV4

TRPV4, originally classified as vanilloid receptor-related osmotically active channel (VR-OAC), is activated by extracellular osmolarity (Liedtke et al. 2000). It is active under isotonic conditions, shows increased activation with decreased extracellular osmolarity, and is inactivated with increased extracellular osmolarity (Strotmann et al. 2000; White et al. 2016). TRPV4 is also mechanosensitive and can respond to cell stretch in a variety of tissue types (Pedersen and Nilius 2007; Ryskamp et al. 2014) (Fig. 7.2). However, the exact mechanism of mechanical activation remains a highly debated topic. One of the proposed mechanisms is activation of the channel by direct membrane tension changes. Evidence supporting this theory includes the channels association with adherence junctions (Janssen et al. 2011; White et al. 2016), and co-expression with stretch-activated channels (Alessandri-Haber et al. 2009; Nilius and Honore 2012). Additional theories include activation by direct mechanical force on cell membrane structures gating the channel (Brohawn et al. 2014; Kung 2005) or membrane deformation-stimulated release of activation products that gate TRPV4, such as cell swelling. Additionally, it has been shown that TRPV4 is activated by phospholipase A2-dependent formation of arachidonic acid

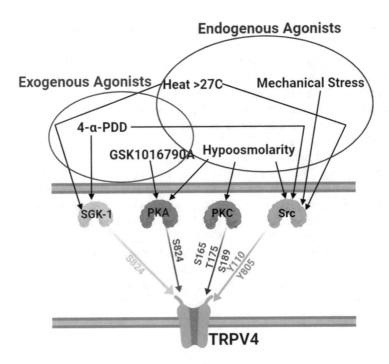

Fig. 7.2 Exogenous and endogenous agonists sensitize TRPV4 through the action of different kinases. SGK-1, serum and glucocorticoid-induced kinase 1; PKA, protein kinase A; PKC, protein kinase C; Src, Src family kinase

metabolites by the cytochrome P-450 epoxygenase-dependent pathway (Watanabe et al. 2003; White et al. 2016; Fernandes et al. 2008).

TRPV4 can also be activated by heat, pH, and inflammation. It is constitutively active at normal body temperatures, but demonstrates temperature activation above 25 °C (Watanabe et al. 2002; Güler et al. 2002; Gao et al. 2003). TRPV4 is sensitized to heat in response to hypotonic stress and reduced in hyperosmotic conditions. The exact purpose of TRPV4 activation due to heat increases remains uncertain and may be species dependent. Likewise, pH has been shown to modulate TRPV4 activity, as demonstrated by a study wherein low pH in in hamster ovarian cells opened the channel (Suzuki et al. 2003). TRPV4 gene disruption also decreased acid nociception in mice (Suzuki et al. 2003). Conversely, in esophageal epithelial cells acidic conditions can inhibit TRPV4 Ca^{2+} cellular influx. Inhibition of TRPV4 with acid is thought to aggravate the condition in gastroesophageal reflux disease (GERD) (Shikano et al. 2011). TRPV4 antagonists have also been shown to reduce inflammation and pain in models of inflammatory bowel disease (Fichna et al. 2012).

Inflammation also can affect TRPV4 in part because it results in temperature and pH changes and inflammatory responses sensitize TRPV4 to hypotonic stimuli (Chen et al. 2007). However, activation of TRPV4 can produce proinflammatory cytokines such as TNF-alpha and interleukins (IL)-1alpha, IL-1β, IL-6, and IL-8 (Pairet et al. 2018; Wang et al. 2019; D'Aldebert et al. 2011). Inhibition of TRPV4 can decrease release of these cytokines (Shi et al. 2013; Wang et al. 2019) or TRPV4 activity can, in turn, be modulated by cytokines (Simpson et al. 2019). Taken together, these results suggest that TRPV4 can be regulated by a variety of physiological conditions including acidic conditions, as well as heat and corresponding inflammation.

7.6 TRPV4 Modulation by GPCRs

TRPV4 has been shown to be modulated by a number of different G-protein coupled receptors via the action of various downstream kinases, as shown in Fig. 7.3. Studies by Xia et al. (2013) demonstrated a role for TRPV4 in enhanced pulmonary

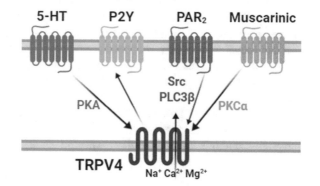

Fig. 7.3 Activation of TRPV4 via G-protein coupled receptors is sensitized by various kinases

vasoreactivity to serotonin (5-HT) in chronic hypoxic pulmonary hypertension indicating that TRPV4 contributes to 5-HT-dependent pharmaco-mechanical coupling in pulmonary arteries. They found that either global knockout of TRPV4 (such as in the *trpv4–/–* mouse) or addition of submicromolar HC-067047, a specific TRPV4 antagonist, decreased the sensitivity of 5-HT-mediated contraction of pulmonary arteries. This effect was independent of endothelin or phenylephrine-mediated contraction. Akiyama et al. (2016) demonstrated a role for TRPV4 in serotonin-mediated itch response in the peripheral nervous system. In this study, approximately 90% of all 5-HT-sensitive DRG cells were found to contain TRPV4. In trpv4–/– mice, they found an overall decrease in serotonin-mediated scratching bouts, and this same result was recapitulated by treating wild-type mice with a TRPV4 antagonist. Additionally, this study utilized calcium imaging of primary DRG cultures to demonstrate that there was a decrease in 5-HT-activated, TRPV4-mediated calcium influx in trpv4–/– cultures, but not in response to other pruritogens.

Rahman et al. (2018) demonstrated that a TRPV4 agonist, 4α-PDD triggers the release of ATP via pannexin channels in human pulmonary fibroblasts. This activation can by attenuated by P2Y purinergic receptor/channel blocker pyridoxalphosphate-6-azophenyl-2′,4′-disulphonic acid (PPADS).

Initial studies (Grant et al. 2007; Zhang et al. 2008) demonstrated that the protease-activated receptor-2 (PAR2) sensitizes TRPV4 to initiate mechanical hyperalgesia in mice through sensitizing calcium signals in DRGs. In HEK cells TRPV4 expression was found to generate sustained intracellular calcium increases, which were attenuated by Src inhibition of PAR2 (Poole et al. 2013). This role for PAR2 activation of TRPV4 was found to be dependent upon the Tyr-110 residue as a binding site for Src kinase (Alessandri-Haber et al. 2008). Additionally, PAR2 was shown to induce long term depression in the hippocampal network through the PKA-dependent, TRPV4-mediated activation of NMDAR (*N*-methyl-D-aspartate receptors) (Shavit-Stein et al. 2017). Lastly, TRPV4 has been shown to activate in response to PAR2 through a PLCB3-dependent mechanism in human respiratory epithelial cells (Li et al. 2011).

TRPV4 activation has been linked to muscarinic receptor mediated endothelium-dependent vasodilation, which is enhanced by the shear stress mechanism. This was demonstrated by Darby et al. (2018), where shear stress was shown to increase endothelial TRPV4 agonist (GSK1016790A) sensitivity. In shear treated arterioles, the acetylcholine response was attenuated by a TRPV4 antagonist (GSK2193874). Additionally, TRPV4 has been found to colocalize with PKC in mesenteric arteries in rats, enabling multiple TRPV4 channels to have a functional effect by opening calcium sensitive potassium channels (Pankey et al. 2014). In summation, this large body of work indicates that GPCR activation can sensitize TRPV4 activity through either direct activity or the interaction with various kinases such as Src, PKA, and PKC.

The multiplicity of possible external stimuli combined with a panoply of intracellular signals has led to the characterization of TRPV4 as a hub protein capable of integrating multiple signals in a tissue-specific manner.

7.7 TRPV4 in the Brain

In the nervous tissue, specifically in neurons and glia, TRPV4 has numerous roles although the mechanisms remain poorly characterized. TRPV4 plays a major role in neurovascular coupling, excitatory glia, and sensory nerve transmission (Cao et al. 2009; Dunn et al. 2013; Shibasaki et al. 2014; Shibasaki and Tominaga 2007; Kumar et al. 2018).

Activation of TRPV4 affects astrocytes, the support cells of the central nervous system, by changing their morphology to an activated state. In astrocytes, activation of TRPV4 causes cell hypertrophy and increases the amount of glial fibrillary activation protein (GFAP), thereby changing the morphology of the cells and increasing their aggregative properties (Wang et al. 2019). TRPV4 is localized predominately in astrocyte endfeet where it regulates vascular permeability and contributes to neurovascular coupling (Dunn et al. 2013). In cases of oxidative stress, TRPV4 can also mediate astrocyte cell death (Bai and Lipski 2010). Furthermore, TRPV4 as an osmosensor has been implicated in regulatory volume decrease (RVD) (Liu et al. 2006). In this mechanism, specific water channels called aquaporins (AQP) have been postulated to interact with TRPV4 either directly or indirectly. In astrocytes, the AQP4 water channel interacts with TRPV4 and this complex is considered to be crucial for RVD (Benfenati et al. 2011).

TRPV4 has been shown to be osmotically activated and expressed in the circumventricular organs (CVOs), where *trpv4−/−* mice showed reduced nociceptive responses to hypotonic and mildly hypertonic stimuli (Benfenati et al. 2007; Liedtke and Friedman 2003; Alessandri-Haber et al. 2003; Kumar et al. 2018). TRPV4 was also shown to significantly increase mEPSC (mini, excitatory postsynaptic current) frequency in hippocampal neurons via the PKC pathway, indicating a role for TRPV4 in excitatory neural signaling in the limbic system (Cao et al. 2009). TRPV4 increases the frequency of mEPSCs in dorsal horn neurons, indicating that TRPV4 also has a role for synaptic transmission in the periphery (Cao et al. 2009). In peripheral nociceptive neurons, TRPV4 acts as a sensor for noxious mechanical and osmotic stimuli (Alessandri-Haber et al. 2003). Furthermore, TRPV4 appears to play a role in the detection and modulation of mechanical hyperalgesia and mechanical allodynia, indicating it as a promising target for a novel class of analgesics (Alessandri-Haber et al. 2003, 2006).

TRPV4 expression was found to increase in many crucial brain and spinal cord regions which are indicated in aged individuals (Lee and Choe 2014). TRPV4 expression has also been shown to be upregulated in regions associated with neurodegenerative pathologies, such as the cerebral cortex, hippocampal formation, basal ganglia, thalamus, cortical pyramidal neurons, and neurons in the spinal cord (Lee and Choe 2014; Lee et al. 2012; Kumar et al. 2018). Increased TRPV4 expression has been observed in a mouse model of amyotrophic lateral sclerosis (ALS) (Lee et al. 2012). In models of stroke, TRPV4 is activated and can be followed by neuronal death (Jie et al. 2016; Li et al. 2013). While intriguing, the significance of these changes remains unknown.

TRPV4 has specific vascular functions which can impact brain tissue including roles in angiogenesis, capillary permeability, and fluid exchange (Dejana and Orsenigo 2013; Pries and Secomb 2014; Wu et al. 1996; Marrelli et al. 2007). Reductions in intravascular pressure as well as flow-induced "shear stress" have been shown to activate endothelial TRPV4 channels (Bagher et al. 2012; Mendoza et al. 2010) and increase TRPV4 membrane localization (Loot et al. 2008). Stimulation of TRPV4 with an agonist, GSK1016790A, causes dilation of blood vessels while antagonist-mediated inhibition of TRPV4 can prevent vasodilation (Mendoza et al. 2010; Köhler et al. 2006). Agonist activation of TRPV4 increases cerebral arteriogenesis (Schierling et al. 2011) and it has been implicated in endothelial proliferation in brain capillaries (Hatano et al. 2013). TRPV4 has been shown to be important in blood-tissue barriers. In pulmonary endothelia, agonist treatment increased permeability of the tissue (Alvarez et al. 2006; Villalta et al. 2014; Wu et al. 2009) but it is not known if this permeability effect is also manifested within the brain vasculature and if so, whether this is as a direct effect or mediated via astrocyte control of the blood brain barrier.

In summary, TRPV4 may play important roles in brain homeostasis via its effects in astrocytes and the vasculature. In addition, a myriad of roles in neuronal function may have important implications for signaling and sensation. Of particular interest in this regard is the increased expression of TRPV4 in neurological disease and during the normal ageing process.

7.8 TRPV4 in the Choroid Plexus

The presence of TRPV4 in choroid plexus epithelia was demonstrated by Liedtke et al. (2000) and subsequent studies have shown both an intracellular and apical membrane distribution of the channel (Willette et al. 2007; Takayama et al. 2014; Narita et al. 2015; Millar et al. 2007). While TRPV4 broadly localizes to the apical membrane of choroid plexus epithelium, it does not localize to the tufts of immotile cilia on the apical surface as it does in the oviduct, cortical collecting duct, mesenchymal stem cells, the eye, and the airway (Narita et al. 2015; Li et al. 2019; Saigusa et al. 2019; Corrigan et al. 2018; Luo et al. 2014; Alenmyr et al. 2014). Though TRPV4 is expressed in astrocytes and neurons, it has been noted with both in situ hybridization and immunofluorescence that the TRPV4 transcript and protein localization are most abundant in the choroid plexus epithelia compared to the rest of the brain tissue (Liedtke et al. 2000; Willette et al. 2007; Hochstetler et al. 2018; Smith et al. 2019).

Functional studies to elucidate the role of TRPV4 in the choroid plexus have used a variety of primary and continuous cells lines. Murine, rat, swine and primate primary cultures have been characterized and used for a range of studies (Klas et al. 2010; Narita et al. 2015; Takayama et al. 2014; Gregoridades et al. 2019; Delery and MacLean 2019). These cultures demonstrate epithelial phenotypes, correct polarization of transporters, and complete junctional complexes, but few exhibit tight barrier function as evidenced by high transepithelial electrical resistance measurements.

TRPV4 has been studied in a subset of the described primary cultures with some interesting results. In primary cultures of swine choroid plexus, the TRPV4 agonist, GSK1016790A (10 nM), decreased barrier integrity by disrupting tight junctions as seen by loss of fluorescent signal when immunostaining for ZO-1 and Claudin-1 (Narita et al. 2015). The agonist also led to a substantial drop in transepithelial electrical resistance as well as a degradation of the filamentous actin cytoskeletal structure and the integrity of the epithelial monolayer. It was suggested that this change may facilitate the infiltration of immune cells into the CSF. Treatment with HC067047, a TRPV4 antagonist, decreased the amount of basolateral to apical transport of α-2-macroglobulin which the authors hypothesize indicates the involvement of TRPV4 in intracellular vesicular trafficking of proteins at the apical membrane. The change in transepithelial electrical resistance, an indicator of barrier function of the epithelium, is in agreement with findings in vascular endothelium where TRPV4 has been shown increase permeability of the tissue.

In terms of channel regulation in the choroid plexus, the well-documented Ca^{2+} influx mediated by TRPV4 activation has the potential to secondarily stimulate Ca^{2+}-activated electrolyte channels. In mouse primary choroid plexus cultures, activation of TRPV4 generated an outward Cl^- current (Takayama et al. 2014) via ANO1 (TMEM16A), a Ca^{2+}-sensitive Cl^- channel. This was activated by either warmth or hypo-osmolarity and was blocked in cells that were deficient in TRPV4. Furthermore, the authors reported a physical interaction between TRPV4 and ANO1. Cell volume changes were dependent on ANO1 activation and the cell volume was decreased at negative membrane potentials by TRPV4-induced Cl^- channel activation. The authors hypothesized the formation of a complex containing TRPV4, ANO1 and AQP1 or AQP4 in the apical membrane.

An important addition to the research armamentarium for studying choroid plexus function was the development of two continuous high resistance choroid plexus cell lines: Human Choroid Plexus Papilloma Cells (HIBCPP) and Porcine Choroid Plexus—Reims (PCP-R) lines (Schwerk et al. 2012; Schroten et al. 2012). These lines have been well characterized and express many of the characteristics of the native tissue including the expression and polarization of key transport proteins.

In contrast to Narita et al. (2015), Preston et al. (2018) demonstrated that in PCP-R cells the TRPV4 agonist GSK1016790A (3 nM) did not affect the expression of tight junctional protein claudin-1 nor did the epithelial cells show morphological changes or loss of monolayer structure. However, the agonist did dramatically increase the conductance of the epithelial monolayer indicating a change in transepithelial barrier function. In agreement with Narita et al. (2015), they found that an antagonist of TRPV4 (RN 1734) inhibited this change in conductance. The differences in the results between the two studies with regard to effects on the tight junctions and cell structure may be the use of a continuous, high resistance cell line (Preston et al. 2018) versus the use of medium to low resistance primary cultures (Narita et al. 2015) and/or the differences in agonist concentration.

Preston et al. (2018) identified a multiphasic change in electrogenic transepithelial ion flux in response to TRPV4 activation. A portion of the stimulated flux was consistent with cation secretion and was dependent on the Ca^{2+}-sensitive intermediate conductance channel IK (KCNN4c; $K_{ca}3.1$).

Simpson et al. (2019) characterized the modulation of TRPV4 by various cytokines in the PCP-R cell line. This study demonstrated that select pro-inflammatory cytokines TNF-a, IL-1B, and TFG-B1 had inhibitory effects on TRPV4-stimulated ion flux and conductance changes. Anti-inflammatory cytokines had no effect on TRPV4 activity. In opposition to published studies in other tissues, this work also demonstrated that arachidonic acid was inhibitory to TRPV4-stimulated transepithelial ion flux. However, inhibition of EET production also inhibited TRPV4 activity suggesting the importance of these arachidonic acid metabolites. Lastly, this study showed that inhibition of transcription factor NF-kB through the use of an inhibitor blocked TRPV4-stimulated activity, thereby proposing a role for TRPV4 in cytokine production.

In a preliminary report, TRPV4 antagonists have been shown to be an effective treatment for hydrocephalus in preclinical studies in rodent models (Smith et al. 2019), presumably via an effect on TRPV4 in the choroid plexus. Although its role is understudied in the choroid plexus, preliminary studies have indicated a role for TRPV4 in fluid secretion, barrier integrity, and immune surveillance within the blood-cerebrospinal fluid barrier.

7.9 Concluding Statement

TRPV4 is a polymodal, ubiquitous hub protein which acts acutely to normalize osmotic imbalances, reestablish membrane potentials, control cellular permeability, modulate immune responses, and activate downstream signaling cascades. Future studies with the potential for clinical implications in many diseases including those that involve cerebrospinal fluid production by the choroid plexus are eagerly awaited.

Acknowledgements The authors are supported by grants from the United States Department of Defense, Investigator Initiated Research Award W81XWH-16-PRMRP-IIRA and the Mayfield Education and Research Foundation.

References

Akiyama T, Ivanov M, Nagamine M et al (2016) Involvement of TRPV4 in serotonin-evoked scratching. J Invest Dermatol 136(1):154–160

Alenmyr L, Uller L, Greiff L, Hogestatt ED, Zygmunt PM (2014) TRPV4-mediated calcium influx and ciliary activity in human native airway epithelial cells. Basic Clin Pharmacol Toxicol 114 (2):210–216

Alessandri-Haber N, Yeh JJ, Boyd AE, Parada CA et al (2003) Hypotonicity induces TRPV4-mediated nociception in rat. Neuron 39(3):497–511

Alessandri-Haber N, Dina OA, Joseph EK, Reichling D, Levine JD (2006) A transient receptor potential vanilloid 4-dependent mechanism of hyperalgesia is engaged by concerted action of inflammatory mediators. J Neurosci 26(14):3864–3874

Alessandri-Haber N, Dina OA, Joseph EK, Reichling D, Levine JD (2008) Interaction of transient receptor potential vanilloid 4, integrin, and SRC tyrosine kinase in mechanical hyperalgesia. J Neurosci 28(5):1046–1057

Alessandri-Haber N, Dina OA, Chen X, Levine JD (2009) TRPC1 and TRPC6 channels cooperate with TRPV4 to mediate mechanical hyperalgesia and nociceptor sensitization. J Neurosci 29:6217–6228

Alvarez DF, King JA, Weber D, Addison E, Liedtke W, Townsley MI (2006) Transient receptor potential vanilloid 4-mediated disruption of the alveolar septal barrier: a novel mechanism of acute lung injury. Circ Res 99:988–995

Arniges M, Fernandez-Fernandez JM, Albrecht N, Shaefer M, Valverde MA (2006) Human TRPV4 channel splice variants revealed a key role of ankyrin domains in multimerization and trafficking. J Biol Chem 281(3):1580–1586

Bagher P, Beleznai T, Kansui Y, Mitchell R, Garland CJ, Dora KA (2012) Low intravascular pressure activates endothelial cell TRPV4 channels, local Ca2 events, and IKCa channels, reducing arteriolar tone. Proc Natl Acad Sci U S A 109:18174–18179

Bai JZ, Lipski J (2010) Differential expression of TRPM2 and TRPV4 channels and their potential role in oxidative stress-induced cell death in organotypic hippocampal culture. Neurotoxicology 31:204–214

Becker D, Muller M, Leuner K, Jendrach M (2008) The C-terminal domain of TRPV4 is essential for plasma membrane localization. Mol Membr Biol 25(2):139–151

Benfenati V, Amiry-Moghaddam M, Caprini M, Mylonakou MN et al (2007) Expression and functional characterization of transient receptor potential vanilloid-related channel 4 (TRPV4) in rat cortical astrocytes. Neuroscience 148(4):876–892

Benfenati V, Caprini M, Dovizio M, Mylonakou MN et al (2011) An aquaporin-4/transient receptor potential vanilloid 4 (AQP4/TRPV4) complex is essential for cell-volume control in astrocytes. Proc Natl Acad Sci U S A 108(6):2563–2568

Brohawn SG, Su Z, MacKinnon R (2014) Mechanosensitivity is mediated directly by the lipid membrane in TRAAK and TREK1 K channels. Proc Natl Acad Sci U S A 111:3614–3619

Cao DS, Yu SQ, Premkumar LS (2009) Modulation of transient receptor potential Vanilloid 4-mediated membrane currents and synaptic transmission by protein kinase C. Mol Pain 5:5

Cao DS, Anishkin A, Zinkevich NS, Nishijima Y et al (2018) Transient receptor potential vanilloid 4 (TRPV4) activation by arachnidonic acid requires protein kindase A-mediated phosphorylation. J Biol Chem 293(14):5307–5322

Chen X, Alessandri-Haber N, Levine JD (2007) Marked attenuation of inflammatory mediator-induced C-fiber sensitization for mechanical and hypotonic stimuli in TRPV4−/−mice. Mol Pain 3:31

Cheng X, Shen D, Samie M, Xu H (2011) Mucolipins: intracellular TRPML1-3 channels. FEBS Lett 584(10):2013–2021

Corrigan MA, Johnson GP, Stavenschi E, Riffault M, Labour MN, Hoey DA (2018) TRPV4-mediates oscillatory fluid shear mechanotransduction in mesenchymal stem cells in part via the primary cilium. Sci Rep 8(1):3824

Cortright DN, Szallasi A (2009) TRP channels and pain. Curr Pharm Des 15:1736–1749

Cosens DJ, Manning A (1969) Abnormal electroretinogram from a Drosophila mutant. Nature 224 (5216):285–287

Cuajungco MP, Grimm C, Oshima K, D'hoedt D et al (2006) PACSINs bind to the TRPV4 cation channel. J Biol Chem 281:18753–18762

D'Aldebert E, Cenac N, Rousset P et al (2011) Transient receptor potential vanilloid 4 activated inflammatory signals by intestinal epithelial cells and colitis in mice. Gastroenterology 140:275–285

Darby WG, Potocnik S, Ramachandran R, Hollenberg MD, Woodman OL, McIntyre P (2018) Shear stress sensitizes TRPV4 in endothelium-dependent vasodilation. Pharmacol Res 133:152–159

Dejana E, Orsenigo F (2013) Endothelial adherens junctions at a glance. J Cell Sci 126:2545–2549

Delery EC, MacLean AG (2019) Culture model for non-human primate choroid plexus. Front Cell Neuro 13:296

Deng Z, Paknejad D, Maksaev G et al (2018) Cryo-EM and X-ray structures of TRPV4 reveal insight into ion permeation and gating mechanisms. Nat Struct Mol Biol 25(3):252–260

Deruyver Y, Weyne E, Dewulf K et al (2018) Intravesical activation of the cation channel TRPV4 improves bladder function in a rat model for detrusor underactivity. Eru Urol 74(3):336–345

D'hoedt D, Owsianik G, Prenen J, Cuajungco MP et al (2007) Stimulus-specific modulation of the cation channel TRPV4 by PACSIN-3. J Biol Chem 283:6272–6280

Du J, Ma X, Shen B, Huang Y, Birnbaumer L, Yao X (2014) TRPV4, TRPC1, and TRPP2 assemble to form a flow-sensitive heteromeric channel. FASEB J 28(11):4677–4685

Dubowitz V, Daniels RJ, Davies KE (1995) Olivopontocerebellar hypoplasia with anterior horn cell involvement (SMA) does not localize to chromosome 5q. Neuromuscul Disord 5(1):25–29

Dunn KM, Hill-Eubanks DC, Liedtke WB, Nelson MT (2013) TRPV4 channels stimulate Ca2+-induced Ca2+ release in astrocytic endfeet and amplify neurovascular coupling responses. Proc Natl Acad Sci U S A 110(15):6157–6162

Everaerts W, Nilius B, Owsianik G (2010) The vanilloid transient receptor potential channel TRPV4: from structure to disease. Prog Biophys Mol Biol 103:2–17

Fan HC, Zhang X, McNaughton PA (2009) Activation of the TRPV4 ion channel is enhanced by phosphorylation. J Biol Chem 284(41):27884–27891

Fernandes J, Lornezo IM, Andrade YN, Garcia-Elias A et al (2008) IP3 sensitizes TRPV4 channel to the mechano- and osmotransducing messenger 5',6' epoxyeicosatrienoic acid. J Cell Biol 181:143–155

Ferrandiz-Hertas C, Mathivanan S, Jakob Wolf C, Devesa I, Ferrer-Montiel A (2014) Trafficking of ThermoTRP channels. Membranes (Basel) 4(3):525–564

Fichna J, Mokrowiecka A, Cygankiewicz AI, Zakrzewski PK et al (2012) Transient receptor potential vanilloid 4 blockade protects against experimental colitis in mice: a new strategy for inflammatory bowel diseases treatment? Neurogastroenterol Motil 24:e557–e560

Fu Y, Subramanya A, Rozansky D, Cohen DM (2006) WNK kinases influence TRPV4 channel function and localization. Am J Physiol Ren Physiol 290:F1305–F1314

Gao X, Wu L, O'Neil RG (2003) Temperature-modulated diversity of TRPV4 channel gating: activation by physical stresses and phorbol ester derivatives through protein kinase C-dependent and -independent pathways. J Biol Chem 278:27129–27137

Garcia-Elias A, Lorenzo IM, Vicente R, Valverde MA (2008) IP3 receptor binds to and sensitizes TRPV4 channel to osmotic stimuli via a calmodulin site. J Biol Chem 283:31284–31288

Garcia-Elias A, Mrkonjic S, Pardo-Pastor C, Inada H et al (2013) Phosphatidylinositol-4,5-biphosphate-dependent rearrangement of TRPV4 cytosolic tails enables channel activation by physiological stimuli. Proc Natl Acad Sci USA 110:9553–9558

Gavaert T, Vriens J, Segal A, Everaerts W et al (2007) Deletion of the transient receptor potential cation channel TRPV4 impairs murine bladder voiding. J Clin Invest 117(11):3453–3462

Gees M, Colsoul B, Nilius B (2010) The role of transient receptor potential cation channels in Ca2+ signaling. Cold Spring Harb Perspect Biol 2(10):a003962

Goswami C, Kuhn J, Heppenstall PA, Hucho T (2010) Importance of non-selective cation channel TRPV4 interaction with cytoskeleton and their reciprocal regulations in cultured cells. PLoSOne 5:e11654

Grant AD, Cottrell GS, Amadesi S, Trevisani M et al (2007) Protease-activated receptor 2 sensitizes the transient receptor potential vanilloid 4 ion channel to cause mechanical hyperalgesia in mice. J Physiol 157(3):715–733

Gregoridades JMC, Madaris A, Alvarez F, Alvarez-Leefmans FJ (2019) Genetic and pharmacological inactivation of apical Na+-K+2Cl- cotransporter 1 in choroid plexus epithelial cells reveals the physiological function of the cotransporter. Am J Physiol Cell Physiol 316: C525–C544

Güler A, Lee H, Iida T, Shimizu I, Tominaga M, Caterina M (2002) Heat-evoked activation of TRPV4 (VR-OAC). J Neurosci 22:6408–6414

Hatano N, Suzuki H, Itoh Y, Muraki K (2013) TRPV4 partially participates in proliferation of human brain capillary endothelial cells. Life Sci 92:317–324

Hochstetler AE, Whitehouse L, Antonellis P, Berbari NF, Blazer-Yost B (2018) Characterizing the expression of TRPV4 in the choroid plexus epithelia as a prospective component in the development of hydrocephalus in the Gas8^{GT} juvenile mutant mouse. FASEB J 32:750.12

Inda H, Procko E, Sotomayor M, Gaudet R (2012) Structural and biochemical consequences of disease-causing mutations in the ankyrin repeat domain of the human TRPV4 channel. Biochemistry 51:6195–6206

Jang Y, Jung J, Kim H, Oh J et al (2012) Axonal neuropathy-associated TRPV4 regulates neurotrophic factor-derived axonal growth. J Biol Chem 287:6014–6024

Janssen DA, Hoenderop JG, Jansen KC, Kemp AW, Heesakkers JP, Schalken JA (2011) The mechanoreceptor TRPV4 is localized in adherence junctions of the human bladder urothelium: a morphological study. J Urol 186:1121–1127

Jie P, Lu Z, Hong Z, Li L et al (2016) Activation of transient receptor potential vanilloid 4 is involved in neuronal injury in middle cerebral artery occlusion in mice. Mol Neurobiol 53:8–17

Klas J, Wolburg H, Terasaki T, Fricker G, Reichel V (2010) Characterization of immortalized choroid plexus epithelial cell lines for studies of transport processes across the blood-cerebrospinal fluid barrier. Cerebrosp Fluid Res 7:11

Köhler R, Heyken WT, Heinau P, Schubert R et al (2006) Evidence for a functional role of endothelial transient receptor potential V4 in shear stress-induced vasodilatation. Arterioscler Thromb Vasc Biol 26:1495–1502

Kornak U, Mundlos S (2003) Genetic disorders of the skeleton: a developmental approach. Am J Hum Genet 73:447–474

Kumar H, Lee SH, Kim KT, Zeng X, Han I (2018) TRPV4: a sensor for homeostasis and pathologcial events in the CNS. Mol Neurobiol 55:8695–8708

Kung C (2005) A possible unifying principle for mechanosensation. Nature 436:647–654

Lamande SR, Yuan Y, Gresshoff IL, Rowley L et al (2011) Mutations in TRPV4 cause an inherited arthropathy of hands and feet. Nat Genet 43:1142–1146

Lanciotti A, Brignone MS, Molinari P, Visentin S et al (2012) Megalencephalic leukoencephalopathy with subcortical cysts protein 1 functionally cooperates with the TRPV4 cation channel to activate the response of astrocytes to osmotic stress: dysregulation by pathological mutations. Hum Mol Genet 21:2166–2180

Lee JC, Choe SY (2014) Age-related changes in the distribution of transient receptor potential vanilloid 4 channel (TRPV4) in the central nervous system of rats. J Mol Histol 45(5):497–505

Lee H, Iida T, Mizuno A, Suzuki M, Caterina MJ (2005) Altered thermal selection behavior in mice lacking transient receptor potential vanilloid 4. J Neurosci 25(5):1304–1310

Lee EJ, Shin SH, Chun J, Hyun S, Kim Y, Kang SS (2010) The modulation of TRPV4 channel activity through its Ser 824 residue phosphorylation by SGK1. Anim Cell Sys 14(2):99–114

Lee JC, Joo KM, Choe SY, Cha CI (2012) Region-specific changes in the immunoreactivity of TRPV4 expression in the central nervous system of SOD1 (G93A) transgenic mice as an in vivo model of amyotrophic lateral sclerosis. J Mol Histol 43(6):625–631

Lei L, Cao X, Yang F et al (2013) A TRPV4 channel C-terminal folding recognition domain critical for trafficking and function. J Biol Chem 288:10427–10439

Li J, Kanju P, Patterson M, Chew WL et al (2011) TRPV4-mediated calcium influx into bronchial epithelia upon exposure to diesel exhaust particles. Eviron Health Perspect 119(6):784–793

Li L, Qu W, Zhou L, Lu Z et al (2013) Activation of transient receptor potential vanilloid 4 increases NMDA-activated current in hippocampal pyramidal neurons. Front Cell Neurosci 7:17

Li M, Fang XZ, Zheng YF et al (2019) Transient receptor potential Vanilloid 4 is a critical mediator in LPS mediated inflammation by mediating calcineurin/NFATc3 signaling. Biochem Biophys Res Commun 513(4):1005–1012

Liedtke W (2008) Molecular mechanisms of TRPV4-mediated neural signaling. Ann N Y Acad Sci 1144:42–52

Liedtke W, Friedman JM (2003) Abnormal osmotic regulation in trpv4−/− mice. Proc Natl Acad Sci USA 100(23):13698–13703

Liedtke W, Choe Y, Marti-Renom MA, Bell AM et al (2000) Vanilloid receptor-related osmotically activated channel (VR-OAC), a candidate vertebrate osmoreceptor. Cell 103(3):525–535

Liu X, Bandyopadhyay BC, Nakamoto T, Singh B et al (2006) A role for AQP5 in activation of TRPV4 by hypotonicity: concerted involvement of AQP5 and TRPV4 in regulation of cell volume recovery. J Biol Chem 281:15485–15495

Loot AE, Popp R, Fisslthaler B, Vriens J, Nilius B, Fleming I (2008) Role of cytochrome P450-dependent transient receptor potential V4 activation in flow-induced vasodilatation. Cardiovasc Res 80:445–452

Luo N, Conwell MD, Chen X et al (2014) Primary cilia signaling mediates intraocular pressure sensation. Proc Natl Acad Sci U S A 111(35):12871–12876

Ma X, Cao J, Luo J, Nilius B et al (2010) Depletion of intracellular Ca^{2+} stores stimulates the translocation of vanilloid transient receptor potential 4-c1 heteromeric channels to the plasma membrane. Arterioscler Thromb Vasc Biol 30:2249–2255

Ma X, Cheng KT, Wong CO, O'Neil RG et al (2011) Heteromeric TRPV4-C1 channels contribute to store-operated Ca(2+) entry in vascular endothelial cells. Cell Calcium 50(6):502–509

Main M, Kairon H, Mercuri E, Muntoni F (2003) The Hammersmith functional motor scale for children with spinal muscular atrophy: a scale to test ability and monitor progress in children with limited ambulation. Eur J Paediatr Neurol 7:155–159

Mamenko M, Zaika OL, Boukelmoune N, Berrout J, O'Neil RG, Pochynyuk O (2013) Discrete control of TRPV4 channel function in the distal nephron by protein kinases A and C. J Biol Chem 288:20306–20314

Marrelli SP, O'Neil RG, Brown RC, Bryan RM Jr (2007) PLA2 and TRPV4 channels regulate endothelial calcium in cerebral arteries. Am J Physiol Heart Circ Physiol 292:H1390–H1397

Masuyama R, Vriens J, Voets T, Karashima Y et al (2008) TRPV4-mediated calcium influx regulates terminal differentiation of osteoclasts. Cell Metab 8:257–265

Masuyama R, Mizuno A, Komori H, Kajiya H et al (2012) Calcium/calmodulin-signaling supports TRPV4 activation in osteoclasts and regulates bone mass. J Bone Miner Res 27:1708–1721

McEntagart M (2012) TRPV4 axonal neuropathy spectrum disorder. J Clin Neurosci 19:927–933

Mendoza SA, Fang J, Gutterman DD, Wilcox DA et al (2010) TRPV4-mediated endothelial Ca2 influx and vasodilation in response to shear stress. Am J Physiol Heart Circ Physiol 298:H466–H476

Millar ID, Bruce Jl, Brown PD (2007) Ion channel diversity, channel expression, and function in the choroid plexuses. Cerebrospinal Fluid Res 4:8

Narita K, Sasamoto S, Koizumi S, Okazaki S et al (2015) TRPV4 regulates the integrity of the blood-cerebrospinal fluid barrier and modulates transepithelial protein transport. FASEB J 29 (6):2247–2259

Nilius B, Honore E (2012) Sensing pressure with ion channels. Trends Neurosci 35:477–486

Nishimura G, Dai J, Lausch E, Unger S et al (2010) Spondylo-epiphyseal dysplasia, Maroteaux type (pseudo-Morpquio syndrome type 2), and parastremmatic dysplasia are caused by TRPV4 mutations. Am J Med Genet 152A:1443–1449

Owsianik G, Cao L, Nilius B (2003) Rescue of functional DeltaF508-CFTR channels by co-expression with truncated CFTR constructs in COS-1 cells. FEBS Lett 554:173–178

Pairet N, Mang S, Fois G, Keck M et al (2018) TRPV4 inhibition attenuates stretch-induced inflammatory cellular responses and lung barrier dysfunction during mechanical ventilation. PLoS One 13:e0196055

Pankey EA, Zsombok A, Lasker GF, Kadowitz PJ (2014) Analysis of responses to the TRPV4 agonist GSK1016790A in the pulmonary vascular bed of the intact-chest rat. Am J Physiol Heart Circ Physiol 306(1):H33–H40

Pedersen SF, Nilius B (2007) Transient receptor potential channels in mechanosensing and cell volume regulation. Methods Enzymol 428:183–207

Poole DP, Amadesi S, Veldhuis NA, Abogadie FC et al (2013) Protease-activated receptor 2 (PAR2) protein and transient receptor potential vanilloid 4 (TRPV4) protein coupling is required for sustained inflammatory signaling. J Biol Chem 288(8):5790–5802

Preston D, Simpson S, Halm D, Hochstetler A et al (2018) Activation of TRPV4 stimulates transepithelial ion flux in a porcine choroid plexus cell line. Am J Physiol Cell Physiol 315 (3):C357–C366

Pries AR, Secomb TW (2014) Making microvascular networks work: angiogenesis, remodeling, and pruning. Physiology 29:446–455

Rahman M, Sun R, Mukherjee S, Nilius B, Janssen LJ (2018) TRPV4 stimulation releases ATP via Pannexin channels in human pulmonary fibroblasts. Am J Respir Cell Mol Biol 59(1):87–95

Ryskamp DA, Jo AO, Frye AM, Vazquez-Chona F et al (2014) Swelling and eicosanoid metabolites differentially gate TRPV4 channels in retinal neurons and glia. J Neurosci 34:15689–15700

Saigusa T, Yue Q, Bunni MA, Bell PD, Eaton DC (2019) Loss of primary cilia increase polycystin-2 and TRPV4 and the appearance of a nonselective cation channel in the mouse cortical collecting duct. Am J Physiol Renal Physiol 317(3):F632–F637

Saliez J, Bouzin C, Rath G, Ghisdal P et al (2008) Role of caveolar compartmentation in endothelium-derived hyperpolarizing factor-mediated relaxation: Ca^{2+} signals and gap junction function are regulated by caveolin in endothelial cells. Circulation 117:1065–1074

Schierling W, Troidl K, Apfelbeck H, Troidi C et al (2011) Cerebral arteriogenesis is enhanced by pharmacological as well as fluid-sheer-stress activation of the TRPV4 calcium channel. Eur J Vasc Endovasc Surg 41:589–596

Schroten M, Hanisch F, Quednau N, Stump C et al (2012) A novel porcine in vitro model of the blood-cerebrospinal fluid barrier with strong barrier function. PloS ONE 7:e39835

Schwerk C, Papandreou T, Schuhmann D, Nickol L et al (2012) Polar invasion and translocation of Neisseria meningitides and Streptococcus suis in a novel human model of the blood-cerebrospinal fluid barrier. Plos ONE 7:e30069

Shavit-Stein E, Artan-Furman A, Feingold E, Ben Shimon M et al (2017) Protease activated receptor 2 (PAR2) induces long-term depression in the hippocampus through transient receptor potential vanilloid 4 (TRPV4). Front Mol Neurosci 10:42

Shi M, Du F, Liu Y, Li L et al (2013) Glial cell-expressed mechanosensitive channel TRPV4 mediates infrasound-induced neuronal impairment. Acta Neuropathol 126:725–739

Shibasaki K, Tominaga M (2007) Implication that TRPV4 activation induces the excitation of astrocytes. In: Proceedings of annual meeting of the Physiological Society of Japan, pp 082–082

Shibasaki K, Ikenaka K, Tamalu F, Tominaga M, Ishizaki Y (2014) A novel subtype of astrocytes expressing TRPV4 (transient receptor potential vanilloid 4) regulates neuronal excitability via release of gliotransmitters. J Biol Chem 289(21):14470–14480

Shikano M, Ueda T, Kamiya T, Ishida Y et al (2011) Acid inhibits TRPV4 mediated Ca^{2+} influx in mouse esophageal epithelial cells. Neurogastroenterol Motil 23:1020–1028

Shin SH, Lee EJ, Hyun S, Chun J, Kim Y, Kang SS (2012) Phosphorylation on the Ser 824 residue of TRPV4 prefers to bind with F-actin than with microtubulues to expand the cell surface area. Cell Signal 24(3):641–651

Simpson S, Preston D, Schwerk C, Schroten H, Blazer-Yost B (2019) Cytokine and inflammatory mediator effects on TRPV4 function in choroid plexus epithelial cells. Am J Physiol Cell Physiol 317(5):C881–C893. Ahead of print

Smith H, Hochstetler A, Preston D, Balzer-Yost BL (2019) Preclinical testing of TRPV4 antagonists for the treatment of hydrocephalus. FASEB J 33:A708.4

Sonkusare SK, Bonev AD, Ledoux J et al (2012) Elementary Ca^{2+} signals through endothelial TRPV4 channels regulate vascular function. Science 336(6081):597–601

Stewart AP, Smith GD, Sandford RN, Edwardson JM (2010) Atomic force microscopy reveals the alternating subunit arrangement of the TRPP2-TRPV4 heterotetramer. Biophys J 99 (3):790–797

Strotmann R, Harteneck C, Nunnenmacher K, Schultz G, Plant TD (2000) OTRPC4, a nonselective cation channel that confers sensitivity to extracellular osmolarity. Nat Cell Biol 2:695–702

Strotmann R, Schultz G, Plant TD (2003) Ca^{2+}-dependent potentiation of the nonselective cation channel TRPV4 is mediated by a C-terminal calmodulin binding site. J Biol Chem 278:26541–26549

Suzuki M, Mizuno A, Kodaira K, Imai M (2003) Impaired pressure sensation with mice lacking TRPV4. J Biol Chem 278:22664–22668

Tabuchi K, Suzuki M, Mizuno A, Hara A (2005) Hearing impairment in TRPV4 knockout mice. Neurosci Lett 382:304–308

Takayama Y, Shibasaki K, Suzuki Y, Yamanaka A, Tominaga M (2014) Modulation of water efflux through functional interaction between TRPV4 and TMEM16A/anoctamin 1. FASEB J 28(5):2238–2248

Teng J, Loukin SH, Anishkin A, Kung C (2015) L596-W733 bond between the start of the S4-S5 linker and the TRP box stabilizes the closed state of the TRPV4 channel. Proc Natl Acad Sci USA 112(11):3386–3391

Thorneloe KS, Cheung M, Bao W, Alsaid H et al (2012) An orally active TRPV4 channel blocker prevents and resolves pulmonary edema induced by heart failure. Sci Transl Med 4:159ra148

Verma P, Kumar A, Goswami C (2010) TRPV4-mediated channelopathies. Channels 4(4):319–328

Villalta PC, Rocic P, Townsley MI (2014) Role of MMP2 and MMP9 in TRPV4-induced lunginjury. Am J Physiol Lung Cell Mol Physiol 307:L652–L659

Vriens J, Owsianik G, Fisslthalter B, Suzuki M et al (2005) Modulation of the Ca^{2+} permeable cation channel TRPV4 by cytochrome P450 epoxygenases in vascular endothelium. Circ Res 97:908–915

Wang Y, Fu X, Gaiser S, Kottgen M et al (2007) OS-9 regulates the transit and polyubiquitination of TRPV4 in the endoplasmic reticulum. J Biol Chem 282:36561–36570

Wang Z, Zhou L, An D, Xu W et al (2019) TRPV4-induced inflammatory response is involved in neuronal death in pilocarpine model of temporal lobe epilepsy in mice. Cell Death Dis 10:386

Watanabe H, Vriens J, Suh SH, Benham CD, Droogmans G, Nilius B (2002) Heat-evoked activation of TRPV4 channels in an HEK293 cell expression system and in native mouse aorta endothelial cells. J Biol Chem 277:47044–47051

Watanabe H, Vriens J, Prenen J, Droogmans G, Voets T, Nilius B (2003) Anandamide and arachidonic acid use epoxyeicosatrienoic acids to activate TRPV4 channels. Nature 424:434–438

Wegierski T, Hill K, Schaefer M, Walz G (2006) The HECT ubiquitin ligase AIP4 regulates the cell surface expression of select TRP channels. EMBO J 25:5659–5669

Wegierski T, Lewandrowski U, Muller B, Sickmann A, Walz G (2009) Tyrosine phosphorylation modulates the activity of TRPV4 in response to defined stimuli. J Biol Chem 284:2923–2933

White JPM, Cibelli M, Urban L, Nilius B, McGeown JG, Nagy I (2016) TRPV4: molecular conductor of a diverse orchestra. Physiol Rev 96(3):911–973

Willette R, Bao W, Nerurkar S, Yue TL et al (2007) Systemic activation of the transient receptor potential vanilloid subtype 4 channel causes endothelial failure and circulatory collapse: part 2. J Pharmacol 326(2):443–452

Wu HM, Huang Q, Yuan Y, Granger HJ (1996) VEGF induces NO-dependent hyperpermeability in coronary venules. Am J Physiol Heart Circ Physiol 271:H2735–H2739

Wu S, Jian MY, Xu YC, Zhou C et al (2009) Ca2 entry via alpha1G and TRPV4 channels differentially regulates surface expression of P-selectin and barrier integrity in pulmonary capillary endothelium. Am J Physiol Lung Cell Mol Physiol 297:L650–L657

Xia Y, Fu Z, Hu J, Huang C et al (2013) TRPV4 channel contributes to serotonin-induced pulmonary vasoconstriction and the enhanced vascular reactivity in chronic hypoxia pulmonary hypertension. Am J Physiol Cell Physiol 305(7):C704–C715

Yu FJ, Yarov-Yarovoy V, Gutman GA, Catterall WA (2005) Overview of molecular relationship in the voltage-gated ion channel superfamily. Pharmacol Rev 57:387–395

Zerangue N, Schwappach B, Jan YN, Jan LY (1999) A new ER trafficking signal regulates the subunit stoichiometry of plasma membrane K(ATP) channels. Neuron 22(3):537–548

Zhang Y, Wang YH, Ge HY, Arendt-Nielsen L, Wang R, Yue SW (2008) A transient receptor potential vanilloid 4 contributes to mechanical allodynia following chronic compression of dorsal root ganglion in rats. Neurosci Lett 432(3):222–227

Chapter 8
Neuroprotective Mechanisms at the Blood-CSF Barrier of the Developing and Adult Brain

Jean-Francois Ghersi-Egea, Alexandre Vasiljevic, Sandrine Blondel, and Nathalie Strazielle

Abstract In addition to CSF secretion, the choroid plexuses fulfill neuroendocrine, neuroimmune and neuroprotective functions. Choroidal neuroprotection results from a combination of tight junctions that prevent the paracellular passage of blood-borne compounds into the CSF, efflux transporters that reduce the CSF bioavailability of numerous potentially toxic drugs and other xenobiotics, and metabolizing enzymes that detoxify reactive organic molecules and reactive oxygen species. The choroid plexuses display developmental stage-specific neuroprotective properties that are reviewed in this chapter.

8.1   Introduction

The choroid plexuses, located in all four brain ventricles, fulfill multiple functions. They secrete the cerebrospinal fluid (CSF) and control inorganic ion concentrations in this fluid. They synthesize and secrete into the CSF a panel of endocrine factors, hormone carriers such as transthyretin, and growth factors. They transport selected micronutrients from the blood into the brain, including vitamin C and manganese. The choroid plexuses also protect the CSF and brain through transporter- and enzyme-based barrier properties. More recently discovered, they participate to the regulation of neuroimmune interactions. These different functions have mainly been investigated and characterized at the adult stage, and are described in details either in

J.-F. Ghersi-Egea (✉) · A. Vasiljevic
FLUID Team, Lyon Neurosciences Research Center, INSERM U1028, CNRS UMR5292, Lyon1 University, Lyon, France
e-mail: Jean-francois.ghersi-egea@inserm.fr

S. Blondel
BIP Facility, Lyon Neurosciences Research Center, Lyon, France

N. Strazielle
FLUID Team, Lyon Neurosciences Research Center, INSERM U1028, CNRS UMR5292, Lyon1 University, Lyon, France

Brain-i, Lyon, France

© The American Physiological Society 2020
J. Praetorius et al. (eds.), *Role of the Choroid Plexus in Health and Disease*, Physiology in Health and Disease, https://doi.org/10.1007/978-1-0716-0536-3_8

the other chapters of this book, or in recent reviews covering these various subjects (Strazielle and Ghersi-Egea 2013; Demeestere et al. 2015; Spector et al. 2015; Benarroch 2016; Kaur et al. 2016; Marques et al. 2017; Praetorius and Damkier 2017; Ghersi-Egea et al. 2018b; Lauer et al. 2018).

During development, choroid plexuses show a strong degree of maturation early on during pre- and postnatal development (Dziegielewska et al. 2001), but they display stage-specific functionalities, which distinguish them from the choroid plexuses in adult. For instance, the choroid plexuses play a specific endocrine role for the developing brain, by secreting factors such as insulin-like growth factor 2 or otx2 homeoprotein, which influence cell differentiation and axonal guidance (reviewed in Prochiantz et al. 2014; Lun et al. 2015; Ghersi-Egea et al. 2018b). The choroid plexus transcriptomic profile of influx transporters, which supply nutrients and micronutrients to the brain, differ between developing and adult stages. Selected SLC transporters, e.g. for amino acids or thyroid hormones, are expressed at higher levels in the fetal choroidal tissue (Saunders et al. 2015). By contrast, in both rodents and humans the rate of CSF secretion is low up to the perinatal period and only increases postnatally (reviewed in Drake et al. 1991; Yasuda et al. 2002; Johanson et al. 2008). All these specificities suggest an adaptation of the blood-CSF interface to the particular needs of the developing brain.

The function of choroid plexuses as a gate of entry into the brain for blood-borne compounds calls for special consideration in the context of brain development. The choroid plexuses develop precociously, as early as embryonic weeks 6 and 7 in human, and present a substantial surface area of exchange between blood and CSF, because of their anatomical organization in villi and the peculiar morphology of their epithelium, which presents basal membrane infoldings as well as an extensive network of microvilli at the apical membrane. In contrast, the density of the microvasculature increases only gradually during pre- and postnatal development. Differential developmental changes in blood flow among cerebral structures result in a choroidal blood-to-parenchymal blood supply ratio higher in the immature brain than in the adult brain. Thus, during pre- and perinatal development, the choroidal blood-CSF barrier is an important gate of entry into the CNS for the solutes circulating in blood (reviewed in Ghersi-Egea et al. 2018a). In addition, the CSF can act as a large reservoir for drugs or nutrients that cross the blood-CSF barrier, because the ratio of CSF-to-brain tissue volume is higher in the developing brain than in the adult, and because the low CSF flow rate in the developing brain compared to the adult brain should favor blood-borne solute accumulation into the CSF. Diffusion of solutes from the CSF into the interstitial spaces is also favored during development, because the extracellular space volume and tortuosity change during postnatal development. Overall, these considerations, discussed in Ghersi-Egea et al. (2018a), point to the choroid plexus-CSF system as an important player in the control of drug and solute concentrations into the parenchyma during the perinatal and postnatal periods.

The blood-CSF barrier is selective in its gatekeeper function, thus preventing potentially noxious molecules to enter the CSF. It also participates in the inactivation or efflux of brain-derived reactive species. These neuroprotective properties result from three distinct features (Fig. 8.1), which include (1) the presence of tight

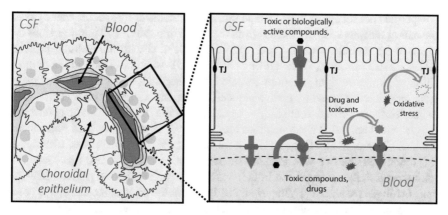

Fig. 8.1 Neuroprotective properties of the choroidal epithelium. The left panel shows a schematic drawing of a choroidal loop depicting the leaky choroidal capillaries and the tight epithelium. The right panel illustrates three characteristics of the choroidal epithelium responsible for the neuroprotective functions of the blood-CSF barrier. The tight junctions linking the epithelial cells together (TJ, black) prevent the paracellular passage of blood-borne hydrophilic compounds. The activity of efflux transporters (blue) either reduces the blood-CSF permeability of various amphiphilic and lipophilic compounds, or increases the CSF efflux of centrally released factors (arrows). Detoxification enzymes (open arrows) inactivate reactive and potentially deleterious molecules such as reactive oxygen species and toxicants

junctions linking neighboring epithelial cells together and preventing the paracellular passage of blood-borne hydrophilic compounds, (2) the activity of efflux transporters that either reduce the blood-CSF permeability of various amphiphilic and lipophilic compounds, or increase the CSF efflux of centrally released factors, and (3) detoxification enzymes that inactivate reactive and potentially deleterious molecules such as reactive oxygen species and toxicants (Strazielle and Ghersi-Egea 2013). The proteins involved in the protective functions of the blood-CSF barrier undergo developmental regulation which needs to be deciphered in view to appreciate the efficiency of this barrier during development. Functional differences between adult and developing stages will critically affect the pharmacological and toxicological consequences of brain exposure to drug and toxins.

The similarities and differences between the developing and adult choroidal features that form the basis of the blood-CSF barrier neuroprotective functions are discussed in the following paragraphs.

8.2 Tight Junctions in the Developing and Adult Choroid Plexuses

8.2.1 Choroidal Tight Junction Proteins

In adult, tight junctions that seal the choroidal epithelial cells forming the blood-CSF barrier greatly restricts diffusion through the intercellular route. Hence polar compounds not recognized by specific transport proteins or by specific transcytosis-mediating receptors, do not diffuse from the blood into the CSF. Tight junctions are made of intercellular claudins and occludin, associated with intracellular proteins including ZO proteins (Tietz and Engelhardt 2015). Analyzing the protein composition of choroidal tight junctions revealed that choroidal epithelial claudins differ from claudins expressed by endothelial cells of the blood-brain barrier (BBB). The main claudin expressed at the BBB is claudin-5, while the main claudins present at the choroidal epithelial tight junctions are Claudin-1, 2, and -3, in both rodents and humans (Kratzer et al. 2012). Evidence for several other well-expressed claudins at the blood-CSF barrier was gathered in rat (Kratzer et al. 2012), and claudin-5 immunoreactivity was reported in human, but not rodent choroid plexus epithelial cells (Mollgard et al. 2017; Virag et al. 2017). It is noteworthy that claudin-2, a pore-forming protein that enables the paracellular passage of selected inorganic cations and possibly water, is a landmark of the choroidal epithelium. The presence of this claudin, and possibly other pore-forming claudins, explains the low transepithelial electric resistance associated with the blood-CSF barrier, in clear contrast with the high transendothelial electric resistance attributed to the BBB. Claudin-2 may also be somehow involved in the mechanisms of CSF secretion, although this has not been formally demonstrated yet (Ghersi-Egea and Damkier 2017). Finally, the identity of proteins that form adherens junctions also differs between the blood-CSF barrier and the BBB (Tietz and Engelhardt 2015).

8.2.2   Developmental Profile of Tight Junction Proteins at the Blood-CSF Barrier

The blood-brain interfaces of the developing brain have conventionally been considered immature and leaky. Nowadays, it is acknowledged that tight junctions form precociously both in the nascent cerebral microvessels and newly formed choroidal epithelium, in rodent as in human. In the former, the influence of the Wnt/β-catenin pathway and of recruited pericytes is crucial to induce the formation of tight junctions and to inhibit the non-specific endocytosis across endothelial cells. The development and differentiation of the choroidal interface result from interactions between the embryonic neuroepithelium and the underlying primitive mesenchyme, and are specified by secreted signaling molecules including bone morphogenic proteins and sonic hedgehog (reviewed in Strazielle and Ghersi-Egea 2013). The mechanism that induces tight junction formation exclusively between choroidal epithelial cells, and not between neuroepithelial/ependymal cells from which the choroid plexus grows, remains to date unknown. Claudin-1, a landmark protein of the blood-CSF barrier tight junction, is present at cell-cell contacts between choroidal epithelial cells early on during development in rodent, and as early as week 8 of gestation and probably sooner in human (Kratzer et al. 2012) (Fig. 8.2). Claudin-5 has been immunolocalized in the human choroidal epithelium at the same developmental age (Mollgard et al. 2017). Early junctions are fully functional and occlude

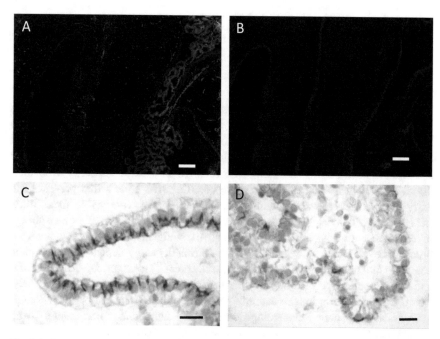

Fig. 8.2 Expression of claudins at the blood-cerebrospinal fluid barrier in rat embryo. (**a**). Claudin-1 immunostaining (in green, honeycomb pattern) continuously localized at the choroidal inter-epithelial junctions in a 19-day-old rat embryo. Claudin-5 (red) is also present at the intercellular junctions of the endothelial cells forming the BBB. (**b**). Negative control by primary antibody omission. (**c**). Claudin-1 immunostaining continuously localized at the inter-epithelial junctions in a lateral ventricle choroid plexus of a human embryo at 8 weeks of gestation. (**d**). Claudin-2 immunostaining displaying only a discontinuous labeling limited to some epithelial cells in a lateral ventricle choroid plexus of a human embryo at 8 weeks of gestation. In Blue: dapi (**a** and **b**) or hematoxylin (**c**, **d**), staining of nuclei. Scale bars: (**a**, **b**): 200 μm, (**c**, **d**): 20 μm

the paracellular pathway, as demonstrated experimentally in rat and opossum fetuses using polar tracers that can be visualized by microscopy (Ek et al. 2006; Liddelow et al. 2013).

There are however, differences in the composition of choroidal epithelial tight junctions between the developing and adult stages. Quantitative expression, Western blot and immunohistochemical data demonstrated that claudin-2 is greatly upregulated from embryonic day 19 to the adult stage in rat, a profile also observed in human choroid plexuses by immunohistochemistry. In contrast, an inverse developmental profile was observed for claudin-3 in rat, being expressed at a higher level during the perinatal period than in the adult (Kratzer et al. 2012). The developmental profile of claudin-2 at the choroidal epithelium, while claudin-3 was already strongly expressed early during development, was later confirmed in mouse (Steinemann et al. 2016). Thus, the peculiar pattern of claudin-2 expression observed during development in choroid plexuses is conserved among species. It may be linked to the possible role of the pore-forming claudin-2 in the secretion of the CSF fluid which increases sharply after birth, but this hypothesis remains to be tested.

8.3   Efflux Transporters in the Developing and Adult Choroid Plexuses

8.3.1   Families and Functions of Efflux Transporters

While tight junctions restrict the permeability of polar compounds across blood-brain interfaces, molecules presenting a more lipophilic nature can diffuse through the lipid bilayers of cell membranes and access the brain parenchyma. This is best illustrated by the easy penetration of alcohol or caffeine into the brain. A large panel of drugs and xenobiotics however do not access the brain to the extent expected from their lipophilicity. The reason is the presence of efflux transporters that prevent drug entry into, or increase drug elimination from the brain. Contrary to most influx transporters, which display a relatively strict substrate selectivity, efflux transporters involved in neuroprotection generally belong to multigenic families of proteins with broad and overlapping substrate specificities. Primary energy-dependent transporters belong to the families of ATP-Binding Cassette (ABC[1]) transporters ABCB, ABCC and ABCG, collectively known for their participation to multidrug resistance. ABC transporters accept various structurally unrelated lipophilic and amphiphilic compounds as substrates, including therapeutic compounds that range from immunosuppressor, antiretroviral, and antitumor drugs, to some antibiotic, antiepileptic, antidepressant and psychotropic agents, and to drug conjugates in the case of ABCCs (references in Strazielle and Ghersi-Egea 2015). Efflux transporters can also be secondary energy-dependent carriers and belong mainly to the solute carrier (SLC) families SLC22 (organic anion and cation transporters) and SLC21/OATP (organic anion transport polypeptides). Certain members of the SLC15 (peptide transporters) and SLC29 (monoamine transporters) families also participate in this barrier function. Substrates transported by these carriers are cationic and anionic compounds including environmental pollutants, nonsteroidal antiinflammatory, antibiotic, nucleoside-based antiviral or antiepileptic agents, as well as peptidomimetic drugs for SLC15 (references in Strazielle and Ghersi-Egea 2015). Some of these ABC and SLC efflux transporters also recognize biologically active endogenous compounds, including steroid hormones and eicosanoids (leukotrienes, prostaglandins), and may play a role in protecting the brain from a potentially deleterious central accumulation of these compounds.

8.3.2   Choroidal Efflux Transporters

In adult, ABCB1, also known as P-glycoprotein, and ABCG2, also known as breast cancer resistant protein BCRP, are landmarks of the BBB across species, including

---

[1]Capitalized characters are used when referring to human transporters or to transporters listed in a general context. Minor characters are used when specifically referring to rodent transporters.

human. Low levels of ABCB1 have been reported by quantitative Western blot in whole choroid plexuses of rats, and the finding was confirmed in human (Gazzin et al. 2008). The choroidal expression of ABCB1 may be attributed in part to the larger penetrating choroidal vessels that maintain a BBB phenotype. The immunohistological studies localizing ABCB1 at the epithelium are controversial. The functional significance of this transporter at the blood-CSF barrier therefore remains elusive. ABCG2 also has been reported in the choroid plexus of several species including human and non-human primate, albeit at lower level than at the BBB. Oddly, its localization appear to be apical at the epithelium, which implies that this carrier would transport its substrates into the CSF. Accordingly, contrary to its function at the BBB, this carrier would not participate to the barrier function of the choroidal epithelium. Indeed, the level of raltegravir, a substrate for ABCG2 which is an integrase inhibitor used in aids treatment, is decreased in the CSF of human individuals with ABCG2 single-nucleotide polymorphisms leading to a reduced expression of the transporter (Tsuchiya et al. 2014). The main ABC transporters active at the blood-CSF barrier belong to the ABCC family and are therefore different from those localized at the BBB. In rodent and in human they include mainly ABCC1 and ABCC4 (Leggas et al. 2004; Gazzin et al. 2008; Zhang et al. 2018). The relative expression of choroidal ABC transporters may be different in other species as one report shows similar protein levels for ABCB1, ABCG2, and ABCC1 in pig (Zhang et al. 2017).

Among SLC efflux transporters identified at the choroid plexus epithelium, SLC22A8/OAT3, Slco1a5/Oatp3, Slc15a2/Pept2, are the most documented in human and/or rodent. Another organic anion transporter, Slco1a4/oatp2 is found at the basolateral membrane of the choroidal epithelium in rat and may also be involved in brain efflux rather than influx processes. In human, two SLCO3A1 variants v1 and v2, of similar and broad substrate specificity, are respectively located at the basolateral and apical membranes of the choroidal epithelium, suggesting that SLCO3A1 may facilitate both influx and efflux transports (reviewed in Strazielle and Ghersi-Egea 2015). More recently the multidrug and toxin extrusion 1 (MATE1/SLC47A1) transporter has been detected in human, but not rat choroid plexus (Uchida et al. 2015). This transporter extrudes various cationic drugs such as metformin in exchange for the cell entry of H^+ (Motohashi and Inui 2013). It may act as an important efflux carrier at the human blood-CSF barrier, providing its localization is basolateral, which has not been confirmed yet. Similarly, SLC29A4, a monoamine transporter accepting cationic drugs and neurotoxins as substrates in addition to biogenic amines, has been detected at the mouse and rat blood-CSF barrier where it may play an important function to protect the CSF from deleterious compounds (Duan and Wang 2013; Usui et al. 2016). Finally, the receptor-mediated transcytotic efflux of polypeptides and proteins was also evidenced at the blood-CSF barrier. This includes the efflux of amyloid beta-peptide, probably involving the low-density lipoprotein receptor-related protein LRP1 which is highly expressed at the choroid plexus epithelium (Crossgrove et al. 2005; Strazielle and Ghersi-Egea 2016; Storck and Pietrzik 2017). The implication of this choroidal receptor in the etiology of Alzheimer's disease is currently investigated. It also includes the efflux of immunoglobulin out of CSF by a mechanism likely involving the Fc neonatal receptor FcRn (Schlachetzki et al. 2002; Strazielle and Ghersi-Egea 2013). This process may decrease the already low cerebral bioavailability of therapeutic immunoglobulins.

8.3.3 Developmental Profile of Efflux Transporters at the Blood-CSF Barrier

Anaysis of data generated from choroid plexuses of laboratory animals and human has highlighted different developmental patterns of expression and function for efflux transporters, which have been reviewed in details previously (Strazielle and Ghersi-Egea 2015). Figure 8.3 summarizes these data. Briefly, among major choroidal efflux transporters, Abccs, Slc15, Slco1a4 carriers are well expressed already during the perinatal period, while Slco1a5 and Slc22a8 increases and decreases, respectively, during development. Overall, the blood-CSF barrier matures early during development with respect to efflux mechanisms, in contrast with the BBB. In the latter, for instance, ABCB1 expression and functional activity is low pre-and postnatally by comparison to the adult stage in various animal species, including non-human primates. This suggests a specific relevance of the blood-CSF barrier for brain protection during development, at a time when the vascularization is still developing and not fully efficient in its neuroprotective function. Of note, the choroidal expression of ABCB1 and ABCG2 carriers, the two main efflux transporters localized at the BBB, is higher in the developing brain than of in the adult brain, suggesting that their functional relevance at the blood-CSF barrier maybe more important during development than in adult. ABCG2 and ABCB1, together with ABCC1, were observed by immunohistochemistry in the choroid plexus epithelium in human fetuses as early as 7 weeks post-conception (Mollgard et al. 2017). Given their unclear membrane localization and transport directionality at this barrier, the nature of their function at the choroid plexus during development remains to be understood. With respect to the developmental profile of transcytotic receptors involved in the efflux of peptides and proteins, FcRn is already expressed before birth in rat, with a predominant expression in the choroid plexuses compared to the BBB during the perinatal period (Strazielle and Ghersi-Egea 2013).

8.4 Detoxification Processes in the Developing and Adult Choroid Plexuses

8.4.1 Choroidal Enzymes Involved in Detoxification

Besides tight junctions and efflux transport mechanisms, enzymatic detoxification contributes to the neuroprotective role of the blood-CSF barrier. Detoxification

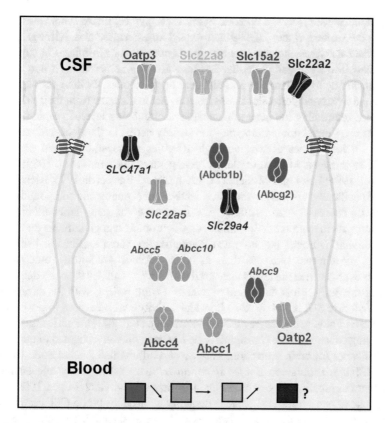

Fig. 8.3 Drug efflux transporters at the blood–cerebrospinal fluid barrier in adult and during development. Both primary ATP-dependent transporters (ABC) and other solute carriers (SLC) identified at the interface are shown. The major transporters are underlined. Transporters whose cell membrane location has not been formally established are named in italic. The expression or function of the transporters between parentheses is debated. Transporters whose expression decreases from the perinatal to the adult stage are noted in blue. Transporters whose expression is stable during development are noted in green. Transporters whose expression increases from the perinatal to the adult stage are noted in orange. Transporters whose developmental profile is unknown appear in grey. Modified from Ghersi-Egea et al. (2018a)

enzymes comprise drug metabolizing enzymes and reactive oxygen species-directed antioxidant enzymes. The former belong to several multigenic families of isoenzymes. They are intracellular enzymes and catalyze either the addition of a functional group to a toxic molecule or drug (phase 1 enzymes), or the conjugation of a polar moiety such as glutathione, sulfate, glucuronic acid to the molecule (phase II enzymes). Phase I and phase II enzymes can act separately or sequentially. The resulting metabolites are usually less toxic than the parent compounds and more hydrophilic, which favors their elimination from the body. Once formed, they are excreted extracellularly by an efflux transporter of the ABCC or OATP families. Typical phase I enzymes include cytochrome P-450 dependent monooxygenases, flavin monooxygenases, and epoxide hydrolases. Typical phase II enzymes include glutathione-S-transferases, UDG-glucuronosyl transferases, and sulfotransferases. Antioxidant enzymes include superoxide dismutases transforming the superoxide free radical into hydrogen peroxide, and also glutathione peroxidases and catalase, which inactivate hydroperoxide and different lipoperoxides. Of note, the superoxide radical and hydrogen peroxide can diffuse and act at distance from their production site, hence reinforcing the need to control their circulating levels in brain fluids.

While drug metabolizing enzymes are mainly active in the liver, pioneer works starting in the eighties discovered high activities of several of them as well as antioxidant enzymes in choroid plexus homogenates (Tayarani et al. 1989; Ghersi-Egea et al. 1994; Leininger-Muller et al. 1994; Ghersi-Egea et al. 1995; Richard et al. 2001). Specifically, epoxide hydrolases inactivating carcinogenic compounds, phase II enzymes decreasing the reactivity of a large range of xeno- and endobiotics by conjugating them to glutathione, sulfate, or glucuronide, and glutathione peroxidases that inactivate hydrogen peroxide and lipoperoxides were described as being more active in the choroid plexuses than in other cerebral structures. Some of these activities even reached the levels measured in the liver, which is the main detoxifying organ of the body. Most data were obtained in adult rodent, with the exception of sulfotransferase SULT1A1 whose choroidal activity was measured in post mortem tissue from human neonates (reviewed in Strazielle et al. 2004). This intriguing high detoxifying capacity of the choroidal tissue has been little investigated to understand its significance for brain protection. The use of a differentiated cellular model of the blood-CSF barrier obtained as primary culture of choroidal epithelial cells enabled to show that conjugation to glucuronic acid (Strazielle and Ghersi-Egea 1999), or to glutathione (Ghersi-Egea et al. 2006) efficiently reduced the blood-CSF permeability of the epithelial monolayer to substrates of these enzymes. These enzyme-based mechanisms confer an apparent metabolic barrier function to the choroidal epithelium.

8.4.2   Developmental Profile of Detoxifying Enzymes at the Blood-CSF Barrier and Functional Relevance for Neonatal Brain Protection

Exposure of the developing brain to toxins, drugs, or deleterious endogenous compounds during the perinatal period can trigger alterations in cell division, migration, differentiation, and synaptogenesis, leading to lifelong neurological impairment. A transcriptomic analysis of drug metabolizing enzymes expressed in the rat choroid plexus showed that a large number of isoforms are expressed at similar or higher levels during prenatal and early postnatal stages of development than in adulthood. This is especially true for the glutathione-S-transferases whose expression levels were the highest during the perinatal period (Kratzer et al. 2013). A combination of enzymatic assays and live tissue fluorescence microscopy (Fig. 8.4) was used to demonstrate that during the early postnatal period the choroid plexus

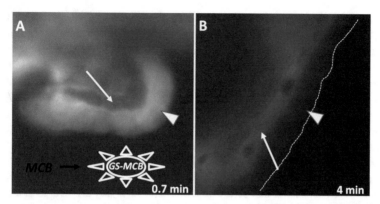

Fig. 8.4 Evidence for the high enzymatic glutathione conjugation capacity of the blood-cerebrospinal fluid barrier. (**a**). GST-mediated conjugation of monochlorobimane, whose glutathione conjugate is fluorescent (white), in freshly isolated live choroid plexuses. Note the intense signal at the choroidal epithelial cells (white arrowhead) a few seconds after exposure to the molecule (2.5 μM). (**b**). Basolateral efflux of the conjugate metabolite. Note the fluorescence shift toward the choroidal vessels (white arrow) a few minute after exposure to the molecule. The dashed line delimits the apical membrane of epithelial cells. Scale bar, 20 μm. Modified from Kratzer et al. (2018)

epithelium and the ependymal cell layer bordering the ventricles harbor a high detoxifying capacity that involves glutathione S-transferases (Kratzer et al. 2018). The spectrum of inactivated molecules is broad and includes drugs, pollutants, and products of oxidative stress. It reflects the multiplicity of glutathione-S-transferase isoenzymes expressed in the choroidal epithelium and the ependyma, and their differential substrate specificities. The use of a rat model functionally knocked down for choroidal glutathione conjugation, enabled to demonstrate in neonates, that this metabolic pathway efficiently prevents the penetration of blood-borne reactive compounds into CSF (Kratzer et al. 2018). The conjugated metabolite formed within the choroidal epithelium was transported back into the blood by efflux systems (Fig. 8.4). These data indicate that the choroid plexuses play an important role in brain protection against noxious compounds at a time of high vulnerability, when the astrocytic network is still immature to protect neuronal cells and liver xenobiotic metabolism is limited.

The conjugating enzyme SULT1A1 catalyzes the sulfation of thyroid hormones as well as of phenolic xenobiotics. Its gene expression level in choroid plexuses is two to three times higher during the perinatal period than in adult in rat (Kratzer et al. 2013). The choroidal protein level and the enzymatic activity are three to twenty times higher in choroid plexuses compared to other brain structures in human embryos of 15–20 weeks of gestation (Richard et al. 2001). This enzyme is therefore likely to reinforce the early detoxification function of the blood-CSF barrier.

The choroid plexuses may also be of special interest in the control of reactive oxygen species circulating within the brain, especially during development. Hydrogen peroxide, when released at a low physiological concentration, is involved in different cell signaling pathways during brain development (Wilson et al. 2018). It can diffuse at distance from its site of production. It is released at supraphysiological concentrations in brain fluids following an inflammatory, hypoxic, or toxic stress. It will then initiate lipid peroxidation, protein and nucleic acid damage unless rapidly inactivated. It may therefore contribute to long-term neurological impairment associated with perinatal diseases (Gitto et al. 2002; Torres-Cuevas et al. 2017). The activity of glutathione peroxidase is much higher in both the lateral and fourth ventricle choroid plexuses than in other brain structures, and the choroid plexus-to-brain activity ratio is maximal a few days after birth in rat. In agreement with this, elevated amounts of glutathione peroxidase 1 and 4 proteins were detected in choroid plexus epithelial cells (Saudrais et al. 2018). Live choroid plexuses isolated from newborn rats were used to show that this tissue is highly efficient in detoxifying H_2O_2 from the bathing fluid mimicking the CSF. Modeling these ex vivo data to in vivo parameters indicated that the lateral ventricle choroid plexus can reduce the half-life of CSF-borne hydrogen peroxide to 10 s. At pathophysiologically relevant hydrogen peroxide concentrations, most of the inactivation proceeds from a glutathione peroxidase activity. Finally, the choroidal cells also form an enzymatic barrier to blood-borne hydroperoxides, preventing them to reach the CSF (Saudrais et al. 2018). Thus, the choroid plexuses seem to play a key function in the control of hydroperoxide levels in the cerebral fluid environment during development, again at a time when the protective glial cell network is still immature.

Although exploration of the detoxifying mechanisms associated with the choroid plexuses remains limited, the data reviewed above argue for an important detoxification function of the blood-CSF barrier during the perinatal period, which ought to compensate for the immaturity of the liver and glial cells at this period of life.

8.5   Conclusion

The neuroprotective functions attributed to the choroid plexuses extend far beyond the presence of tight junctions that seal the paracellular pathway across the choroidal epithelium. They involve diverse efflux transporters and detoxifying enzymes that collectively prevent a large range of potentially noxious compounds from reaching the CSF and brain from the blood circulation, or favor the elimination of active metabolites from the brain.

Our knowledge of the developmental expression profiles of the carriers and enzymes involved in neuroprotection at brain interfaces is currently limited. Lessons learned from tight junction protein analysis and from the few papers on the "biochemical" barrier properties of these interfaces point to an early maturation of the choroid plexus during development, and to developmental stage-specific protective properties of the choroid plexuses that remain to be investigated in details. In adult, the glial network controls the neuronal environment, and the liver controls the overall exposition of the body to toxicants. In the perinatal period, both are immature. Deciphering the protective functions of the blood-CSF barrier at that stage will lead to a clear understanding of the mechanisms that protect the brain during this sensitive period of development. It may also lead to the design of new strategies to improve or restore neuroprotection in the context of perinatal injuries.

Acknowledgments   This work was supported by ANR-10-IBHU-0003 CESAME.

References

Benarroch EE (2016) Choroid plexus--CSF system: recent developments and clinical correlations. Neurology 86:286–296

Crossgrove JS, Li GJ, Zheng W (2005) The choroid plexus removes beta-amyloid from brain cerebrospinal fluid. Exp Biol Med (Maywood) 230:771–776

Demeestere D, Libert C, Vandenbroucke RE (2015) Therapeutic implications of the choroid plexus-cerebrospinal fluid interface in neuropsychiatric disorders. Brain Behav Immun 50:1–13

Drake JM, Sainte-Rose C, DaSilva M, Hirsch JF (1991) Cerebrospinal fluid flow dynamics in children with external ventricular drains. Neurosurgery 28:242–250

Duan H, Wang J (2013) Impaired monoamine and organic cation uptake in choroid plexus in mice with targeted disruption of the plasma membrane monoamine transporter (Slc29a4) gene. J Biol Chem 288:3535–3544

Dziegielewska KM, Ek J, Habgood MD, Saunders NR (2001) Development of the choroid plexus. Microsc Res Tech 52:5–20

Ek CJ, Dziegielewska KM, Stolp H, Saunders NR (2006) Functional effectiveness of the blood-brain barrier to small water-soluble molecules in developing and adult opossum (*Monodelphis domestica*). J Comp Neurol 496:13–26

Gazzin S, Strazielle N, Schmitt C, Fevre-Montange M, Ostrow JD, Tiribelli C, Ghersi-Egea JF (2008) Differential expression of the multidrug resistance-related proteins ABCb1 and ABCc1 between blood-brain interfaces. J Comp Neurol 510:497–507

Ghersi-Egea JF, Damkier HH (2017) Blood–brain interfaces organization in relation to inorganic ion transport, CSF secretion, and circulation. In: Badaut J, Plesnila N (eds) Brain edema. Academic, Oxford, pp 29–48

Ghersi-Egea JF, Leninger-Muller B, Suleman G, Siest G, Minn A (1994) Localization of drug-metabolizing enzyme activities to blood-brain interfaces and circumventricular organs. J Neurochem 62:1089–1096

Ghersi-Egea JF, Leininger-Muller B, Cecchelli R, Fenstermacher JD (1995) Blood-brain interfaces: relevance to cerebral drug metabolism. Toxicol Lett 82-83:645–653

Ghersi-Egea JF, Strazielle N, Murat A, Jouvet A, Buenerd A, Belin MF (2006) Brain protection at the blood-cerebrospinal fluid interface involves a glutathione-dependent metabolic barrier mechanism. J Cereb Blood Flow Metab 26:1165–1175

Ghersi-Egea JF, Saudrais E, Strazielle N (2018a) Barriers to drug distribution into the perinatal and postnatal brain. Pharm Res 35:84

Ghersi-Egea JF, Strazielle N, Catala M, Silva-Vargas V, Doetsch F, Engelhardt B (2018b) Molecular anatomy and functions of the choroidal blood-cerebrospinal fluid barrier in health and disease. Acta Neuropathol 135:337–361

Gitto E, Reiter RJ, Karbownik M, Tan DX, Gitto P, Barberi S, Barberi I (2002) Causes of oxidative stress in the pre- and perinatal period. Biol Neonate 81:146–157

Johanson CE, Duncan JA 3rd, Klinge PM, Brinker T, Stopa EG, Silverberg GD (2008) Multiplicity of cerebrospinal fluid functions: new challenges in health and disease. Cerebrospinal Fluid Res 5:10

Kaur C, Rathnasamy G, Ling EA (2016) The choroid plexus in healthy and diseased brain. J Neuropathol Exp Neurol 75:198–213

Kratzer I, Vasiljevic A, Rey C, Fevre-Montange M, Saunders N, Strazielle N, Ghersi-Egea JF (2012) Complexity and developmental changes in the expression pattern of claudins at the blood-CSF barrier. Histochem Cell Biol 138:861–879

Kratzer I, Liddelow SA, Saunders NR, Dziegielewska KM, Strazielle N, Ghersi-Egea JF (2013) Developmental changes in the transcriptome of the rat choroid plexus in relation to neuroprotection. Fluids Barriers CNS 10:25

Kratzer I, Strazielle N, Saudrais E, Monkkonen K, Malleval C, Blondel S, Ghersi-Egea JF (2018) Glutathione conjugation at the blood-CSF barrier efficiently prevents exposure of the developing brain fluid environment to blood-borne reactive electrophilic substances. J Neurosci 38:3466–3479

Lauer AN, Tenenbaum T, Schroten H, Schwerk C (2018) The diverse cellular responses of the choroid plexus during infection of the central nervous system. Am J Physiol Cell Physiol 314: C152–C165

Leggas M, Adachi M, Scheffer GL, Sun D, Wielinga P, Du G, Mercer KE, Zhuang Y, Panetta JC, Johnston B, Scheper RJ, Stewart CF, Schuetz JD (2004) Mrp4 confers resistance to topotecan and protects the brain from chemotherapy. Mol Cell Biol 24:7612–7621

Leininger-Muller B, Ghersi-Egea JF, Siest G, Minn A (1994) Induction and immunological characterization of the uridine diphosphate-glucuronosyltransferase conjugating 1-naphthol in the rat choroid plexus. Neurosci Lett 175:37–40

Liddelow SA, Dziegielewska KM, Ek CJ, Habgood MD, Bauer H, Bauer HC, Lindsay H, Wakefield MJ, Strazielle N, Kratzer I, Mollgard K, Ghersi-Egea JF, Saunders NR (2013) Mechanisms that determine the internal environment of the developing brain: a transcriptomic, functional and ultrastructural approach. PLoS One 8:e65629

Lun MP, Monuki ES, Lehtinen MK (2015) Development and functions of the choroid plexus-cerebrospinal fluid system. Nat Rev Neurosci 16:445–457

Marques F, Sousa JC, Brito MA, Pahnke J, Santos C, Correia-Neves M, Palha JA (2017) The choroid plexus in health and in disease: dialogues into and out of the brain. Neurobiol Dis 107:32–40

Mollgard K, Dziegielewska KM, Holst CB, Habgood MD, Saunders NR (2017) Brain barriers and functional interfaces with sequential appearance of ABC efflux transporters during human development. Sci Rep 7:11603

Motohashi H, Inui K (2013) Multidrug and toxin extrusion family SLC47: physiological, pharmacokinetic and toxicokinetic importance of MATE1 and MATE2-K. Mol Asp Med 34:661–668

Praetorius J, Damkier HH (2017) Transport across the choroid plexus epithelium. Am J Physiol Cell Physiol 312:C673–C686

Prochiantz A, Fuchs J, Di Nardo AA (2014) Postnatal signalling with homeoprotein transcription factors. Philos Trans R Soc Lond Ser B Biol Sci 369(1652):20130518. https://doi.org/10.1098/rstb.2013.0518

Richard K, Hume R, Kaptein E, Stanley EL, Visser TJ, Coughtrie MW (2001) Sulfation of thyroid hormone and dopamine during human development: ontogeny of phenol sulfotransferases and arylsulfatase in liver, lung, and brain. J Clin Endocrinol Metab 86:2734–2742

Saudrais E, Strazielle N, Ghersi-Egea JF (2018) Choroid plexus glutathione peroxidases are instrumental in protecting the brain fluid environment from hydroperoxides during postnatal development. Am J Physiol Cell Physiol 315:C445–C456

Saunders NR, Dziegielewska KM, Mollgard K, Habgood MD, Wakefield MJ, Lindsay H, Stratzielle N, Ghersi-Egea JF, Liddelow SA (2015) Influx mechanisms in the embryonic and adult rat choroid plexus: a transcriptome study. Front Neurosci 9:123

Schlachetzki F, Zhu C, Pardridge WM (2002) Expression of the neonatal fc receptor (FcRn) at the blood-brain barrier. J Neurochem 81:203–206

Spector R, Keep RF, Robert Snodgrass S, Smith QR, Johanson CE (2015) A balanced view of choroid plexus structure and function: focus on adult humans. Exp Neurol 267:78–86

Steinemann A, Galm I, Chip S, Nitsch C, Maly IP (2016) Claudin-1, -2 and -3 are selectively expressed in the epithelia of the choroid plexus of the mouse from early development and into adulthood while Claudin-5 is restricted to endothelial cells. Front Neuroanat 10:16

Storck SE, Pietrzik CU (2017) Endothelial LRP1 - a potential target for the treatment of Alzheimer's disease : theme: drug discovery, development and delivery in Alzheimer's disease guest editor: Davide Brambilla. Pharm Res 34:2637–2651

Strazielle N, Ghersi-Egea JF (1999) Demonstration of a coupled metabolism-efflux process at the choroid plexus as a mechanism of brain protection toward xenobiotics. J Neurosci 19:6275–6289

Strazielle N, Ghersi-Egea JF (2013) Physiology of blood-brain interfaces in relation to brain disposition of small compounds and macromolecules. Mol Pharm 10:1473–1491

Strazielle N, Ghersi-Egea JF (2015) Efflux transporters in blood-brain interfaces of the developing brain. Front Neurosci 9:21

Strazielle N, Ghersi-Egea JF (2016) Potential pathways for CNS drug delivery across the blood-cerebrospinal fluid barrier. Curr Pharm Des 22:5463–5476

Strazielle N, Khuth ST, Ghersi-Egea JF (2004) Detoxification systems, passive and specific transport for drugs at the blood-CSF barrier in normal and pathological situations. Adv Drug Deliv Rev 56:1717–1740

Tayarani I, Cloez I, Clement M, Bourre JM (1989) Antioxidant enzymes and related trace elements in aging brain capillaries and choroid plexus. J Neurochem 53:817–824

Tietz S, Engelhardt B (2015) Brain barriers: crosstalk between complex tight junctions and adherens junctions. J Cell Biol 209:493–506

Torres-Cuevas I, Parra-Llorca A, Sanchez-Illana A, Nunez-Ramiro A, Kuligowski J, Chafer-Pericas C, Cernada M, Escobar J, Vento M (2017) Oxygen and oxidative stress in the perinatal period. Redox Biol 12:674–681

Tsuchiya K, Hayashida T, Hamada A, Kato S, Oka S, Gatanaga H (2014) Low raltegravir concentration in cerebrospinal fluid in patients with ABCG2 genetic variants. J Acquir Immune Defic Syndr 66:484–486

Uchida Y, Zhang Z, Tachikawa M, Terasaki T (2015) Quantitative targeted absolute proteomics of rat blood-cerebrospinal fluid barrier transporters: comparison with a human specimen. J Neurochem 134:1104–1115

Usui T, Nakazawa A, Okura T, Deguchi Y, Akanuma SI, Kubo Y, Hosoya KI (2016) Histamine elimination from the cerebrospinal fluid across the blood-cerebrospinal fluid barrier: involvement of plasma membrane monoamine transporter (PMAT/SLC29A4). J Neurochem 139:408–418

Virag J, Haberler C, Baksa G, Piurko V, Hegedus Z, Reiniger L, Balint K, Chocholous M, Kiss A, Lotz G, Glasz T, Schaff Z, Garami M, Hegedus B (2017) Region specific differences of claudin-5 expression in pediatric intracranial ependymomas: potential prognostic role in Supratentorial cases. Pathol Oncol Res 23:245–252

Wilson C, Munoz-Palma E, Gonzalez-Billault C (2018) From birth to death: a role for reactive oxygen species in neuronal development. Semin Cell Dev Biol 80:43–49

Yasuda T, Tomita T, McLone DG, Donovan M (2002) Measurement of cerebrospinal fluid output through external ventricular drainage in one hundred infants and children: correlation with cerebrospinal fluid production. Pediatr Neurosurg 36:22–28

Zhang Z, Uchida Y, Hirano S, Ando D, Kubo Y, Auriola S, Akanuma SI, Hosoya KI, Urtti A, Terasaki T, Tachikawa M (2017) Inner blood-retinal barrier dominantly expresses breast cancer resistance protein: comparative quantitative targeted absolute proteomics study of CNS barriers in pig. Mol Pharm 14:3729–3738

Zhang Z, Tachikawa M, Uchida Y, Terasaki T (2018) Drug clearance from cerebrospinal fluid mediated by organic anion transporters 1 (Slc22a6) and 3 (Slc22a8) at arachnoid membrane of rats. Mol Pharm 15:911–922

Chapter 9
Roles of the Choroid Plexus in Aging

Caroline Van Cauwenberghe, Nina Gorlé,
and Roosmarijn E. Vandenbroucke

Abstract The choroid plexus comprises of a monolayer of tightly connected epithelial cells that form an important physical, enzymatic, and immunologic barrier, called the blood–cerebrospinal fluid (CSF) barrier. It is a highly vascularized structure located in the brain ventricles and plays a key role in maintaining brain homeostasis by producing CSF.

During aging, the morphology and normal function of the choroid plexus is compromised. Different alterations of the choroid plexus have been reported such as atrophy of the choroid plexus epithelial cells, decreased CSF production and secretion, decreased CSF clearance and absorption resulting in reduced clearance of toxic compounds, reduced enzymatic and metabolic activity, loss of barrier integrity, and insufficient distribution of nutrients. The described degeneration of the structure and function of the choroid plexus can result in multiple brain deficits and contribute to cognitive deterioration. In fact, these alterations of the choroid plexus are even more prominent in age-related neurodegenerative diseases including late-onset Alzheimer's disease. A better understanding of the alterations in structure, activity, and function of the choroid plexus epithelial cells during aging and how the choroid plexus is implicated in aging and age-associated neurological diseases might reveal novel strategies to combat age-related cognitive decline and age-related neurological disorders.

9.1 Introduction

Aging is a complex, multifactorial process influenced by many unknown genetic and environmental factors. It is associated with progressive decline in normal cell and organ functioning. An important hallmark of aging is 'inflamm-aging', a state of

C. Van Cauwenberghe · N. Gorlé · R. E. Vandenbroucke (✉)
VIB Center for Inflammation Research, Ghent, Belgium

Barriers in Inflammation Lab, Department of Biomedical Molecular Biology, Ghent University, Ghent, Belgium
e-mail: Roosmarijn.Vandenbroucke@irc.vib-UGent.be

© The American Physiological Society 2020
J. Praetorius et al. (eds.), *Role of the Choroid Plexus in Health and Disease*,
Physiology in Health and Disease, https://doi.org/10.1007/978-1-0716-0536-3_9

chronic, low-grade inflammation, caused by an elevated concentration of inflammatory markers in the circulation (Calder et al. 2017). The balance between pro- and anti-inflammatory cytokines in the healthy adult brain is shifted with aging towards a pro-inflammatory state (Franceschi 2007). This immunological fragile state makes the aged brain more susceptible to diseases, infection, and stress, which might even influence the onset of age-related neurodegenerative brain diseases (Franceschi and Campisi 2014; Gorle et al. 2016).

With life expectancy exponentially increasing, age-related diseases will become an emerging epidemic and a tremendous public health issue due to the high costs of dementia care. Numbers are predicted to increase to 152 million in 2050 and there are over 9.9 million new cases of dementia each year worldwide (Report 2018), in addition no treatment to reverse or halt disease progression exists. Increasing evidence indicates that degeneration of the choroid plexus can result in brain deficits and contribute to cognitive impairment. Therefore, extensive insights in the aging choroid plexus are essential in understanding age-associated neurodegenerative and neuroinflammatory disorders and pave new ways for therapy.

The role of the choroid plexus in health and disease is being increasingly recognized and it has been reported to play a central role during aging (Gorle et al. 2016; Vandenbroucke 2016; Marques et al. 2017). The choroid plexus is a highly vascularized brain structure, consisting of a monolayer of choroid plexus epithelial cells firmly interconnected by tight junctions (De Bock et al. 2014), that form one of the brain barriers, called the blood-cerebrospinal fluid (CSF) barrier. Together with other brain barriers, the blood-CSF barrier assures a balanced and well-controlled micro-environment in the central nervous system (CNS), providing protection against external insults such as toxins, infectious agents, and peripheral blood fluctuations (Gorle et al. 2016). The choroid plexus produces CSF and receives input from both circulatory, autonomic and immune system. It can respond as a key regulator to local changes of different physiological signals by changing its secretome, including proteins (Marques and Sousa 2015; Silva-Vargas et al. 2016) and extracellular vesicles (EVs) (Balusu et al. 2016b). The normal functioning of the choroid plexus is severely affected during aging and this dysfunction is even aggravated in age-related neurodegenerative brain diseases like Alzheimer's disease (Balusu et al. 2016a). Understanding how the function and activity of the choroid plexus is altered in aging might lead to the identification of strategies to attenuate aging-associated cognitive decline and related diseases (Baruch et al. 2014; Vandenbroucke 2016; Gorle et al. 2016).

9.2 Morphological Changes of the Choroid Plexus Epithelium Upon Aging

Several reports have been published describing the morphological alterations of the choroid plexus epithelium upon aging (Fig. 9.1), which is comparable to other secretory epithelia. Across species epithelial atrophy and weight increase have

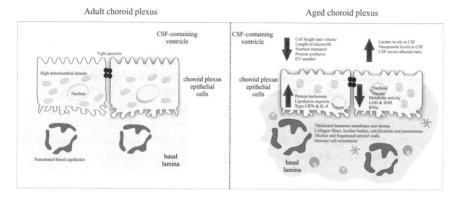

Fig. 9.1 Schematic representation of the changes at the choroid plexus during aging. Several morphological changes are observed at the choroid plexus: the choroid plexus epithelial cells are flattened with an irregular nucleus and shortened microvilli, more Biondi rings and lipofuscin are present, and the basement membrane is thickened and contains fragmented vessels, collagen fibers, hyaline bodies, calcifications, and psammomas. Functionally multiple alterations were shown: increased cerebrospinal fluid (CSF)/serum albumin ratio, reduced metabolic activity, decreased extracellular vesicles (EVs) in the CSF, increased levels of lactate and vasopressin, and altered immune cell recruitment (linked with changes in the interferon (IFN) balance). Key: CSF: cerebrospinal fluid, EV: extracellular vesicles, IFN: interferon, IL: interleukin, LDH: lactate dehydrogenase, SDH: succinate dehydrogenase

been observed (Wen et al. 1999), however slightly different modifications have been described according to species.

In humans, the height of the epithelial cells decreases with approximately 11% during life and the cells become more flattened (Serot et al. 2000). The aged cell cytoplasm contains protein inclusions called Biondi ring tangles. In addition, the presence of lipofuscin deposits can be found. Since this age pigment is a product of lipid peroxidation by free oxygen radicals it will probably alter the cell functioning (ZS-Nagy et al. 1995). The nuclei become more irregular in elderly and have a flattened shape (Serot et al. 2000, 2001). Moreover, the epithelial basement membrane has been reported to become thicker with aging. Also the stroma of the aged choroid plexus is thicker and contains collagen fibers, hyaline bodies, calcifications, and psammomas (i.e. dystrophic calcifications) (Eriksson and Westermark 1986; Jovanovic et al. 2004; Sturrock 1988; Wen et al. 1999). An age-associated increase in size and volume density of the psammoma bodies has been described (Zivkovic et al. 2017). The arterial walls become thicker, especially the media and adventitia, while the blood vessel volume density decreases and elastic fibers are fragmented (Serot et al. 2000; Shuangshoti and Netsky 1970; Zivkovic et al. 2017), resulting in a reduced contact area between the blood and the choroid plexus epithelium.

Rodent models show similar epithelial disruptions of the choroid plexus epithelium compared to humans (Serot et al. 2001; Sturrock 1988). In elderly rats the epithelial cells lose height, approximately 15%, and become more flattened. The cells show an irregular, elongated nucleus and shortened microvilli, causing a decrease in the choroid plexus epithelium-CSF contact area. Lipid vacuoles are present in the cytoplasm of the choroidal epithelial cells. Irregular fibrosis has been described in the stroma of elderly rats together with thickening of the basement membranes (Serot et al. 2001; Sturrock 1988).

Age-associated reduction in contact area between blood-choroid plexus epithelium and choroid plexus epithelium-CSF due to morphological alterations, together with the changes in choroidal proteins involved in CSF production (Masseguin et al. 2005), negatively influence the CSF production (Vandenbroucke 2016). These morphological changes will result in functional alterations, which may consequently have an impact on brain homeostasis.

9.3   Functional Alterations of the Choroid Plexus in Aging

9.3.1   CSF Dynamics

9.3.1.1   CSF Production and Secretion

One of the major functions of the choroid plexus is CSF production and secretion. CSF flows from the choroid plexus through the ventricular system to the subarachnoid space and continues to the spinal column. The classic theory suggests that CSF flow is pulsatile and generated by cardiac pulsations and pulmonary respiration (Khasawneh et al. 2018; Sakka et al. 2011). CSF not only provides mechanical support to the brain (Segal 2000) but also helps to remove toxic catabolites of the brain metabolism (Brown et al. 2004). Furthermore, CSF can be considered as a route of communication within the brain as it carries hormones, growth factors, and neurotransmitters between different areas of the brain (Kaur et al. 2016; Marques et al. 2011; Silva-Vargas et al. 2016; Preston 2001; Brown et al. 2004; Strazielle and Ghersi-Egea 2000).

The adult brain contains a constant volume of 150 ml CSF, of which 25 ml in the brain ventricles and 125 ml in the subarachnoid compartments. The total CSF production is about 500 ml per day in healthy individuals at a rate of about 0.3–0.4 ml per minute and is completely replaced about four times a day (Brown et al. 2004; Khasawneh et al. 2018). CSF is for 99% composed of water with the remaining 1% accounted for by proteins, ions, neurotransmitters, and glucose. Ion concentrations of Na^+, Cl^-, and Mg^{2+} are higher in CSF than the levels in plasma, while K^+ and Ca^{2+} concentrations are lower (Bulat and Klarica 2011; Sakka et al. 2011). The majority of the total CSF volume (60–90%) is being produced and secreted by the choroid plexus epithelium, the remaining CSF originates from the brain interstitial fluid, ependyma, and cerebral capillaries (Redzic and Segal 2004; Sakka et al. 2011). CSF secretion by the choroid plexus is dependent on active translocation of ions and water from the basolateral membrane to the cytoplasm and subsequently across the apical membrane into the brain ventricles (Brown et al. 2004). Transport of Na^+, Cl^-, K^+, and HCO_3^- takes place via different transporters

Table 9.1 Changes in CSF secretion

Method	Change	Species	Reference
Radiotracer dilution	↓	Human	Cutler et al. (1968), May et al. (1990)
MRI	↔	Human	Barkhof et al. (1994), Gideon et al. (1994)
	↓	Human	Stoquart-ElSankari et al. (2007)
Ventriculo-cisternal perfusion	↓	Rat	Preston (2001)
In-situ perfusion	↓	Sheep	Chen et al. (2009)

Key: MRI: magnetic resonance imaging; CSF: cerebrospinal fluid

present on the choroid plexus epithelium. Na^+ and Cl^- are transported into the epithelial cells by Cl^-/HCO_3^- and Na^+ linked Cl^-/HCO_3^- transporters present on the basolateral surface (Lindsey et al. 1990). Translocation of these ions creates an osmotic gradient which drives water transport facilitated by aquaporins (AQP) on the epithelial surface (Liddelow 2015). At the apical surface, the Na^+, K^+-ATPase plays an important role creating an osmotic gradient which facilitates transfer of various molecules in and out the choroid plexus epithelium (Pershing and Johanson 1982; Plotkin et al. 1997; Redzic and Segal 2004; Speake et al. 2001; Johanson et al. 2008). Besides the Na^+, K^+-ATPase transporter, also the electrogenic sodium-bicarbonate cotransporter (NBCe2) facilitates Na^+ ion transport into the CSF. Interestingly, a knockout of this NBCe2 cotransporter resulted in significant remodeling of choroid plexus epithelium including abnormal mitochondrial distribution, cytoskeletal protein expression, CSF electrolyte imbalance, and neurological impairment (Kao et al. 2011), reflecting the importance of cotransporter in the normal physiology of the nervous system (Christensen et al. 2018).

In elderly, the CSF production is reduced as shown in multiple studies in human, rat, and sheep (Table 9.1). The reduced expression of choroidal proteins involved in CSF secretion such as carbonic anhydrase II and AQP1 have been described in aging rat models and sheep. In addition, Na^+, K^+-ATPase mRNA levels decrease with age (Chen et al. 2009; Kvitnitskaia-Ryzhova and Shkapenko 1992; Masseguin et al. 2005). Next to a decreased CSF production rate, the mean CSF pressure declines steadily after the age of 50. In comparison to a 20–49 year old group, the 50–54 age group showed a reduction of 2.5% and this even increased to 13% in individuals older than 70 years of age (Fleischman et al. 2012).

9.3.1.2 CSF Absorption

The site of CSF absorption is still a point of discussion in the research field. For decades it was believed that CSF returns to the venous blood in the brain sinuses through the arachnoid villi and granulations (Kida et al. 1988). These arachnoid granulations are projections of the arachnoid membrane into the dural venous sinuses. The driving force of the absorption of CSF into the venous bloodstream is a difference in fluid pressure between the subarachnoid space and the venous system. As such, fluid is driven out of the granulations into the circulation (Damkier et al.

2013). In addition, CSF absorption sites have been identified on meningeal recesses of spinal and cranial nerve roots, particularly the trigeminal and cochlear nerve (Sakka et al. 2011). Recently, dynamic imaging suggested that lymphatic outflow might be the major outflow route for CSF. Using non-invasive imaging techniques, the authors were able to demonstrate that tracers added to the CSF rapidly reach the lymph nodes using perineural routes through the foramina of the skull to finally reach the peripheral blood (Ma et al. 2017; Proulx et al. 2017). Interestingly, this lymphatic outflow system showed significant decline in aged mice (Ma et al. 2017).

9.3.1.3   CSF Turnover and Circulation

Moderate brain tissue atrophy that occurs during healthy aging, leading to an increase in total cranial CSF compartment and volume, affects the turnover or replacement time of CSF (Table 9.2) (Preston 2001). The decreased production and secretion of CSF together with increased CSF volume results in a longer CSF turnover with age. These observations have been confirmed by reduced clearance of radio-iodinated human serum albumin from the brain in individuals around 62 years of age (Henriksson and Voigt 1976). Similarly, reduced clearance of ^{3}H-polyethylene glycol and 125I-Amyloid beta (Aβ) (1–40) was observed in older rats (Preston 2001). Cross-sectional studies in healthy humans show a doubling of CSF volume between the age of 30 and 70 years (Foundas et al. 1998; Matsumae et al. 1996a). In elderly humans, CSF turnover is reduced to two times daily in comparison to three to four times in young adults (Chiu et al. 2012; Johanson et al. 2008). In humans, additional factors have been described that contribute to the reduced CSF turnover, including increased resistance for CSF drainage by fibrosis present in the arachnoid membranes and an increase in central venous pressure, both noted to be increased in normal aging (Preston 2001; Bellur et al. 1980; Rubenstein 1998).

The diminished CSF production, secretion, and reduced CSF turnover rate might have serious complications. In senescence, alterations in CSF composition due to reduced turnover could bring about inadequate distribution of nutritive components

Table 9.2 CSF volume, turnover and clearance

Observation	Change	Species	Reference
CSF volume	↑	Human	Foundas et al. (1998), Matsumae et al. (1996b), Silverberg et al. (2001, 2003), Wahlund et al. (1996)
	↑	Rat	Preston (2001)
Resistance to CSF drainage	↑	Human	Albeck et al. (1998)
CSF turnover	↓	Rat	Preston (2001)
	↓	Human	Rubenstein (1998), Silverberg et al. (2003)
Albumin clearance	↓	Human	Henriksson and Voigt (1976)
Aβ clearance	↓	Rats	Preston (2001)

Key: CSF: cerebrospinal fluid; Aβ: beta-amyloid

and trophic factors, and the diminished CSF clearance leads to accumulation of toxic compounds and waste products from the brain (Marques et al. 2017; Preston 2001). Both will result in increased cellular stress and changes in the cerebral metabolism and blood flow, disrupting cognitive and motor functions (Rubenstein 1998), eventually influencing age-related cognitive decline and development of age-associated neurological diseases (Emerich et al. 2005). In addition, adult neural stem cells contact the CSF in the ventricular-subventricular stem cell niche of which the lateral choroid plexus is an important component. This implicates that secreted factors and toxic compounds accumulated due to diminished CSF clearance can impact the neural stem cells, which are especially sensitive to age-related changes (Silva-Vargas et al. 2016).

CSF production, clearance rate, and CSF flow are altered during aging, thereby affecting brain homeostasis. Interestingly, a highly organized pattern of ependymal cilia is responsible for the transport of CSF in the ventricles of the mouse brain. Coordinated cilia beating patterns collectively give rise to a network of fluid flows that allow for precise CSF directional flow, which may control substance distribution in the ventricle. A cilia-based switch was discovered that reliably and periodically alters the flow pattern and may control substance distribution in the ventricle (Faubel et al. 2016). However, it remains to be determined whether changes in beating patterns occur in aging and whether this affects the distribution of components throughout the brain. Peak CSF volume flow is altered in aging and higher aqueductal peak CSF flow velocities were described in elderly healthy volunteers (Gideon et al. 1994). In addition, the total cerebral blood flow decreases with aging and consequently CSF stroke volumes (i.e. mean volume of CSF passing through the aqueduct during both systole and diastole) and pulsations were significantly reduced in elderly (Stoquart-Elsankari et al. 2007).

9.3.1.4 Choroid Plexus Biochemistry

Choroid plexus functioning and metabolism are largely energy-dependent. All the homeostatic and secretory functions of the choroid plexus are linked to energy dependent mechanisms, explaining the huge number of mitochondria in the choroid plexus epithelial cells. Morphometric studies in different model organisms indicate that mitochondria constitute 10–14% of the choroid plexus cytoplasm (Cornford et al. 1997). In addition, proteome analysis in rats identified a total of 1400 proteins in the choroid plexus of which a high percentage (33.5%) are mapped to metabolism, e.g. several enzymes like hydrolases, oxidoreductases, and transferases. The presence of a substantial number of mitochondrial proteins in the proteome analyses suggests a high mitochondrial density (Sathyanesan et al. 2012). Aging however leads to a reduced metabolic activity of the choroid plexus epithelial cells, as demonstrated by in vitro choroid plexus cultures (Emerich et al. 2007). Additionally, the number of epithelial cells deficient in cytochrome C oxidase has been shown to be increased with age (Cottrell et al. 2001).

The mammalian brain depends on glucose as main energy source and a continuous supply is essential to sustain neural activity (Siesjo 1978; Simpson et al. 2007). Glucose provides energy for physiological brain functioning (biosynthesis of neurotransmitters, maintenance of action potentials, information processing) by oxidative metabolism and tight regulation of the glucose metabolism is necessary (Mergenthaler et al. 2013). Glucose transporter proteins transfer glucose from the blood circulation to the brain. The blood-CSF barrier expresses GLUT1 (Redzic 2011; Serot et al. 2003; Simpson et al. 2007). Disruption of the glucose metabolism and the pathways involved in glucose delivery can have pathophysiological consequences and lead to brain diseases. There is compelling evidence that the aging tissue is unable to maintain appropriate energy output. During aging, the expression of enzymes necessary for anaerobic respiration and oxidative phosphorylation, such as lactate dehydrogenase (LDH) and succinate-dehydrogenase (SDH), are diminished and consequently energy production in choroid plexus epithelial cells decreases (Fig. 9.1) (Emerich et al. 2005; Ferrante and Amenta 1987; Gorle et al. 2016). Both LDH and SDH play a key role in glucose metabolism and show a major reduction with age, respectively 9 and 26% (Ferrante and Amenta 1987; Preston 2001). Impairment of glucose dependent energy transduction mechanisms may influence the functional activity of the choroid plexus epithelial cells (Ferrante and Amenta 1987). In addition, in humans CSF levels of lactate increase with age (Fig. 9.1). Since CSF lactate and brain lactate concentration correlate closely, this might suggest a decline in the efficiency of glucose metabolism in brain tissue (Yesavage et al. 1982).

Different imaging methods have been developed to allow non-invasive brain measurements. Functional magnetic resonance spectroscopy (fMRS) is used for the measurement of metabolite concentrations in the human brain (Jahng et al. 2016). In addition, alterations of the glucose metabolism in the choroid plexus can be visualized and measured in vivo with dynamic fluorodeoxyglucose positron emission tomography (dynamic ^{18}F-FDG-PET). By using this technique, the dynamic uptake of FDG in the choroid plexus and CSF can be measured over time. A recent study showed the presence of decreased glucose metabolism in Alzheimer's disease patients. Conversely, dynamic uptake was higher in CSF for Alzheimer's disease patients. The activity of the choroid plexus gradually decreases in patients with cognitive decline. This results in the disturbance of the glucose exchange at the blood-CSF barrier and alters the CSF-choroid plexus glucose equilibrium (Daouk et al. 2016).

9.3.1.5   Iron Metabolism

Iron is an essential element for different metabolic processes, tissue homeostasis, and brain functioning. However, in excessive amounts, iron becomes toxic for cells. Therefore, the iron delivery in the brain is strictly regulated through receptor mediated endocytosis of iron-bound transferrin by the blood-CSF barrier and the blood-brain barrier (Morris et al. 1992; Deane et al. 2004; Rouault et al. 2009;

Hubert et al. 2019). During aging, decreased metabolic activity, increased oxidative stress, impaired barrier functioning, impaired protein secretion, and diminution of the CSF flow might all affect the iron metabolism and iron-mediated toxicity (Marques et al. 2009, 2007; Chen et al. 2012b). Moreover, pro-inflammatory cytokines like IL-6, which are increased in the blood with age, can influence the secretion of hepcidin by choroid plexus epithelial cells. Hepcidin is a central regulator of iron homeostasis and secretion is influenced through the Stat3 signal transduction pathway (Chongbin et al. 2014; Chen et al. 2008; Villeda et al. 2011; Rouault et al. 2009; Hubert et al. 2019; Leitner and Connor 2012; Lu et al. 1995).

9.3.2 Growth Factors and Hormones Secreted by the Choroid Plexus

The choroid plexus is uniquely located at the interface between blood and CSF. It expresses many receptors for growth factors and hormones, such as growth hormone (GH), prolactin, corticotrophin-releasing hormone, vasopressin, and leptin, in order to respond to local and peripheral signals (Kaur et al. 2016; Marques et al. 2011; Silva-Vargas et al. 2016). In this way, the choroid plexus is a key component in neuroendocrine regulation, having an impact on hormonal signaling and in addition, also the choroid plexus functioning is regulated by a variety of hormones (Preston 2001). Several studies revealed that different hormones and neuropeptides might be actively processed by the choroid plexus, namely GH, nerve growth factor (NGF), brain-derived neurotrophic factor (BDNF), vascular endothelial growth factor (VEGF), insulin-like growth factor (IGF1 and 2), and insulin-like growth factor binding protein 2 (IGFBP2) (Emerich et al. 2005; Holm et al. 1994; Nilsson et al. 1996; Vega et al. 1992). IGF2 plays an important role in cell growth, in development, and maintenance of the nervous system and regulates the functional plasticity of the adult brain (Lenoir and Honegger 1983; Mill et al. 1985; Mozell and Mcmorris 1991). The production of IGF partially depends on the presence of GH (Cohen et al. 1992). In elderly humans, reduced binding of GH to the choroid plexus has been reported, resulting in reduced activity of IGF2, which might have an impact on epithelial cell growth and repair (Nilsson et al. 1992; Preston 2001). Next to changes in IGF, in vitro studies showed that VEGF secretion in aged choroid plexus epithelial cells was reduced compared to young epithelial cells (Emerich et al. 2007).

Interestingly, the growth factors secreted by the choroid plexus into the CSF may be involved in the proliferation, differentiation, and survival of neural progenitor cells in the subventricular zone (Falcao et al. 2012; Lun et al. 2015). More recently, it was shown that the lateral ventricle choroid plexus affects the behavior of neural stem cells of the ventricular-subventricular zone by the secretion of several factors promoting for colony formation and proliferation (Silva-Vargas et al. 2016). The functional effect of the lateral ventricle secretome changes throughout life, with activated neural stem cells being especially sensitive to age-related changes. The lateral ventricle choroid plexus is an important compartment that contributes to the age-related changes of the ventricular-subventricular zone stem cells. Transcriptome analysis revealed two proteins, BMP5 and IGF1, that might play an important role in these age-dependent effects of the choroid plexus (Silva-Vargas et al. 2016). The expression of both proteins decreases with aging, BMP5 levels are lower in aged human CSF and systemic IGF levels decrease with aging (Baird et al. 2012; Bartke et al. 2013). Furthermore, a study revealed that implants of young choroid plexus in rats were potently neuroprotective, whereas the choroid plexus implants from aged rats were only modestly effective and less potent. This study links aging with a diminished neuroprotective capacity of the choroid plexus epithelial cells (Emerich et al. 2007).

Vasopressin is a neurohormone produced by the hypothalamus, involved in the regulation of blood pressure. A high density of Vasopressin receptors (V1) is present at the choroid plexus. The activation of the V1 receptors regulates CSF production by decreased choroidal blood flow or by the effect on the choroidal epithelial cells' ion channels (Chodobski and Szmydynger-Chodobska 2001; Faraci et al. 1988). Vasopressin is able to reduce the efflux of Cl^- ions by regulating the Na^+, K^+, $2Cl^-$ cotransporter, and maintains the volume of the choroidal epithelial cells. Vasopressin levels in blood and CSF can vary substantially, and elevated levels of vasopressin have been found in the CSF of old rats and in the plasma of elderly humans (Fig. 9.1) (Frolkis et al. 1999). In addition to vasopressin, also angiotensin II and endothelin-1, secreted by the choroid plexus, can affect the choroidal blood flow and CSF secretion (Kaur et al. 2016). A reduced production and secretion of CSF could influence the delivery of many components to the brain and may interfere with the normal physiological pathways (Kaur et al. 2016; Preston 2001).

The Klotho protein is a transmembrane protein that was identified as aging-suppressor (Kuro et al. 1997). A defect in Klotho gene expression in mice accelerates aging-like phenotypes and results in a syndrome that resembles human aging including a short lifespan, impaired cognition (Uchida et al. 2001; Shiozaki et al. 2008), abnormal brain pathology, infertility, arteriosclerosis (Arking et al. 2003), skin atrophy, osteoporosis (Ogata et al. 2002), and emphysema (Kuro et al. 1997). Vice versa, the overexpression of Klotho in mice extends life span and improves memory (Kurosu et al. 2005; Li et al. 2019). Moreover, gene expression analysis of brain white matter in rhesus monkeys also indicated the implication of Klotho in the regulation of brain aging (Duce et al. 2008). In humans, a functional variant of the *KLOTHO* (*KL*) gene showed to be associated with high-density cholesterol, blood pressure, stroke, and longevity (Arking et al. 2002, 2005). The Klotho protein functions as a circulating hormone that represses intracellular signals of insulin and IGF1, an evolutionarily conserved mechanism for extending life span. In Klotho-deficient mice the disruption of insulin and IGF1 signaling lead to the improvement of aging-like phenotypes, suggesting that Klotho-mediated inhibition of insulin and IGF1 signaling contributes to its anti-aging properties (Kurosu et al. 2005). Klotho is predominantly secreted by the choroid plexus, the distal tubule cells of the kidney, and parathyroid glands; high levels of Klotho are expressed in the choroid plexus of juvenile and adult mice, humans, and mammals (Kuro 2010).

Soluble Klotho has been demonstrated to be present in human CSF and serum (Imura et al. 2004; Semba et al. 2014). Aging is associated in mice with decreased klotho expression in the choroid plexus (Zhu et al. 2018). Moreover, CSF klotho concentrations are lower in older versus younger cognitive healthy individuals and in addition, CSF klotho concentrations are significantly lower in Alzheimer's disease patients compared to adults without cognitive problems (Semba et al. 2014). Selective depletion of klotho in the choroid plexus triggered the expression of multiple proinflammatory factors and macrophage infiltration into the choroid plexus. Furthermore, experimental reduction of klotho in the choroid plexus demonstrated enhanced microglial activation in the hippocampus following peripheral stimulation with lipopolysaccharide (Zhu et al. 2018). These results suggest that klotho depletion from the choroid plexus could contribute to the age-dependent priming of microglia for activation by peripheral infections (Henry et al. 2009; Zhu et al. 2018). In primary macrophage cultures, Klotho suppressed the activation of the NLRP3 inflammasome by enhancing fibroblast growth factor (FGF)23 (Zhu et al. 2018). This suggests that Klotho controls the brain-immune systems interface in the choroid plexus. Moreover, Klotho depletion in aging or disease may weaken this barrier and promote immune-mediated neuropathogenesis (Zhu et al. 2018). In addition, Klotho secreted by the choroid plexus might enhance oligodendrocyte maturation and myelination of the CNS (Chen et al. 2013). In this way it may play a role in the prevention of myelin degeneration in the aging brain (Chen et al. 2013; Semba et al. 2014).

9.3.3 Barrier Permeability and Transport by the Choroid Plexus

The blood-CSF barrier ensures a stable, balanced, and well-controlled micro-environment of the brain, which is necessary for proper functioning of the CNS. Transport across the barrier is restricted by tight junctions between the choroid plexus epithelial cells and require transporter and receptor systems in a directional way (Redzic 2011; Saunders et al. 2013). The choroid plexus produces CSF by passive filtration of fluid across the fenestrated capillaries and regulated secretion of molecules across the choroid plexus epithelial cells (Brinker et al. 2014), together with active production of molecules by the choroid plexus epithelial cells (Thouvenot et al. 2006). Dysregulation of choroid plexus transporters and tight junction complexes subsequently reflects into CSF compositional changes. Several studies have reported a compromised blood-CSF barrier in response to inflammatory signals (Brkic et al. 2015; Marques and Sousa 2015; Vandenbroucke et al. 2012). As described previously, aging is associated with morphological changes of the choroid plexus epithelial cells and is in addition associated with a state of low grade, chronic inflammation or inflamm-aging which might lead to the loss of the barrier function at the choroid plexus. Loss of barrier integrity might result in leakage of components from the blood circulation into the CSF, thus changing CSF composition. In agreement with this, a study performed by Chen et al., showed increased blood-CSF permeability for proteins upon aging in sheep. However, no complete disruption of the barrier is present since the passage of larger molecules was still prevented (>109.51–120 kDa) (Chen et al. 2009, 2012a). Studies conducted in healthy elderly individuals report only small changes in CSF composition: the concentration of molecules including Transthyretin (TTR) (Serot et al. 2003; Kleine et al. 1993b), alpha2-macroglobulin (Garton et al. 1991; Kleine et al. 1993b), and IgG (Blennow et al. 1993a; Garton et al. 1991; Kleine et al. 1993b; Chen et al. 2018) increases slightly with age. The CSF versus serum albumin ratio is used to evaluate blood-CSF barrier functioning and an increased variability in the CSF/serum albumin ratio has been observed from the age of 45 years, indicating that the blood-CSF barrier is compromised in elderly humans (Fig. 9.1) (Blennow et al. 1993a, b, c). However, it is difficult to determine whether this is the result of increased blood-CSF barrier permeability or altered clearance of the proteins (Preston 2001; Serot et al. 2003, 1997). Often these elevated levels are interpreted as blood-CSF barrier integrity loss.

Choroid plexus epithelial cells are able to secrete EVs, including exosomes, into the CSF as a mechanism of blood-brain communication (Balusu et al. 2016b). EVs are membrane-derived vesicles that can enclose specific repertoires of proteins, lipids, and RNA molecules (Van Niel et al. 2018; Mathieu et al. 2019) and are able to transport these molecules both to adjacent and distant cells (Baixauli et al. 2014; Mittelbrunn and Sanchez-Madrid 2012; Paolicelli et al. 2018). In the CNS, EVs have shown to mediate intercellular communication over long range distances and are believed to be important for the cross-talk between neurons and glial cells in the brain (Paolicelli et al. 2018). Inflammation, which is also present in the aging brain, was shown to induce an increase in EV production, together with an altered EV content (Balusu et al. 2016b). The number of EVs present in the CSF declines in elderly humans and their miRNA content changes throughout life (Fig. 9.1) (Tietje et al. 2014). However, the size of the EVs and their size distribution did not change during aging (Tietje et al. 2014; Yang et al. 2015). EVs in the CSF can be produced by different cell types and no data is currently available on EV production by the choroid plexus epithelial cells during aging. However, if affected, this might have consequences for the nutrient delivery to the brain. As an example, exosomes, a specific type of EVs, which are secreted via the fusion of multivesicular bodies with the plasma membrane, are important for the delivery of folate, an important vitamin for the brain, across the choroid plexus epithelial cells into the CSF (Grapp et al. 2013).

9.4 Immune Cell Trafficking at the Choroid Plexus

Migration of immune cells into brain tissue and (limited) inflammatory reactions are fundamental mechanisms to sustain normal physiology, immune surveillance, host defense, and learning processes (Engelhardt and Coisne 2011; Garner et al. 2006; Ransohoff and Engelhardt 2012; Galea et al. 2007). However, these mechanisms are tightly controlled by the presence of different brain barriers. The blood-CSF barrier is perfectly located at the interface between blood and CSF to provide active immune surveillance and serves as active and selective gate for immune cell trafficking (Demeestere et al. 2015). The tightly connected choroidal epithelium limits paracellular transport of not only molecules, but also immune cells. Additionally, the choroid plexus contains fenestrated capillaries, allowing free communication between the stroma and peripheral blood (Demeestere et al. 2015). The choroid plexus stroma contains a large population of macrophages, dendritic cells, CD3+ and CD4+ T cells, and $CX3CR1^{hi}$ $Ly6C^{low}$ monocytes (Shechter et al. 2013). Macrophages at the apical side of the choroid plexus are called epiplexus or Kolmer cells (Maslieieva and Thompson 2014) and are thought to contribute to the immune component of the blood-CSF barrier. In healthy conditions, the CSF contains CD4 + T cells, natural killer cells, and B cells (Ransohoff and Engelhardt 2012). Although leukocytes enter the CSF, in steady state conditions they do not invade the brain parenchyma (Shechter et al. 2013). Leukocyte infiltration however is modulated in response to disease or trauma like meningitis, multiple sclerosis, or peripheral inflammation. The cells can transmigrate from the blood across the fenestrated endothelium to enter the stroma matrix. After travelling through the stroma of the choroidal cells, the immune cells can, in response to specific triggers, cross the choroid plexus epithelium and enter the CSF where they are able to skew toward specific effector responses, including regulatory T cells, T helper 2 cells, and alternatively activated macrophages (Shechter et al. 2013). The CD4+ T cells present in the CSF are distinct from the T cell populations in the blood circulation and brain parenchyma, indicating that the influx of T cells via the choroid plexus into the CSF is highly regulated (Engelhardt and Ransohoff 2012). After entering the CSF, leukocytes might be able to cross the ependymal cell layer and migrate further into the brain parenchyma under inflammatory conditions or might travel to the arachnoidea via the CSF flow.

The expression of adhesion molecules, chemokines, and chemokine receptors control leukocyte trafficking across the blood-CSF barrier. Inflammation causes the upregulation of adhesion molecules in the choroid plexus epithelial cells, such as intercellular adhesion molecule 1 (ICAM-1), vascular cellular adhesion molecule 1 (VCAM-1), and mucosal vascular addressin cell adhesion molecule 1 (MADCAM-1) (Endo et al. 1998). The expression of cell adhesion molecules at the brain barriers could possibly be increased during aging because of the pro-inflammatory state of the brain. The choroid plexus also produces cytokines (e.g. interleukin-1β (Il-1β) and tumor necrosis factor α (TNFα)) and chemokines (e.g. C-X-C motif chemokine ligand (CXCL10) and monocyte chemoattractant protein 1 (MCP1)). These cyto- and chemokines are necessary for the activation and/or recruitment of immune cells during systemic inflammation (Demeestere et al. 2015). Interestingly, aging leads to an increased expression of Il-1β in the choroid plexus (Silva-Vargas et al. 2016).

In the healthy adult brain, a balance is present between pro- and anti-inflammatory cytokines, but Baruch and colleagues observed in the choroid plexus a shift towards a Th2-like pro-inflammatory state with increasing age (Baruch et al. 2013; Sparkman and Johnson 2008). This pro-inflammatory state is reflected by the reduced production of interferon (IFN)-γ and increased production of IL-4, which negatively affect brain functioning (Baruch et al. 2013). Mice lacking the IFNγ receptor show a decreased number of leukocytes in the CSF and premature cognitive decline (Baruch et al. 2013). A type I IFN signature was described in the aged choroid plexus (Baruch et al. 2014). In both mouse models and human samples, the choroid plexus showed an increased type I and decreased type II IFN dependent gene expression profile (Fig. 9.1). This type I IFN signature negatively influences type II IFN signaling, leading to a reduced expression of homing and trafficking molecules (*Cd34*, *Madcam1*, *Ccl2*, *Cx3cr1*, *Cxcl13*, *Il2*) that are required for leukocyte entry in the CSF during aging, eventually leading to increased brain inflammation and cognitive decline (Baruch et al. 2014; Kunis et al. 2013). Interestingly, blocking IFN type I signaling restored cognitive functioning and hippocampal neurogenesis and in addition was able to diminish astrogliosis and microgliosis, and increase the anti-inflammatory cytokine IL-10 in the hippocampus (Baruch et al. 2014). It was suggested that this IFN I signaling is a mechanism to attenuate neuroinflammation which eventually becomes detrimental to brain plasticity resulting in age-associated cognitive decline (Baruch et al. 2014). However, therapeutically targeting the IFN pathway could influence the immune surveillance at the choroid plexus, since a tight balance between type I IFNs and IFNγ is central in leukocyte entry and cognition (Deczkowska et al. 2016). The importance of this balance is reflected in the IFNβ treatment, which is used in the clinic to reduce clinical relapses in multiple sclerosis (Wingerchuk and Carter 2014). Patients treated with IFN experience several adverse effects including increased risk for developing depression, cognitive decline, and they develop Parkinson like symptoms (Manouchehrinia and Constantinescu 2012). Similarly, type I IFNs aggravate disease in multiple mouse models of Parkinson's disease (Main et al. 2017).

It remains to be determined whether loss of barrier integrity is responsible for immune cell trafficking across the blood-CSF barrier. Independent of barrier impairment, other mechanisms might determine the leukocyte migration across the blood-CSF barrier. Nitric oxide, a negative regulator of leukocyte trafficking has been found to be upregulated at the choroid plexus during aging (Baruch et al. 2015). Additionally, transcellular migration events of leukocytes occurring in close proximity of the tight junctions have been undervalued and might have been mistaken for paracellular migration (Phillipson et al. 2008; Wewer et al. 2011; Wolburg et al. 2005). Steinmann et al. were able to demonstrate transcellular migration of leukocytes across the blood-CSF barrier after bacterial infection as well as T-cell transmigration after viral stimulation (Steinmann et al. 2013; Wewer et al. 2011). Moreover, polymorphonuclear (PMN) and monocytes differentially migrate in a human blood-CSF barrier model (Steinmann et al. 2013).

9.5   Choroid Plexus and Neurodegenerative Diseases

Interestingly, all the age-related morphological changes in choroid plexus structure described above, such as the flattening of epithelium, thickening of basement membrane, and lipofuscin deposits, are significantly more prominent in neurodegenerative diseases.

The most prevalent neurodegenerative disease, Alzheimer's disease, is characterized by the decline of memory and other cognitive functions. It is a progressive deteriorating disease, eventually leading to loss of autonomy and ultimately patients require full-time medical care (Jost and Grossberg 1995). Pathologically, Alzheimer's disease is defined by severe neuronal loss, resulting in the loss of brain volume, which is most pronounced around the medial temporal lobe areas, and particularly in the hippocampus. Furthermore, the aggregation of beta-amyloid (Aβ) in extracellular senile plaques and formation of intraneuronal neurofibrillary tangles consisting of hyperphosphorylated tau protein have been identified to play a major role in Alzheimer's disease pathogenesis (Braak and Braak 1991).

Ultrastructural changes are similar to those described in aging namely epithelial cell atrophy. In humans, the cells decrease in height with approximately 22% compared to healthy controls (Serot et al. 2000, 2003). The cytoplasm of the epithelial cells contains multiple lipofuscin and Biondi tangles (Miklossy et al. 1998). The basement membrane of the epithelium is thickened and irregular. Apical microvilli become irregular and fibrotic (Serot et al. 2000; Jellinger 1976). The stroma contains calcifications and psammomas, hyaline bodies, and thickened vessel walls (Serot et al. 2003).

TTR, a highly expressed protein at the choroid plexus, is a carrier for the thyroid hormones, but also has the ability to bind to Aβ and prevents the aggregation and deposition of Aβ plaques in the brain (Marques et al. 2013; Schwarzman et al. 1994). Studies reporting the production and secretion of TTR by the choroid plexus during aging show conflicting data: both an increase and decrease of TTR in the CSF has been described (Kleine et al. 1993a; Redzic et al. 2005; Serot et al. 1997). The blood-CSF barrier expresses several transporter systems including low-density lipoprotein receptor-related protein 1 (LRP1), receptor for advanced glycation end products (RAGE), receptor glycoprotein 330/megalin (LRP2), and Pgp. The LRP and Pgp receptors are responsible for the receptor-mediated efflux of Aβ from the brain, while RAGE mediates the influx of Aβ into the brain (Marques et al. 2013; Storck et al. 2016). Expression of LRP2 in the choroid plexus is decreased during aging, but an increase of LRP1 and Pgp is observed as well as no difference in RAGE expression (Gorle et al. 2016; Pascale et al. 2011).

Several studies have reported the beneficial effect of the choroid plexus on the rejuvenation of damaged brain regions because of the production of neurotrophic factors (Thanos et al. 2010; Borlongan et al. 2004a, b; Bolos et al. 2014). Choroid plexus epithelial cells treated in vitro with Aβ peptide lead to increased proliferation and differentiation of neuronal progenitor cells (Bolos et al. 2014). Moreover, transplantation of healthy choroid plexus epithelial cells into the brain of an Alzheimer's disease mouse model induced a significant reduction in brain Aβ and tau levels and improved memory of the animals (Bolos et al. 2014). Also in other neurodegenerative diseases, such as Huntington's disease choroid plexus cell transplantation studies showed successful results (Emerich and Borlongan 2009).

9.6 Conclusions

The choroid plexus, that contains the blood-CSF barrier, accomplishes important functions in the CNS and actively contributes to brain homeostasis. The choroid plexus is able to respond to changes both in the periphery and the brain parenchyma. However, during aging, the morphology and normal functioning of the choroid plexus is severely compromised. Alterations in brain barrier transport mechanisms, CSF production and clearance, receptor-mediated signaling, enzymatic and metabolic activity, loss of barrier integrity, and insufficient distribution of nutrients have an effect on brain functioning and might influence cognitive performance. Understanding how the blood-CSF barrier is altered in aging and how it can contribute to these age-associated diseases, might lead to novel strategies to attenuate aging-associated cognitive decline and related diseases.

References

Albeck MJ, Skak C, Nielsen PR, Olsen KS, Borgesen SE, Gjerris F (1998) Age dependency of resistance to cerebrospinal fluid outflow. J Neurosurg 89:275–278

Arking DE, Krebsova A, Macek M Sr, Macek M Jr, Arking A, Mian IS, Fried L, Hamosh A, Dey S, Mcintosh I, Dietz HC (2002) Association of human aging with a functional variant of klotho. Proc Natl Acad Sci U S A 99:856–861

Arking DE, Becker DM, Yanek LR, Fallin D, Judge DP, Moy TF, Becker LC, Dietz HC (2003) KLOTHO allele status and the risk of early-onset occult coronary artery disease. Am J Hum Genet 72:1154–1161

Arking DE, Atzmon G, Arking A, Barzilai N, Dietz HC (2005) Association between a functional variant of the KLOTHO gene and high-density lipoprotein cholesterol, blood pressure, stroke, and longevity. Circ Res 96:412–418

Baird GS, Nelson SK, Keeney TR, Stewart A, Williams S, Kraemer S, Peskind ER, Montine TJ (2012) Age-dependent changes in the cerebrospinal fluid proteome by slow off-rate modified aptamer array. Am J Pathol 180:446–456

Baixauli F, Lopez-Otin C, Mittelbrunn M (2014) Exosomes and autophagy: coordinated mechanisms for the maintenance of cellular fitness. Front Immunol 5:403

Balusu S, Brkic M, Libert C, Vandenbroucke RE (2016a) The choroid plexus-cerebrospinal fluid interface in Alzheimer's disease: more than just a barrier. Neural Regen Res 11:534–537

Balusu S, Van Wonterghem E, De Rycke R, Raemdonck K, Stremersch S, Gevaert K, Brkic M, Demeestere D, Vanhooren V, Hendrix A, Libert C, Vandenbroucke RE (2016b) Identification of a novel mechanism of blood-brain communication during peripheral inflammation via choroid plexus-derived extracellular vesicles. EMBO Mol Med 8:1162–1183

Barkhof F, Kouwenhoven M, Scheltens P, Sprenger M, Algra P, Valk J (1994) Phase-contrast cine MR imaging of normal aqueductal CSF flow. Effect of aging and relation to CSF void on modulus MR. Acta Radiol 35:123–130

Bartke A, Sun LY, Longo V (2013) Somatotropic signaling: trade-offs between growth, reproductive development, and longevity. Physiol Rev 93:571–598

Baruch K, Ron-Harel N, Gal H, Deczkowska A, Shifrut E, Ndifon W, Mirlas-Neisberg N, Cardon M, Vaknin I, Cahalon L, Berkutzki T, Mattson MP, Gomez-Pinilla F, Friedman N, Schwartz M (2013) CNS-specific immunity at the choroid plexus shifts toward destructive Th2 inflammation in brain aging. Proc Natl Acad Sci U S A 110:2264–2269

Baruch K, Deczkowska A, David E, Castellano JM, Miller O, Kertser A, Berkutzki T, Barnett-Itzhaki Z, Bezalel D, Wyss-Coray T, Amit I, Schwartz M (2014) Aging. Aging-induced type I interferon response at the choroid plexus negatively affects brain function. Science 346:89–93

Baruch K, Kertser A, Porat Z, Schwartz M (2015) Cerebral nitric oxide represses choroid plexus NFkappaB-dependent gateway activity for leukocyte trafficking. EMBO J 34:1816–1828

Bellur SN, Chandra V, Mcdonald LW (1980) Arachnoidal cell hyperplasia. Its relationship to aging and chronic renal failure. Arch Pathol Lab Med 104:414–416

Blennow K, Fredman P, Wallin A, Gottfries CG, Karlsson I, Langstrom G, Skoog I, Svennerholm L, Wikkelso C (1993a) Protein analysis in cerebrospinal fluid. II. Reference values derived from healthy individuals 18–88 years of age. Eur Neurol 33:129–133

Blennow K, Fredman P, Wallin A, Gottfries CG, Langstrom G, Svennerholm L (1993b) Protein analyses in cerebrospinal fluid. I. Influence of concentration gradients for proteins on cerebrospinal fluid/serum albumin ratio. Eur Neurol 33:126–128

Blennow K, Fredman P, Wallin A, Gottfries CG, Skoog I, Wikkelso C, Svennerholm L (1993c) Protein analysis in cerebrospinal fluid. III. Relation to blood-cerebrospinal fluid barrier function for formulas for quantitative determination of intrathecal IgG production. Eur Neurol 33:134–142

Bolos M, Antequera D, Aldudo J, Kristen H, Bullido MJ, Carro E (2014) Choroid plexus implants rescue Alzheimer's disease-like pathologies by modulating amyloid-beta degradation. Cell Mol Life Sci 71:2947–2955

Borlongan CV, Skinner SJ, Geaney M, Vasconcellos AV, Elliott RB, Emerich DF (2004a) CNS grafts of rat choroid plexus protect against cerebral ischemia in adult rats. Neuroreport 15:1543–1547

Borlongan CV, Skinner SJ, Geaney M, Vasconcellos AV, Elliott RB, Emerich DF (2004b) Intracerebral transplantation of porcine choroid plexus provides structural and functional neuroprotection in a rodent model of stroke. Stroke 35:2206–2210

Braak H, Braak E (1991) Neuropathological stageing of Alzheimer-related changes. Acta Neuropathol 82:239–259

Brinker T, Stopa E, Morrison J, Klinge P (2014) A new look at cerebrospinal fluid circulation. Fluids Barriers CNS 11:10

Brkic M, Balusu S, Van Wonterghem E, Gorle N, Benilova I, Kremer A, Van Hove I, Moons L, De Strooper B, Kanazir S, Libert C, Vandenbroucke RE (2015) Amyloid beta oligomers disrupt blood-CSF barrier integrity by activating matrix metalloproteinases. J Neurosci 35:12766–12778

Brown PD, Davies SL, Speake T, Millar ID (2004) Molecular mechanisms of cerebrospinal fluid production. Neuroscience 129:957–970

Bulat M, Klarica M (2011) Recent insights into a new hydrodynamics of the cerebrospinal fluid. Brain Res Rev 65:99–112

Calder PC, Bosco N, Bourdet-Sicard R, Capuron L, Delzenne N, Dore J, Franceschi C, Lehtinen MJ, Recker T, Salvioli S, Visioli F (2017) Health relevance of the modification of low grade inflammation in ageing (inflammageing) and the role of nutrition. Ageing Res Rev 40:95–119

Chen J, Buchanan JB, Sparkman NL, Godbout JP, Freund GG, Johnson RW (2008) Neuroinflammation and disruption in working memory in aged mice after acute stimulation of the peripheral innate immune system. Brain Behav Immun 22:301–311

Chen RL, Kassem NA, Redzic ZB, Chen CP, Segal MB, Preston JE (2009) Age-related changes in choroid plexus and blood-cerebrospinal fluid barrier function in the sheep. Exp Gerontol 44:289–296

Chen CP, Chen RL, Preston JE (2012a) The influence of ageing in the cerebrospinal fluid concentrations of proteins that are derived from the choroid plexus, brain, and plasma. Exp Gerontol 47:323–328

Chen X, Guo C, Kong J (2012b) Oxidative stress in neurodegenerative diseases. Neural Regen Res 7:376–385

Chen CD, Sloane JA, Li H, Aytan N, Giannaris EL, Zeldich E, Hinman JD, Dedeoglu A, Rosene DL, Bansal R, Luebke JI, Kuro OM, Abraham CR (2013) The antiaging protein Klotho enhances oligodendrocyte maturation and myelination of the CNS. J Neurosci 33:1927–1939

Chen CPC, Preston JE, Zhou S, Fuller HR, Morgan DGA, Chen R (2018) Proteomic analysis of age-related changes in ovine cerebrospinal fluid. Exp Gerontol 108:181–188

Chiu C, Miller MC, Caralopoulos IN, Worden MS, Brinker T, Gordon ZN, Johanson CE, Silverberg GD (2012) Temporal course of cerebrospinal fluid dynamics and amyloid accumulation in the aging rat brain from three to thirty months. Fluids Barriers CNS 9:3

Chodobski A, Szmydynger-Chodobska J (2001) Choroid plexus: target for polypeptides and site of their synthesis. Microsc Res Tech 52:65–82

Chongbin L, Rui W, Chunyan W, Hu C, Qifeng D (2014) Altered hepcidin expression is part of the choroid plexus response to IL-6/Stat3 signaling pathway in normal aging rats. Bioenergetics 3:2

Christensen HL, Barbuskaite D, Rojek A, Malte H, Christensen IB, Fuchtbauer AC, Fuchtbauer EM, Wang T, Praetorius J, Damkier HH (2018) The choroid plexus sodium-bicarbonate cotransporter NBCe2 regulates mouse cerebrospinal fluid pH. J Physiol 596:4709–4728

Cohen P, Ocrant I, Fielder PJ, Neely EK, Gargosky SE, Deal CI, Ceda GP, Youngman O, Pham H, Lamson G et al (1992) Insulin-like growth factors (IGFs): implications for aging. Psychoneuroendocrinology 17:335–342

Cornford EM, Varesi JB, Hyman S, Damian RT, Raleigh MJ (1997) Mitochondrial content of choroid plexus epithelium. Exp Brain Res 116:399–405

Cottrell DA, Blakely EL, Johnson MA, Ince PG, Borthwick GM, Turnbull DM (2001) Cytochrome c oxidase deficient cells accumulate in the hippocampus and choroid plexus with age. Neurobiol Aging 22:265–272

Cutler RW, Page L, Galicich J, Watters GV (1968) Formation and absorption of cerebrospinal fluid in man. Brain 91:707–720

Damkier HH, Brown PD, Praetorius J (2013) Cerebrospinal fluid secretion by the choroid plexus. Physiol Rev 93:1847–1892

Daouk J, Bouzerar R, Chaarani B, Zmudka J, Meyer ME, Baledent O (2016) Use of dynamic (18) F-fluorodeoxyglucose positron emission tomography to investigate choroid plexus function in Alzheimer's disease. Exp Gerontol 77:62–68

De Bock M, Vandenbroucke RE, Decrock E, Culot M, Cecchelli R, Leybaert L (2014) A new angle on blood-CNS interfaces: a role for connexins? FEBS Lett 588:1259–1270

Deane R, Zheng W, Zlokovic BV (2004) Brain capillary endothelium and choroid plexus epithelium regulate transport of transferrin-bound and free iron into the rat brain. J Neurochem 88:813–820

Deczkowska A, Baruch K, Schwartz M (2016) Type I/II interferon balance in the regulation of brain physiology and pathology. Trends Immunol 37:181–192

Demeestere D, Libert C, Vandenbroucke RE (2015) Clinical implications of leukocyte infiltration at the choroid plexus in (neuro)inflammatory disorders. Drug Discov Today 20:928–941

Duce JA, Podvin S, Hollander W, Kipling D, Rosene DL, Abraham CR (2008) Gene profile analysis implicates Klotho as an important contributor to aging changes in brain white matter of the rhesus monkey. Glia 56:106–117

Emerich DF, Borlongan CV (2009) Potential of choroid plexus epithelial cell grafts for neuroprotection in Huntington's disease: what remains before considering clinical trials. Neurotox Res 15:205–211

Emerich DF, Skinner SJ, Borlongan CV, Vasconcellos AV, Thanos CG (2005) The choroid plexus in the rise, fall and repair of the brain. BioEssays 27:262–274

Emerich DF, Schneider P, Bintz B, Hudak J, Thanos CG (2007) Aging reduces the neuroprotective capacity, VEGF secretion, and metabolic activity of rat choroid plexus epithelial cells. Cell Transplant 16:697–705

Endo H, Sasaki K, Tonosaki A, Kayama T (1998) Three-dimensional and ultrastructural ICAM-1 distribution in the choroid plexus, arachnoid membrane and dural sinus of inflammatory rats induced by LPS injection in the lateral ventricles. Brain Res 793:297–301

Engelhardt B, Coisne C (2011) Fluids and barriers of the CNS establish immune privilege by confining immune surveillance to a two-walled castle moat surrounding the CNS castle. Fluids Barriers CNS 8:4

Engelhardt B, Ransohoff RM (2012) Capture, crawl, cross: the T cell code to breach the blood-brain barriers. Trends Immunol 33:579–589

Eriksson L, Westermark P (1986) Intracellular neurofibrillary tangle-like aggregations. A constantly present amyloid alteration in the aging choroid plexus. Am J Pathol 125:124–129

Falcao AM, Marques F, Novais A, Sousa N, Palha JA, Sousa JC (2012) The path from the choroid plexus to the subventricular zone: go with the flow! Front Cell Neurosci 6:34

Faraci FM, Mayhan WG, Farrell WJ, Heistad DD (1988) Humoral regulation of blood flow to choroid plexus: role of arginine vasopressin. Circ Res 63:373–379

Faubel R, Westendorf C, Bodenschatz E, Eichele G (2016) Cilia-based flow network in the brain ventricles. Science 353:176–178

Ferrante F, Amenta F (1987) Enzyme histochemistry of the choroid plexus in old rats. Mech Ageing Dev 41:65–72

Fleischman D, Berdahl JP, Zaydlarova J, Stinnett S, Fautsch MP, Allingham RR (2012) Cerebrospinal fluid pressure decreases with older age. PLoS One 7:e52664

Foundas AL, Zipin D, Browning CA (1998) Age-related changes of the insular cortex and lateral ventricles: conventional MRI volumetric measures. J Neuroimaging 8:216–221

Franceschi C (2007) Inflammaging as a major characteristic of old people: can it be prevented or cured? Nutr Rev 65:S173–S176

Franceschi C, Campisi J (2014) Chronic inflammation (inflammaging) and its potential contribution to age-associated diseases. J Gerontol A Biol Sci Med Sci 69(Suppl 1):S4–S9

Frolkis VV, Kvitnitskaya-Ryzhova TY, Dubiley TA (1999) Vasopressin, hypothalamo-neurohypophyseal system and aging. Arch Gerontol Geriatr 29:193–214

Galea I, Bechmann I, Perry VH (2007) What is immune privilege (not)? Trends Immunol 28:12–18

Garner CC, Waites CL, Ziv NE (2006) Synapse development: still looking for the forest, still lost in the trees. Cell Tissue Res 326:249–262

Garton MJ, Keir G, Lakshmi MV, Thompson EJ (1991) Age-related changes in cerebrospinal fluid protein concentrations. J Neurol Sci 104:74–80

Gideon P, Thomsen C, Stahlberg F, Henriksen O (1994) Cerebrospinal fluid production and dynamics in normal aging: a MRI phase-mapping study. Acta Neurol Scand 89:362–366

Gorle N, Van Cauwenberghe C, Libert C, Vandenbroucke RE (2016) The effect of aging on brain barriers and the consequences for Alzheimer's disease development. Mamm Genome 27:407–420

Grapp M, Wrede A, Schweizer M, Huwel S, Galla HJ, Snaidero N, Simons M, Buckers J, Low PS, Urlaub H, Gartner J, Steinfeld R (2013) Choroid plexus transcytosis and exosome shuttling deliver folate into brain parenchyma. Nat Commun 4:2123

Henriksson L, Voigt K (1976) Age-dependent differences of distribution and clearance patterns in normal RIHSA cisternograms. Neuroradiology 12:103–107

Henry CJ, Huang Y, Wynne AM, Godbout JP (2009) Peripheral lipopolysaccharide (LPS) challenge promotes microglial hyperactivity in aged mice that is associated with exaggerated induction of both pro-inflammatory IL-1beta and anti-inflammatory IL-10 cytokines. Brain Behav Immun 23:309–317

Holm NR, Hansen LB, Nilsson C, Gammeltoft S (1994) Gene expression and secretion of insulin-like growth factor-II and insulin-like growth factor binding protein-2 from cultured sheep choroid plexus epithelial cells. Brain Res Mol Brain Res 21:67–74

Hubert V, Chauveau F, Dumot C, Ong E, Berner LP, Canet-Soulas E, Ghersi-Egea JF, Wiart M (2019) Clinical imaging of choroid plexus in health and in brain disorders: a mini-review. Front Mol Neurosci 12:34

Imura A, Iwano A, Tohyama O, Tsuji Y, Nozaki K, Hashimoto N, Fujimori T, Nabeshima Y (2004) Secreted Klotho protein in sera and CSF: implication for post-translational cleavage in release of Klotho protein from cell membrane. FEBS Lett 565:143–147

Jahng GH, Oh J, Lee DW, Kim HG, Rhee HY, Shin W, Paik JW, Lee KM, Park S, Choe BY, Ryu CW (2016) Glutamine and glutamate complex, as measured by functional magnetic resonance spectroscopy, alters during face-name association task in patients with mild cognitive impairment and Alzheimer's disease. J Alzheimers Dis 52:145–159

Jellinger K (1976) Neuropathological aspects of dementias resulting from abnormal blood and cerebrospinal fluid dynamics. Acta Neurol Belg 76:83–102

Johanson CE, Duncan JA 3rd, Klinge PM, Brinker T, Stopa EG, Silverberg GD (2008) Multiplicity of cerebrospinal fluid functions: new challenges in health and disease. Cerebrospinal Fluid Res 5:10

Jost BC, Grossberg GT (1995) The natural history of Alzheimer's disease: a brain bank study. J Am Geriatr Soc 43:1248–1255

Jovanovic I, Stefanovic N, Antic S, Ugrenovic S, Djindjic B, Vidovic N (2004) Morphological and morphometric characteristics of choroid plexus psammoma bodies during the human aging. Ital J Anat Embryol 109:19–33

Kao L, Kurtz LM, Shao X, Papadopoulos MC, Liu L, Bok D, Nusinowitz S, Chen B, Stella SL, Andre M, Weinreb J, Luong SS, Piri N, Kwong JM, Newman D, Kurtz I (2011) Severe neurologic impairment in mice with targeted disruption of the electrogenic sodium bicarbonate cotransporter NBCe2 (Slc4a5 gene). J Biol Chem 286:32563–32574

Kaur C, Rathnasamy G, Ling EA (2016) The choroid plexus in healthy and diseased brain. J Neuropathol Exp Neurol 75:198–213

Khasawneh AH, Garling RJ, Harris CA (2018) Cerebrospinal fluid circulation: what do we know and how do we know it? Brain Circ 4:14–18

Kida S, Yamashima T, Kubota T, Ito H, Yamamoto S (1988) A light and electron microscopic and immunohistochemical study of human arachnoid villi. J Neurosurg 69:429–435

Kleine TO, Hackler R, Lutcke A, Dauch W, Zofel P (1993a) Transport and production of cerebrospinal fluid (CSF) change in aging humans under normal and diseased conditions. Z Gerontol 26:251–255

Kleine TO, Hackler R, Zofel P (1993b) Age-related alterations of the blood-brain-barrier (bbb) permeability to protein molecules of different size. Z Gerontol 26:256–259

Kunis G, Baruch K, Rosenzweig N, Kertser A, Miller O, Berkutzki T, Schwartz M (2013) IFN-gamma-dependent activation of the brain's choroid plexus for CNS immune surveillance and repair. Brain 136:3427–3440

Kuro OM (2010) Klotho. Pflugers Arch 459:333–343

Kuro OM, Matsumura Y, Aizawa H, Kawaguchi H, Suga T, Utsugi T, Ohyama Y, Kurabayashi M, Kaname T, Kume E, Iwasaki H, Iida A, Shiraki-Iida T, Nishikawa S, Nagai R, Nabeshima YI (1997) Mutation of the mouse klotho gene leads to a syndrome resembling ageing. Nature 390:45–51

Kurosu H, Yamamoto M, Clark JD, Pastor JV, Nandi A, Gurnani P, Mcguinness OP, Chikuda H, Yamaguchi M, Kawaguchi H, Shimomura I, Takayama Y, Herz J, Kahn CR, Rosenblatt KP, Kuro OM (2005) Suppression of aging in mice by the hormone Klotho. Science 309:1829–1833

Kvitnitskaia-Ryzhova T, Shkapenko AL (1992) A comparative ultracytochemical and biochemical study of the ATPases of the choroid plexus in aging. Tsitologiia 34:81–87

Leitner DF, Connor JR (2012) Functional roles of transferrin in the brain. Biochim Biophys Acta 1820:393–402

Lenoir D, Honegger P (1983) Insulin-like growth factor I (IGF I) stimulates DNA synthesis in fetal rat brain cell cultures. Brain Res 283:205–213

Li D, Jing D, Liu Z, Chen Y, Huang F, Behnisch T (2019) Enhanced expression of secreted alpha-klotho in the hippocampus alters nesting behavior and memory formation in mice. Front Cell Neurosci 13:133

Liddelow SA (2015) Development of the choroid plexus and blood-CSF barrier. Front Neurosci 9:32

Lindsey AE, Schneider K, Simmons DM, Baron R, Lee BS, Kopito RR (1990) Functional expression and subcellular localization of an anion exchanger cloned from choroid plexus. Proc Natl Acad Sci U S A 87:5278–5282

Lu J, Kaur C, Ling EA (1995) Expression and upregulation of transferrin receptors and iron uptake in the epiplexus cells of different aged rats injected with lipopolysaccharide and interferon-gamma. J Anat 187(Pt 3):603–611

Lun MP, Johnson MB, Broadbelt KG, Watanabe M, Kang YJ, Chau KF, Springel MW, Malesz A, Sousa AM, Pletikos M, Adelita T, Calicchio ML, Zhang Y, Holtzman MJ, Lidov HG, Sestan N, Steen H, Monuki ES, Lehtinen MK (2015) Spatially heterogeneous choroid plexus transcriptomes encode positional identity and contribute to regional CSF production. J Neurosci 35:4903–4916

Ma Q, Ineichen BV, Detmar M, Proulx ST (2017) Outflow of cerebrospinal fluid is predominantly through lymphatic vessels and is reduced in aged mice. Nat Commun 8:1434

Main BS, Zhang M, Brody KM, Kirby FJ, Crack PJ, Taylor JM (2017) Type-I interferons mediate the neuroinflammatory response and neurotoxicity induced by rotenone. J Neurochem 141:75–85

Manouchehrinia A, Constantinescu CS (2012) Cost-effectiveness of disease-modifying therapies in multiple sclerosis. Curr Neurol Neurosci Rep 12:592–600

Marques F, Sousa JC (2015) The choroid plexus is modulated by various peripheral stimuli: implications to diseases of the central nervous system. Front Cell Neurosci 9:136

Marques F, Sousa JC, Correia-Neves M, Oliveira P, Sousa N, Palha JA (2007) The choroid plexus response to peripheral inflammatory stimulus. Neuroscience 144:424–430

Marques F, Falcao AM, Sousa JC, Coppola G, Geschwind D, Sousa N, Correia-Neves M, Palha JA (2009) Altered iron metabolism is part of the choroid plexus response to peripheral inflammation. Endocrinology 150:2822–2828

Marques F, Sousa JC, Coppola G, Gao F, Puga R, Brentani H, Geschwind DH, Sousa N, Correia-Neves M, Palha JA (2011) Transcriptome signature of the adult mouse choroid plexus. Fluids Barriers CNS 8:10

Marques F, Sousa JC, Sousa N, Palha JA (2013) Blood-brain-barriers in aging and in Alzheimer's disease. Mol Neurodegener 8:38

Marques F, Sousa JC, Brito MA, Pahnke J, Santos C, Correia-Neves M, Palha JA (2017) The choroid plexus in health and in disease: dialogues into and out of the brain. Neurobiol Dis 107:32–40

Maslieieva V, Thompson RJ (2014) A critical role for pannexin-1 in activation of innate immune cells of the choroid plexus. Channels (Austin) 8:131–141

Masseguin C, Lepanse S, Corman B, Verbavatz JM, Gabrion J (2005) Aging affects choroidal proteins involved in CSF production in Sprague-Dawley rats. Neurobiol Aging 26:917–927

Mathieu M, Martin-Jaular L, Lavieu G, Thery C (2019) Specificities of secretion and uptake of exosomes and other extracellular vesicles for cell-to-cell communication. Nat Cell Biol 21:9–17

Matsumae M, Kikinis R, Morocz I, Lorenzo AV, Albert MS, Black PM, Jolesz FA (1996a) Intracranial compartment volumes in patients with enlarged ventricles assessed by magnetic resonance-based image processing. J Neurosurg 84:972–981

Matsumae M, Kikinis R, Morocz IA, Lorenzo AV, Sandor T, Albert MS, Black PM, Jolesz FA (1996b) Age-related changes in intracranial compartment volumes in normal adults assessed by magnetic resonance imaging. J Neurosurg 84:982–991

May C, Kaye JA, Atack JR, Schapiro MB, Friedland RP, Rapoport SI (1990) Cerebrospinal fluid production is reduced in healthy aging. Neurology 40:500–503

Mergenthaler P, Lindauer U, Dienel GA, Meisel A (2013) Sugar for the brain: the role of glucose in physiological and pathological brain function. Trends Neurosci 36:587–597

Miklossy J, Kraftsik R, Pillevuit O, Lepori D, Genton C, Bosman FT (1998) Curly fiber and tangle-like inclusions in the ependyma and choroid plexus--a pathogenetic relationship with the cortical Alzheimer-type changes? J Neuropathol Exp Neurol 57:1202–1212

Mill JF, Chao MV, Ishii DN (1985) Insulin, insulin-like growth factor II, and nerve growth factor effects on tubulin mRNA levels and neurite formation. Proc Natl Acad Sci U S A 82:7126–7130

Mittelbrunn M, Sanchez-Madrid F (2012) Intercellular communication: diverse structures for exchange of genetic information. Nat Rev Mol Cell Biol 13:328–335

Morris CM, Keith AB, Edwardson JA, Pullen RG (1992) Uptake and distribution of iron and transferrin in the adult rat brain. J Neurochem 59:300–306

Mozell RL, Mcmorris FA (1991) Insulin-like growth factor I stimulates oligodendrocyte development and myelination in rat brain aggregate cultures. J Neurosci Res 30:382–390

Nilsson C, Lindvall-Axelsson M, Owman C (1992) Neuroendocrine regulatory mechanisms in the choroid plexus-cerebrospinal fluid system. Brain Res Brain Res Rev 17:109–138

Nilsson C, Hultberg BM, Gammeltoft S (1996) Autocrine role of insulin-like growth factor II secretion by the rat choroid plexus. Eur J Neurosci 8:629–635

Ogata N, Matsumura Y, Shiraki M, Kawano K, Koshizuka Y, Hosoi T, Nakamura K, Kuro OM, Kawaguchi H (2002) Association of klotho gene polymorphism with bone density and spondylosis of the lumbar spine in postmenopausal women. Bone 31:37–42

Paolicelli RC, Bergamini G, Rajendran L (2018) Cell-to-cell communication by extracellular vesicles: focus on microglia. Neuroscience 405:148–157

Pascale CL, Miller MC, Chiu C, Boylan M, Caralopoulos IN, Gonzalez L, Johanson CE, Silverberg GD (2011) Amyloid-beta transporter expression at the blood-CSF barrier is age-dependent. Fluids Barriers CNS 8:21

Pershing LK, Johanson CE (1982) Acidosis-induced enhanced activity of the Na-K exchange pump in the in vivo choroid plexus: an ontogenetic analysis of possible role in cerebrospinal fluid pH homeostasis. J Neurochem 38:322–332

Phillipson M, Kaur J, Colarusso P, Ballantyne CM, Kubes P (2008) Endothelial domes encapsulate adherent neutrophils and minimize increases in vascular permeability in paracellular and transcellular emigration. PLoS One 3:e1649

Plotkin MD, Kaplan MR, Peterson LN, Gullans SR, Hebert SC, Delpire E (1997) Expression of the Na(+)-K(+)-2Cl- cotransporter BSC2 in the nervous system. Am J Phys 272:C173–C183

Preston JE (2001) Ageing choroid plexus-cerebrospinal fluid system. Microsc Res Tech 52:31–37

Proulx ST, Ma Q, Andina D, Leroux JC, Detmar M (2017) Quantitative measurement of lymphatic function in mice by noninvasive near-infrared imaging of a peripheral vein. JCI Insight 2: e90861

Ransohoff RM, Engelhardt B (2012) The anatomical and cellular basis of immune surveillance in the central nervous system. Nat Rev Immunol 12:623–635

Redzic Z (2011) Molecular biology of the blood-brain and the blood-cerebrospinal fluid barriers: similarities and differences. Fluids Barriers CNS 8:3

Redzic ZB, Segal MB (2004) The structure of the choroid plexus and the physiology of the choroid plexus epithelium. Adv Drug Deliv Rev 56:1695–1716

Redzic ZB, Preston JE, Duncan JA, Chodobski A, Szmydynger-Chodobska J (2005) The choroid plexus-cerebrospinal fluid system: from development to aging. Curr Top Dev Biol 71:1–52

Report WA (2018) World Alzheimer report

Rouault TA, Zhang DL, Jeong SY (2009) Brain iron homeostasis, the choroid plexus, and localization of iron transport proteins. Metab Brain Dis 24:673–684

Rubenstein E (1998) Relationship of senescence of cerebrospinal fluid circulatory system to dementias of the aged. Lancet 351:283–285

Sakka L, Coll G, Chazal J (2011) Anatomy and physiology of cerebrospinal fluid. Eur Ann Otorhinolaryngol Head Neck Dis 128:309–316

Sathyanesan M, Girgenti MJ, Banasr M, Stone K, Bruce C, Guilchicek E, Wilczak-Havill K, Nairn A, Williams K, Sass S, Duman JG, Newton SS (2012) A molecular characterization of the choroid plexus and stress-induced gene regulation. Transl Psychiatry 2:e139

Saunders NR, Daneman R, Dziegielewska KM, Liddelow SA (2013) Transporters of the blood-brain and blood-CSF interfaces in development and in the adult. Mol Asp Med 34:742–752

Schwarzman AL, Gregori L, Vitek MP, Lyubski S, Strittmatter WJ, Enghilde JJ, Bhasin R, Silverman J, Weisgraber KH, Coyle PK et al (1994) Transthyretin sequesters amyloid beta protein and prevents amyloid formation. Proc Natl Acad Sci U S A 91:8368–8372

Segal MB (2000) The choroid plexuses and the barriers between the blood and the cerebrospinal fluid. Cell Mol Neurobiol 20:183–196

Semba RD, Moghekar AR, Hu J, Sun K, Turner R, Ferrucci L, O'Brien R (2014) Klotho in the cerebrospinal fluid of adults with and without Alzheimer's disease. Neurosci Lett 558:37–40

Serot JM, Christmann D, Dubost T, Couturier M (1997) Cerebrospinal fluid transthyretin: aging and late onset Alzheimer's disease. J Neurol Neurosurg Psychiatry 63:506–508

Serot JM, Bene MC, Foliguet B, Faure GC (2000) Morphological alterations of the choroid plexus in late-onset Alzheimer's disease. Acta Neuropathol 99:105–108

Serot JM, Foliguet B, Bene MC, Faure GC (2001) Choroid plexus and ageing in rats: a morphometric and ultrastructural study. Eur J Neurosci 14:794–798

Serot JM, Bene MC, Faure GC (2003) Choroid plexus, aging of the brain, and Alzheimer's disease. Front Biosci 8:s515–s521

Shechter R, London A, Schwartz M (2013) Orchestrated leukocyte recruitment to immune-privileged sites: absolute barriers versus educational gates. Nat Rev Immunol 13:206–218

Shiozaki M, Yoshimura K, Shibata M, Koike M, Matsuura N, Uchiyama Y, Gotow T (2008) Morphological and biochemical signs of age-related neurodegenerative changes in klotho mutant mice. Neuroscience 152:924–941

Shuangshoti S, Netsky MG (1970) Human choroid plexus: morphologic and histochemical alterations with age. Am J Anat 128:73–95

Siesjo BK (1978) Brain metabolism and anaesthesia. Acta Anaesthesiol Scand Suppl 70:56–59

Silva-Vargas V, Maldonado-Soto AR, Mizrak D, Codega P, Doetsch F (2016) Age-dependent niche signals from the choroid plexus regulate adult neural stem cells. Cell Stem Cell 19:643–652

Silverberg GD, Heit G, Huhn S, Jaffe RA, Chang SD, Bronte-Stewart H, Rubenstein E, Possin K, Saul TA (2001) The cerebrospinal fluid production rate is reduced in dementia of the Alzheimer's type. Neurology 57:1763–1766

Silverberg GD, Mayo M, Saul T, Rubenstein E, Mcguire D (2003) Alzheimer's disease, normal-pressure hydrocephalus, and senescent changes in CSF circulatory physiology: a hypothesis. Lancet Neurol 2:506–511

Simpson IA, Carruthers A, Vannucci SJ (2007) Supply and demand in cerebral energy metabolism: the role of nutrient transporters. J Cereb Blood Flow Metab 27:1766–1791

Sparkman NL, Johnson RW (2008) Neuroinflammation associated with aging sensitizes the brain to the effects of infection or stress. Neuroimmunomodulation 15:323–330

Speake T, Whitwell C, Kajita H, Majid A, Brown PD (2001) Mechanisms of CSF secretion by the choroid plexus. Microsc Res Tech 52:49–59

Steinmann U, Borkowski J, Wolburg H, Schroppel B, Findeisen P, Weiss C, Ishikawa H, Schwerk C, Schroten H, Tenenbaum T (2013) Transmigration of polymorphnuclear neutrophils and monocytes through the human blood-cerebrospinal fluid barrier after bacterial infection in vitro. J Neuroinflammation 10:31

Stoquart-Elsankari S, Baledent O, Gondry-Jouet C, Makki M, Godefroy O, Meyer ME (2007) Aging effects on cerebral blood and cerebrospinal fluid flows. J Cereb Blood Flow Metab 27:1563–1572

Storck SE, Meister S, Nahrath J, Meissner JN, Schubert N, Di Spiezio A, Baches S, Vandenbroucke RE, Bouter Y, Prikulis I, Korth C, Weggen S, Heimann A, Schwaninger M, Bayer TA, Pietrzik CU (2016) Endothelial LRP1 transports amyloid-beta(1-42) across the blood-brain barrier. J Clin Invest 126:123–136

Strazielle N, Ghersi-Egea JF (2000) Choroid plexus in the central nervous system: biology and physiopathology. J Neuropathol Exp Neurol 59:561–574

Sturrock RR (1988) An ultrastructural study of the choroid plexus of aged mice. Anat Anz 165:379–385

Thanos CG, Bintz B, Emerich DF (2010) Microencapsulated choroid plexus epithelial cell transplants for repair of the brain. Adv Exp Med Biol 670:80–91

Thouvenot E, Lafon-Cazal M, Demettre E, Jouin P, Bockaert J, Marin P (2006) The proteomic analysis of mouse choroid plexus secretome reveals a high protein secretion capacity of choroidal epithelial cells. Proteomics 6:5941–5952

Tietje A, Maron KN, Wei Y, Feliciano DM (2014) Cerebrospinal fluid extracellular vesicles undergo age dependent declines and contain known and novel non-coding RNAs. PLoS One 9:e113116

Uchida A, Komiya Y, Tashiro T, Yorifuji H, Kishimoto T, Nabeshima Y, Hisanaga S (2001) Neurofilaments of Klotho, the mutant mouse prematurely displaying symptoms resembling human aging. J Neurosci Res 64:364–370

Van Niel G, D'Angelo G, Raposo G (2018) Shedding light on the cell biology of extracellular vesicles. Nat Rev Mol Cell Biol 19:213–228

Vandenbroucke RE (2016) A hidden epithelial barrier in the brain with a central role in regulating brain homeostasis. Implications for aging. Ann Am Thorac Soc 13(Suppl 5):S407–S410

Vandenbroucke RE, Dejonckheere E, Van Lint P, Demeestere D, Van Wonterghem E, Vanlaere I, Puimege L, Van Hauwermeiren F, De Rycke R, Mc Guire C, Campestre C, Lopez-Otin C, Matthys P, Leclercq G, Libert C (2012) Matrix metalloprotease 8-dependent extracellular matrix cleavage at the blood-CSF barrier contributes to lethality during systemic inflammatory diseases. J Neurosci 32:9805–9816

Vega JA, Del Valle ME, Calzada B, Bengoechea ME, Perez-Casas A (1992) Expression of nerve growth factor receptor immunoreactivity in the rat choroid plexus. Cell Mol Biol 38:145–149

Villeda SA, Luo J, Mosher KI, Zou B, Britschgi M, Bieri G, Stan TM, Fainberg N, Ding Z, Eggel A, Lucin KM, Czirr E, Park JS, Couillard-Despres S, Aigner L, Li G, Peskind ER, Kaye JA, Quinn JF, Galasko DR, Xie XS, Rando TA, Wyss-Coray T (2011) The ageing systemic milieu negatively regulates neurogenesis and cognitive function. Nature 477:90–94

Wahlund LO, Almkvist O, Basun H, Julin P (1996) MRI in successful aging, a 5-year follow-up study from the eighth to ninth decade of life. Magn Reson Imaging 14:601–608

Wen GY, Wisniewski HM, Kascsak RJ (1999) Biondi ring tangles in the choroid plexus of Alzheimer's disease and normal aging brains: a quantitative study. Brain Res 832:40–46

Wewer C, Seibt A, Wolburg H, Greune L, Schmidt MA, Berger J, Galla HJ, Quitsch U, Schwerk C, Schroten H, Tenenbaum T (2011) Transcellular migration of neutrophil granulocytes through the blood-cerebrospinal fluid barrier after infection with Streptococcus suis. J Neuroinflammation 8:51

Wingerchuk DM, Carter JL (2014) Multiple sclerosis: current and emerging disease-modifying therapies and treatment strategies. Mayo Clin Proc 89:225–240

Wolburg H, Wolburg-Buchholz K, Engelhardt B (2005) Diapedesis of mononuclear cells across cerebral venules during experimental autoimmune encephalomyelitis leaves tight junctions intact. Acta Neuropathol 109:181–190

Yang Y, Keene CD, Peskind ER, Galasko DR, Hu SC, Cudaback E, Wilson AM, Li G, Yu CE, Montine KS, Zhang J, Baird GS, Hyman BT, Montine TJ (2015) Cerebrospinal fluid particles in Alzheimer disease and parkinson Disease. J Neuropathol Exp Neurol 74:672–687

Yesavage JA, Holman CA, Sarnquist FH, Berger PA (1982) Elevation of cerebrospinal fluid lactate with aging in subjects with normal blood oxygen saturations. J Gerontol 37:313–315

Zhu L, Stein LR, Kim D, Ho K, Yu GQ, Zhan L, Larsson TE, Mucke L (2018) Klotho controls the brain-immune system interface in the choroid plexus. Proc Natl Acad Sci U S A 115:E11388–E11396

Zivkovic VS, Stanojkovic MM, Antic MM (2017) Psammoma bodies as signs of choroid plexus ageing - a morphometric analysis. Vojnosanit Pregl 74:1054–1059

ZS-Nagy I, Steiber J, Jeney F (1995) Induction of age pigment accumulation in the brain cells of young male rats through iron-injection into the cerebrospinal fluid. Gerontology 41(Suppl 2):145–158

Chapter 10
Choroid Plexus Tumors

Daniel H. Fulkerson, Adam Leibold, David Priemer, and Karl Balsara

Abstract Choroid plexus tumors are rare tumors, predominantly found in pediatric patients. Lower grade tumors histologically resemble native choroid plexus while higher grade lesions are progressively more disordered and less differentiated. They are commonly associated with hydrocephalus and usually present with symptoms of that condition. Complete resection is typically curative in the case of low grade lesions. Higher grade lesions may be treated with adjuvant chemotherapy or radiotherapy. The hydrocephalus often resolves but in some cases may persist after resection and may require additional treatment.

10.1   Introduction

Tumors arising from the choroid plexus are relatively rare. They comprise less than 1% of all intracranial tumors (Cannon et al. 2015). Guerard made the first detailed notes of a choroid plexus tumor in his autopsy report of a 3 year-old female in 1833. His notes were recorded in a publication by Davis and Cushing (1925). The first recorded attempt of surgical removal occurred in 1906 and the first known surgery resulting in long-term survival of the patient occurred in 1919 (Unger 1906). Over the subsequent few decades, surgical resection of these intraventricular tumors was

D. H. Fulkerson
Indiana University School of Medicine, Indianapolis, IN, USA

Beacon Children's Hospital, Beacon Medical Group, South Bend, IN, USA

A. Leibold
Indiana University School of Medicine, Indianapolis, IN, USA

D. Priemer
Department of Pathology and Laboratory Medicine, Indiana University School of Medicine, Indianapolis, IN, USA

K. Balsara (✉)
Department of Neurological Surgery, Indiana University School of Medicine, Indianapolis, IN, USA
e-mail: kbalsara@iu.edu

© The American Physiological Society 2020
J. Praetorius et al. (eds.), *Role of the Choroid Plexus in Health and Disease*,
Physiology in Health and Disease, https://doi.org/10.1007/978-1-0716-0536-3_10

refined, with Dandy pioneering the transcallosal route to the third ventricle in 1922 and Masson developing the transfrontal approach in 1933 (Dandy 1922; Masson 1934). Until the past two decades, the surgical mortality of these often vascular tumors remained high (Guidetti and Spallone 1981).

10.2 Epidemiology

Choroid plexus tumors are divided into the World Health Organization (WHO) Grade I choroid plexus papilloma (CPP), the Grade II atypical choroid plexus papilloma (aCPP), and the malignant Grade III choroid plexus carcinoma (CPC). Approximately 80% of all choroid plexus tumors are CPPs (Borja et al. 2013).

Choroid plexus tumors may occur at any age, however they are more common in the pediatric population. They comprise 2–4% of intracranial tumors in children, but up to 20% of tumors present within the first year of life (Ostrom et al. 2014; Thomas et al. 2015). The overall incidence of choroid plexus tumors is approximately 0.22–0.3 per 100,000 children (Cannon et al. 2015; Lafay-Cousin et al. 2011). Over 80% of all CPCs occur in patients less than 18 years; the median age of patients with CPCs is 3 years (Sun et al. 2014a). While most tumors are sporadic, there may be predisposing genetic factors. CPPs have been found in siblings and may be associated with specific genetic syndromes (Gozali et al. 2012).

The anatomic location of the tumor varies with age; tumors in children are most commonly located within the lateral ventricles. Large series do not show a predilection for one side; although there was a commonly held belief that left-sided tumors were more common (Boyd and Steinbok 1987; Knierim 1990; McGirr et al. 1988; Spallone et al. 1990). The most common tumor location in adults is within the fourth ventricle. The median age of all choroid plexus tumors is 3.5 years. The median age for tumors found in the fourth ventricle is 22.5 years (Wolff et al. 2002). There are rare reports of choroid plexus tumors in the cerebellopontine angle (the anatomic area defined by the junction of the pons and the cerebellum), third ventricle, or cerebral parenchyma.

CPPs are generally solitary. Multifocal or disseminated CPPs have been reported in adults, but are rare (Abdulkader et al. 2016; Doglietto et al. 2005; Jagielski et al. 2001; Karim et al. 2006; Morshed et al. 2017; Peyre et al. 2012; Scholsem et al. 2012; Serifoglu et al. 2016; Zachary et al. 2014). Multifocal or metastatic lesions are more common in higher grade tumors. Multiple lesions occur in approximately 17% of patients with aCPP (Jinhu et al. 2007; Wrede et al. 2009). Multifocal choroid plexus tumors have been rarely reported in children, often in patients with either high-grade tumors or genetic conditions such as Aicardi syndrome, an X-linked dominant disorder characterized by agenesis of the corpus callosum, infantile spasms, and lacunar chorioretinopathy (Pianetti Filho et al. 2002; Taggard and Menezes 2000; Trifiletti et al. 1995).

10.3   Clinical Presentation and Evaluation

Choroid plexus tumors often present with manifestations of hydrocephalus including headache, vomiting, lethargy, or papilledema (swelling of the optic disc noted on ophthalmologic examination) (Lena et al. 1990). The hydrocephalus may cause an abrupt decline with rapid deterioration and herniation signs including bradycardia, hypertension, and coma (Picht et al. 2006). Herniation signs indicate a dangerously elevated level of intracranial pressure that, if uncorrected, is rapidly fatal. Young children without fused cranial sutures may present with a tense fontanelle or macrocephaly. Other documented presentations include irritability, failure to thrive, developmental delay, cranial nerve palsy, endocrine disturbances, "bobble-head" phenomenon, anorexia, schizophrenia, spontaneous cerebrospinal fluid leak from the nose or ears, and titubation, a characteristic uncontrollable rhythmic tremor of the head (Arasappa et al. 2013; Lechanoine et al. 2017; Nagib and O'Fallon 2000; Singh et al. 2017). Rarely, patients present with rapid decline from tumor hemorrhage (Pandey et al. 2016).

10.3.1   Radiographic Evaluation

A computerized tomography (CT) scan is often the first imaging modality employed in patients with choroid plexus tumors. A CT scan clearly shows the ventricle size and allows assessment of the degree of hydrocephalus. Choroid plexus tumors may be calcified, a feature best appreciated on CT scan. CPPs are isodense to hyperdense on non-contrasted CT scanning (Fig. 10.1a). They appear lobulated and often have a "cauliflower" appearance. Tumors often brightly enhance with administration of iodinated contrast agents. Patients may have diffuse enhancement of the meninges surrounding the brain, suggesting infection or tumor spread. However, meningeal enhancement often resolves completely after removal of the primary tumor (Scala et al. 2017). CPPs are generally contained within a ventricle and show a clear distinction from surrounding brain parenchyma. Aggressive aCPPs may have partially blurred borders with the parenchyma and CPCs may show frank invasion. Images of all choroid plexus tumors may show mild to moderate peritumoral edema (Shi et al. 2017a).

Choroid plexus tumors have characteristic features with magnetic resonance imaging (MRI). MRI provides greater detail than CT scanning and is the imaging modality of choice. They are hyperintense on T2-weighted images (Fig. 10.1b). Choroid plexus tumors are iso- to hypointense to gray matter on T1-weighted images (Fig. 10.1c). Choroid plexus tumors often show homogenous and intense enhancement with gadolinium contrast (Fig. 10.1d) (Kim et al. 2012; Shi et al. 2017b). Rarely, tumors exhibit heterogeneous enhancement. It is difficult to distinguish CPPs from CPCs based on imaging alone, however CPCs often show heterogenous contrast enhancement, invasion of the parenchyma, and robust peritumoral edema (Salunke et al. 2014; Taylor et al. 2001).

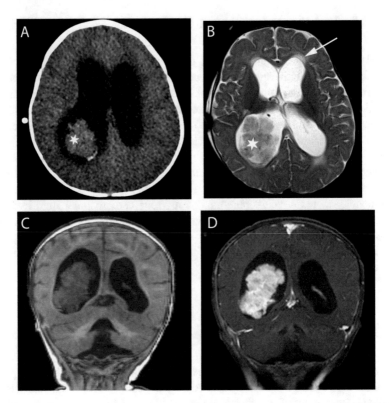

Fig. 10.1 (**a**) Axial image in a non-contrasted CT scan of the head in a 5 month-old child shows hydrocephalus with a choroid plexus papilloma in the right lateral ventricle that is isodense to surrounding brain and shows a small amount of bright calcification on the posterior border (star). (**b**) Axial T2-weighted MRI in the same patient demonstrates the tumor (star) and transependymal flow (arrow), the bright area around the frontal horn of the ventricle indicative of elevated intraventricular pressure from hydrocephalus. (**c**) Coronal non-contrasted T1-weighted MRI demonstrates the large tumor in the lateral ventricle. (**d**) The tumor brightly and homogenously enhances with gadolinium contrast

Diffusion-weighted images (DWI) and apparent diffusion coefficient (ADC) images are MRI sequences that generate pictures based on diffusion of water in cells. These imaging modalities are complementary to the standard imaging techniques. In pediatric tumors, the cellular density correlates with DWI and ADC findings. This may be used to predict the degree of malignancy of a tumor, as higher-grade tumors are often more cellular than lower grade ones. However, choroid plexus tumors may be any density on DWI. There is no difference of these values between grades of tumor and thus the prognostic value of these sequences is limited (Shi et al. 2017a).

MR spectroscopy measures cellular metabolites and is helpful in evaluating the diagnosis and grade of brain tumors. The general finding in brain tumors is an elevated choline peak and decreased *N*-acetyl aspartate (NAA) peak compared to normal brain. Choroid plexus tumors are characterized by an elevated choline peak and absence of NAA (Borja et al. 2013). A higher peak of the metabolites Cho and myoinositol may be useful in differentiating CPCs from CPPs (Borja et al. 2013; Horska et al. 2001; Krieger et al. 2005).

The radiologic differential diagnosis of intraventricular tumors includes ependymomas, meningiomas, central neurocytomas, subependymal giant cell astrocytomas, colloid cysts, astrocytomas, and pineal region tumors. A rare intraventricular tumor is the choroid plexus adenoma. This tumor is distinct from the standard choroid plexus tumor. An adenoma is characterized by well-differentiated tubular glands and lacks the characteristic findings of the standard choroid plexus tumors (Prendergast et al. 2018).

10.4   Histopathology

10.4.1   Gross Pathology

Choroid plexus tumors appear as cauliflower-shaped fronds, with a greyish or brownish-tan appearance. CPPs are well circumscribed and generally separate easily from the ventricular ependymal lining. They are typically soft, but may develop calcifications, hemorrhages, or cystic features (Gaudio et al. 1998). aCPPs appear grossly like CPPs but may have irregular and possibly invasive margins (Safaee et al. 2013a). As opposed to CPCs, aCPPs lack *necrosis*, the finding of intratumoral cellular death that indicates a rapidly progressing, aggressive tumor whose expansion exceeds its blood supply. CPCs are less circumscribed, invade normal brain parenchyma, and may contain areas of necrosis (Gopal et al. 2008).

10.4.2   Microscopic Pathology

Normal human choroid plexus is shown in Fig. 10.2. This structure is characterized by a single layer of cuboidal epithelial cells lining finger-like papillary structures with fibrovascular cores. The epithelial cells have central, uniform round-to-oval nuclei and eosinophilic cytoplasm. Physiologic calcifications are commonly encountered and increase in frequency with patient age (Fig. 10.2, arrow).

Choroid plexus tumors are graded based on histological findings. Low grade tumors are very similar to normal choroid plexus. Higher grade tumors are distinguished by elevated cellular mitotic activity, histological necrosis, loss of papillary architecture, increased cellularity, and nuclear asymmetry or pleomorphism. Brain invasion may occur with all three grades, although it is rare with CPPs and common with CPCs.

CPPs (WHO Grade I) are microscopically similar to normal choroid plexus. They are characterized by well-formed papillary structures containing fibrovascular cores

Fig. 10.2 Normal human choroid plexus (200× magnification) consists of fingerlike fibrous papillary structures lined by a single layer of uniform cuboidal cells. Physiologic calcifications (arrow) are common and increase in frequency with age

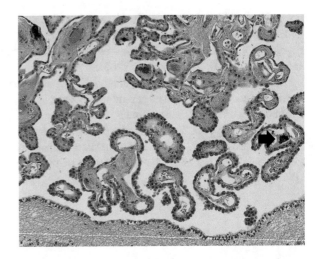

Fig. 10.3 Choroid plexus papilloma (400× magnification) is a WHO Grade I tumor which closely resembles normal choroid plexus. By definition, the mitotic rate is <2 mitoses/ high-powered field

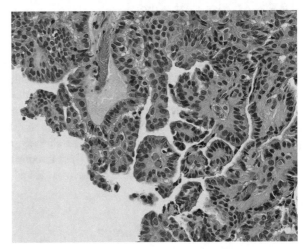

and lined by a single layer of cuboidal to columnar epithelial cells (Fig. 10.3). However, the cells are more elongated and crowded in a higher density than normal choroid plexus. CPPs have a well-defined basement membrane and low mitotic activity (<2 mitosis per 10 high-powered fields). CPPs may have small areas of higher-grade features, but they are rare. Brain invasion is typically absent (Louis et al. 2016). There are rare cases of CPP progressing to CPC (Dhillon et al. 2013; Jeibmann et al. 2007; Niikawa et al. 1993).

CPCs (WHO Grade III) have frankly malignant features (Fig. 10.4). As opposed to the benign variants, CPCs do not have organized papillary structures. The nuclei are not homogenously round, as in normal choroid plexus. Nuclei are elongated, differ from cell to cell, and may form bizarre, irregular shapes. There is a dramatically increased density or crowding of cells compared to the benign variants. There

Fig. 10.4 Choroid plexus carcinoma (400× magnification) is a WHO Grade III tumor defined by the presence of several high-grade features, including a high rate of mitoses (arrow), irregularly and heterogeneously shaped nuclei, loss of papillary architecture, and necrosis

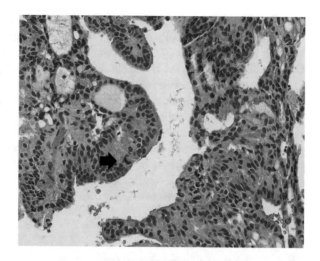

Fig. 10.5 Atypical choroid plexus papilloma (400× magnification) is a WHO Grade II tumor with is defined by a higher amount of mitoses (arrows) than a choroid plexus papilloma, but does not meet the criteria for a choroid plexus carcinoma

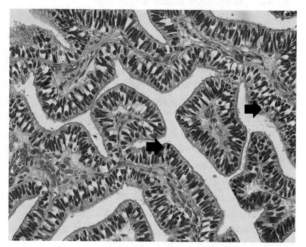

is a high mitotic rate (>5 mitoses per 10 high-powered fields); the arrow in Fig. 10.4 indicates a mitotic figure. CPCs are characterized by at least four of five histologic features: high mitotic rage (>5 mitoses per 10 high-powered fields), increased cellularity, nuclear pleomorphism, necrosis, and loss of papillary architecture (Fig. 10.4) (Ellenbogen et al. 1989; Ogiwara et al. 2012; St Clair et al. 1991). Brain invasion is not necessary for the diagnosis of CPC, but it is a common finding (Louis et al. 2016).

In 2007, the WHO Working group recognized an intermediate Grade II entity (Fig. 10.5). This tumor was considered an "atypical" choroid plexus papilloma and is separated from the standard CPP by an increased mitotic rate of (≥2 mitoses/10 high power fields). aCPPs retain a papillary architecture, but the structures are often more complex and irregular than CPPs. The cells show crowding and may have elongated nuclei. The arrows in Fig. 10.5 indicate mitotic figures. The increased mitotic rate is the defining factor of aCPP compared to CPP, as the other histological factors are variable (Louis et al. 2007). These tumors may have overlapping features with CPCs, without reaching the threshold for a Grade III diagnosis.

The diagnosis of aCPP depends on the patient's age. Patients over 3 years of age with aCPPs have a higher probability of tumor recurrence and an overall worse prognosis compared to patients with CPPs. However, an evaluation of 149 patients in the choroid plexus tumor registry of the International Society of Pediatric Oncology (CPT-SIOP) showed that the clinical outcomes and the progression free survival of children less than 3 years of age was the same in patients with CPP or aCPP (Thomas et al. 2015). Therefore, increased mitotic activity is only considered to be of prognostic value in children over 3 years of age.

10.4.3 Immunohistochemistry

Immunohistochemical staining of a choroid plexus tumor is similar to normal choroid plexus, although CPCs have significant variability (Louis et al. 2016). The normal choroid plexus reacts positively with cytokeratin (CK) stains, particularly CK7. It also reacts with vimentin, S100, and transthyretin (prealbumin). Normal choroid plexus cells also reactive positively for stains directed toward the potassium channel KIR7.1. Stains for epithelial membrane antigen (EMA) and glial fibrillary acidic protein (GFAP) are weak or negative (Louis et al. 2016).

Most CPPs are positive for cytokeratins, vimentin, and podoplanin (Safaee et al. 2013a). KIR7.1 and stanniocalcin-1 are sensitive and specific markers for CPPs (Hasselblatt et al. 2006). CPPs may show any combination of CK7 and 20, but are most commonly CK7 positive and CK20 negative (74%) (Gyure and Morrison 2000). The Ki-67 and mouse intestinal bacteria (MIB-1) indices are also useful for differentiating CPPs from higher grade choroid plexus tumors by defining the mitotic activity (Safaee et al. 2013a).

Aquaporins (AQP) are selective water channel proteins that are integral to human homeostasis and may be integral in cerebrospinal fluid production. Normal choroid plexus heavily expresses AQP1 in the apical portion of cuboidal cells. CPPs also express AQP1, although in a variable, heterogenous degree. The intensity of AQP1 expression in CPPs correlates with the presence of hydrocephalus (Longatti et al. 2006; Paul et al. 2011). However, CPCs do not stain for AQP1.

Immunohistochemistry is an important method to distinguish choroid plexus tumors from other primary or metastatic neoplasms. Choroid plexus tumors may be differentiated from metastatic lesions by their lack of human epithelial antigen (HEA)-125 and Ber-EP4 expression (Safaee et al. 2013a). These are tumor markers sensitive to cancer of the skin, ovarium, colon, prostate, stomach, and lung. Another intraventricular tumor is an ependymoma. Choroid plexus tumors may be differentiated from ependymomas with the tumor markers E-cadherin and neural cell adhesion marker (NCAM). Choroid plexus tumors are E-cadherin+/NCAM- and ependymomas are E-cadherin-/NCAM+. Tumors of the pineal gland may also present in the same anatomic area as choroid plexus tumors. The lack of NCAM and lack of microtubule-associated protein 2 differentiates choroid plexus tumors from primary tumors of the pineal gland (Safaee et al. 2013a).

10.4.4 Molecular Genetics

The most established genetic connection associated with choroid plexus tumors is alteration of the p53 tumor suppressor gene. Alteration in this gene increases the risk of choroid plexus tumors (Gozali et al. 2012). Patients with Li-Fraumeni syndrome, a hereditary condition with p53 gene mutations, develop many types of neoplasms, including choroid plexus tumors. Mutations in the p53 gene are found in 50% of CPCs and 5% of CPPs (Tabori et al. 2010). Tumors with p53 alterations show increased aggressiveness and patients have a worse prognosis (Merino et al. 2015). Polymorphisms in mouse double minute 2 homolog (MDM2), a negative regulator of p53, cause defects in the tumor suppressor function. Defects in MDM2 are also associated with choroid plexus tumors (Tabori et al. 2010).

"Notch" is another gene associated with choroid plexus tumors. The Notch pathway is a central regulator during neural development, promoting proliferation and inhibiting differentiating of embryonic progenitor cells in the central nervous system (Dang et al. 2006). These mechanisms are parasitized by tumors to promote their own growth and survival. The distribution of Notch proteins differs between normal choroid plexus and tumors. In normal choroid plexus, Notch 1, 2, and 4 are distributed in the cell membrane and cytoplasm. In neoplastic cells, these proteins are found in the nuclei; this enhances their signaling capabilities (Beschorner et al. 2013). Notch 3 has been shown to be instrumental in the formation choroid plexus tumors in experimental mice (Dang et al. 2006). The mechanism through which Notch signaling promotes choroid plexus tumor formation is thought to be through the Sonic Hedgehog (SHH) gene (Li et al. 2016).

Multiple other genes are associated with choroid plexus tumors (Merino et al. 2015; Losi-Guembarovski et al. 2007; Tong et al. 2015). Genes associated with tumorigenesis include: Twist-related protein 1 (TWIST1), Wnt inhibitor factor 1 (WIF1), transmembrane protein Shrew-1 (AJAP1), transient receptor protein, M2 channel (TRPM2), BCL2-associated transcription factor (BCLAF1), and IL-6 signal transducer (IL6ST) (Hasselblatt et al. 2009). Platelet derived growth factor (PDGF) is a known oncogene in multiple tumors, including choroid plexus tumors. PDGF receptors are phosphorylated at abnormally high levels in CPCs (Koos et al. 2009; Nupponen et al. 2008). Other genes that are influential in CPC progression include TAF12, NFYC, and RAD54L (Tong et al. 2015). TAF12 and NFYC are epigenomic regulators; RAD54L plays a role in DNA repair.

10.5   Clinical Treatment

Patients often present with signs and symptoms of hydrocephalus. Hydrocephalus may result from mechanical obstruction of the normal CSF pathways or by hypersecretion of CSF from the tumor (Kahn and Luros 1952). The degree of hydrocephalus is not correlated to the anatomic location, size, or pathology of the tumor (Pencalet et al. 1998; Safaee et al. 2013b). Choroid plexus tumors may produce up to 800 ml of CSF per day, well above the normal physiologic production of approximately 450 ml per day (Pencalet et al. 1998; Eisenberg et al. 1974; Ghatak and McWhorter 1976). Approximately 25–50% of all patients with choroid plexus tumors will require permanent CSF diversion (Ogiwara et al. 2012; Pencalet et al. 1998; Tacconi et al. 1996). Complete surgical resection of the tumor reduces the overproduction of CSF and will cure hydrocephalus in approximately 70–80% of tumors. However, 20–30% of patients will still require treatment of hydrocephalus despite tumor removal (Bettegowda et al. 2012). Possible reasons for hydrocephalus after tumor excision include arachnoiditis, hemorrhage, elevated CSF protein, debris from the tumor, or mechanical obstruction of the cerebral aqueduct. Pre-operative endovascular embolization of feeding arteries reduces the production of CSF, however this alone does not alter the overall need for CSF diversion (Haliasos et al. 2013a).

The treatment of choice for choroid plexus tumors is surgical resection. Gross total resection (GTR) of the tumor is the most important factor determining the risk of recurrence and the rate of long-term survival (Wolff et al. 2002; Bettegowda et al. 2012; Bahar et al. 2017; Koh et al. 2014; Krishnan et al. 2004).

Complete resection of CPPs is often curative (McGirr et al. 1988; Bostrom et al. 2011). The 10-year survival of patients with a GTR of CPPs approaches 100% (McGirr et al. 1988; Ogiwara et al. 2012; Pencalet et al. 1998; Safaee et al. 2013b; Bettegowda et al. 2012; Menon et al. 2010). Patients do not require adjuvant therapy after surgical resection. The overall survival in patients with a subtotal resection remains positive. Patients with a small residual should be observed closely, but do not require radiation or chemotherapy. CPPs are often quite indolent, and residual tumor has been observed without signs of growth for years. A second surgery is indicated if residual tumor grows. Adjuvant therapy is considered for inoperable lesions that demonstrate clear signs of growth or malignancy (McGirr et al. 1988; Safaee et al. 2013b; Krishnan et al. 2004; Menon et al. 2010).

Complete resection of aCPPs may be curative (Koh et al. 2014). However, recurrences are more common in aCPPs than CPPs. Jeibmann reported a series of patients with gross total resection of choroid plexus tumors. Six of 103 (6%) patients with CPPs suffered a recurrence. Six of 21 (29%) of patients with aCPPs suffered a recurrence. The difference was statistically significant (Jeibmann et al. 2007). In two of the patients in this series, the pathology of the recurrent tumor had progressed to CPC.

The survival of patients with CPCs is significantly worse than those with CPPs. In a large meta-analysis performed in 2002 by Wolff et al., the 5- and 10- year survival for CPPs was 81% and 77% respectively. For CPCs, the survival dropped to 41% and 35% respectively (Wolff et al. 2002). Similarly, Pencalet et al. reported a 5-year survival of 100% in patients with CPPs and 40% in patients with CPCs in a single institution experience of 38 tumors (Pencalet et al. 1998). Overall, the 5-year survival of patients with CPC ranges from 26–41% (Lam et al. 2013). Choroid plexus carcinomas are approximately 20 times more likely to recur and metastasize compared to CPPs (Bettegowda et al. 2012). Gross total resection significantly improves survival in patients with CPCs (Ogiwara et al. 2012; Gupta 2003; Sun et al. 2014b; Wolff et al. 1999). In a population based study using data from the SEER (surveillance, epidemiology, and end results) database, the 5-year survival of patients with a gross total resection of a CPC was 70.9%, compared to 35.9% after subtotal resection (Lam et al. 2013). GTR may be technically difficult due to brain invasion or anatomic features. In older series, GTR was achieved in less than 50% of patients (Boyd and Steinbok 1987; Gupta 2003; Berger et al. 1998; Chow et al. 1999; Hawkins 3rd 1980). If tumor resection is incomplete, a "second-look" surgery is beneficial. In a series reported by Sun et al., the 2 year survival of patients with CPC and a second resection was 69%, compared to 30% in patients without a second surgery (Sun et al. 2014a).

CPCs are generally treated with adjuvant therapy after surgery. Chemotherapy improves survival. Chemotherapy is of particular importance in children under three years of age, as radiation may be devastating to the developing brain. Multiple chemotherapeutic agents have been employed, including carboplatin, bevacizumab, vincristine, etoposide, and cyclophosphamide. Due to the rarity of these tumors, a large scale, randomized trial has not been performed. However, platinum-based therapies and etoposide are most often used based on the established result in other pediatric brain tumors (Mallick et al. 2017). Etoposide may be the most efficacious single agent (Mallick et al. 2017). There are a few encouraging case reports of high-dose chemotherapy and autologous stem cell transplant in younger children (Bostrom et al. 2011).

The use of radiotherapy is controversial. There are studies that indicate a better overall survival rate in patients treated with craniospinal radiation (Wolff et al. 1999; Mazloom et al. 2010). Other reports failed to find any survival benefit with radiotherapy, including an analysis of the SEER database (Lam et al. 2013; Gupta 2003; Packer et al. 1992; Sampath et al. 2008).

While gross total resection is key to treatment, surgery may be technically challenging. Choroid plexus tumors may be extremely vascular. The major cause of surgical morbidity or mortality in children is perioperative blood loss (St Clair et al. 1991; Pencalet et al. 1998; Due-Tonnessen et al. 2001). Older series cite a surgical mortality rate of up to 30% because of hemorrhage (Guidetti and Spallone 1981). The vascular supply for the tumor generally includes a major feeding vessel from the choroidal arteries (Ogiwara et al. 2012). Unfortunately, this vessel is generally in the deepest part of the surgical bed and may not be encountered until the later stages of the surgery.

Modern surgical techniques have improved the feasibility of resection of tumors. The operating microscope and neuroendoscope provide significant improvements in

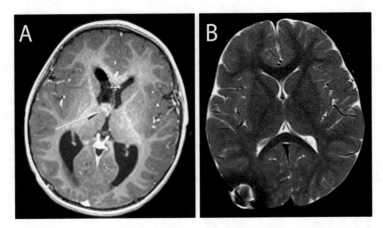

Fig. 10.6 A 35-month old child presented with rapid progression of hydrocephalus. A T1-weighted MRI with gadolinium contrast shows an obstructive lesion at the level of the foramen of Monroe (**a**). This lesion was removed with a right frontal neuroendoscopic approach. The endoscope allowed removal of the lesion with minimal trauma to the surrounding brain. The lesion was a choroid plexus papilloma, WHO Grade I. There is no residual or recurrent tumor at the level of the foramen seen on a follow-up T2 weighted axial MRI one year after surgery (**b**). The child did not receive adjuvant therapy. However, despite complete tumor removal, the child required a cerebrospinal fluid shunt to treat hydrocephalus

visualization of these deep, intraventricular lesions. Figure 10.6a shows a choroid plexus papilloma located at the Foramen of Monro. This tumor was removed through a small, minimally-invasive approach with a neuroendoscope (Fig. 10.6b). In experienced hands, minimally invasive approaches mitigate cortical trauma and intraoperative blood loss. Endovascular embolization may significantly decrease the vascularity of the tumor and also reduce surgical blood loss (Haliasos et al. 2013b). There is a report of tumor regression in a 3-month-old child after embolization alone (Wind et al. 2010).

With the increasing availability of imaging, incidental CPPs may be discovered in patients without symptoms or hydrocephalus. There is no clear evidence to recommend surgery or observation in these patients (Laarakker et al. 2017). The natural history of incidentally discovered choroid plexus tumors is unknown. There are reports of rapid progression of tumors in younger children (Fig. 10.7a, b) (Gorelyshev et al. 2013; Jamjoom et al. 2009). However, there are also reports of very indolent behavior of the residual of subtotally resected tumors. With no clear, scientific evidence, decisions on incidentally found CPPs should be made on a "case-by-case" basis with a thorough discussion of the surgical risk with the patient and the family. The advantage of observation includes the possibility that the tumor will remain indolent and the avoidance of surgical risk. The advantage of surgery is that it may be curative, will provide specimen for pathologic diagnosis, and may avoid subsequent development of symptoms.

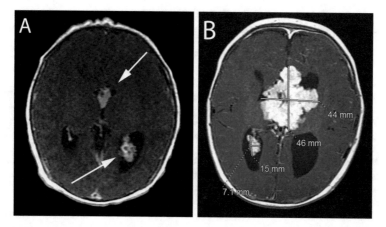

Fig. 10.7 A rapidly progressing multifocal tumor in an infant is shown. (**a**) Axial T1-weighted MRI image with gadolinium contrast in a newborn shows a tumor in the third ventricle and another in the left lateral ventricle. The left ventricular tumor was removed at 3 months of age. By 6 months of age, the third ventricular tumor grew significantly (**b**)

10.6   Conclusions

Choroid plexus tumors may be benign, malignant, or have an intermediate grade. Benign CPPs are the most common, representing 80% of all choroid plexus tumors. The tumors tend to be located within the ventricles and may be very vascular. The tumors are more common in the pediatric population, and are one of the most common brain tumors encountered in children under 1 year of age. Choroid tumors have a very characteristic "cauliflower" appearance on imaging studies. They are often accompanied by hydrocephalus, which may persist despite removal of the tumor. Surgical resection is the treatment of choice; gross total resection improves the outcomes of patients with all three grades of tumor. Adjuvant therapy is reserved for malignant or inoperable tumors. There are encouraging results with chemotherapy for malignant tumors. The role of radiation therapy is controversial. The clinical outcome in patients with CPPs is very good with surgery alone. The long-term survival in patients with CPCs is less encouraging, and multimodal therapy is required.

References

Abdulkader MM, Mansour NH, Van Gompel JJ, Bosh GA, Dropcho EJ, Bonnin JM et al (2016) Disseminated choroid plexus papillomas in adults: a case series and review of the literature. J Clin Neurosci 32:148–154

Arasappa R, Danivas V, Venkatasubramanian G (2013) Choroid plexus papilloma presenting as schizophrenia: a case report. J Neuropsychiatry Clin Neurosci 25:E26–E27

Bahar M, Hashem H, Tekautz T, Worley S, Tang A, de Blank P et al (2017) Choroid plexus tumors in adult and pediatric populations: the Cleveland Clinic and University Hospitals experience. J Neuro-Oncol 132:427–432

Berger C, Thiesse P, Lellouch-Tubiana A, Kalifa C, Pierre-Kahn A, Bouffet E (1998) Choroid plexus carcinomas in childhood: clinical features and prognostic factors. Neurosurgery 42:470–475

Beschorner R, Waidelich J, Trautmann K, Psaras T, Schittenhelm J (2013) Notch receptors in human choroid plexus tumors. Histol Histopathol 28:1055–1063

Bettegowda C, Adogwa O, Mehta V, Chaichana KL, Weingart J, Carson BS et al (2012) Treatment of choroid plexus tumors: a 20-year single institutional experience. J Neurosurg Pediatr 10:398–405

Borja MJ, Plaza MJ, Altman N, Saigal G (2013) Conventional and advanced MRI features of pediatric intracranial tumors: supratentorial tumors. AJR Am J Roentgenol 200:W483–W503

Bostrom A, Bostrom JP, von Lehe M, Kandenwein JA, Schramm J, Simon M (2011) Surgical treatment of choroid plexus tumors. Acta Neurochir 153:371–376

Boyd MC, Steinbok P (1987) Choroid plexus tumors: problems in diagnosis and management. J Neurosurg 66:800–805

Cannon DM, Mohindra P, Gondi V, Kruser TJ, Kozak KR (2015) Choroid plexus tumor epidemiology and outcomes: implications for surgical and radiotherapeutic management. J Neuro-Oncol 121:151–157

Chow E, Reardon DA, Shah AB, Jenkins JJ, Langston J, Heideman RL et al (1999) Pediatric choroid plexus neoplasms. Int J Radiat Oncol Biol Phys 44:249–254

Dandy WE (1922) Diagnosis, localization, and removal of tumours of the third ventricle. Bull Johns Hopkins Hosp 33:188–189

Dang L, Fan X, Chaudhry A, Wang M, Gaiano N, Eberhart CG (2006) Notch3 signaling initiates choroid plexus tumor formation. Oncogene 25:487–491

Davis LE, Cushing H (1925) Papillomas of the choroid plexus: with the report of six cases. Arch Neurol Psychiatr 13:681–710

Dhillon RS, Wang YY, McKelvie PA, O'Brien B (2013) Progression of choroid plexus papilloma. J Clin Neurosci 20:1775–1778

Doglietto F, Lauretti L, Tartaglione T, Gessi M, Fernandez E, Maira G (2005) Diffuse craniospinal choroid plexus papilloma with involvement of both cerebellopontine angles. Neurology 65:842

Due-Tonnessen B, Helseth E, Skullerud K, Lundar T (2001) Choroid plexus tumors in children and young adults: report of 16 consecutive cases. Childs Nerv Syst 17:252–256

Eisenberg HM, McComb JG, Lorenzo AV (1974) Cerebrospinal fluid overproduction and hydrocephalus associated with choroid plexus papilloma. J Neurosurg 40:381–385

Ellenbogen RG, Winston KR, Kupsky WJ (1989) Tumors of the choroid plexus in children. Neurosurgery 25:327–335

Gaudio RM, Tacconi L, Rossi ML (1998) Pathology of choroid plexus papillomas: a review. Clin Neurol Neurosurg 100:165–186

Ghatak NR, McWhorter JM (1976) Ultrastructural evidence for CSF production by a choroid plexus papilloma. J Neurosurg 45:409–415

Gopal P, Parker JR, Debski R, Parker JC Jr (2008) Choroid plexus carcinoma. Arch Pathol Lab Med 132:1350–1354

Gorelyshev SK, Matuev KB, Medvedev OA (2013) Migrating choroid plexus papilloma of the lateral ventricle in infant--modern approaches to surgical treatment. Zh Vopr Neirokhir Im N N Burdenko 77:45–49; discussion 49–50

Gozali AE, Britt B, Shane L, Gonzalez I, Gilles F, McComb JG et al (2012) Choroid plexus tumors; management, outcome, and association with the Li-Fraumeni syndrome: the Children's Hospital Los Angeles (CHLA) experience, 1991-2010. Pediatr Blood Cancer 58:905–909

Guidetti B, Spallone A (1981) The surgical treatment of choroid plexus papillomas: the results of 27 years experience. Neurosurg Rev 4:129–137

Gupta N (2003) Choroid plexus tumors in children. Neurosurg Clin N Am 14:621–631

Gyure KA, Morrison AL (2000) Cytokeratin 7 and 20 expression in choroid plexus tumors: utility in differentiating these neoplasms from metastatic carcinomas. Mod Pathol 13:638–643

Haliasos N, Brew S, Robertson F, Hayward R, Thompson D, Chakraborty A (2013a) Pre-operative embolisation of choroid plexus tumours in children. Part II. Observations on the effects on CSF production. Childs Nerv Syst 29:71–76

Haliasos N, Brew S, Robertson F, Hayward R, Thompson D, Chakraborty A (2013b) Preoperative embolisation of choroid plexus tumours in children: Part I-Does the reduction of perioperative blood loss affect the safety of subsequent surgery? Childs Nerv Syst 29:65–70

Hasselblatt M, Bohm C, Tatenhorst L, Dinh V, Newrzella D, Keyvani K et al (2006) Identification of novel diagnostic markers for choroid plexus tumors: a microarray-based approach. Am J Surg Pathol 30:66–74

Hasselblatt M, Mertsch S, Koos B, Riesmeier B, Stegemann H, Jeibmann A et al (2009) TWIST-1 is overexpressed in neoplastic choroid plexus epithelial cells and promotes proliferation and invasion. Cancer Res 69:2219–2223

Hawkins JC 3rd (1980) Treatment of choroid plexus papillomas in children: a brief analysis of twenty years' experience. Neurosurgery 6:380–384

Horska A, Ulug AM, Melhem ER, Filippi CG, Burger PC, Edgar MA et al (2001) Proton magnetic resonance spectroscopy of choroid plexus tumors in children. J Magn Reson Imaging 14:78–82

Jagielski J, Zabek M, Wierzba-Bobrowicz T, Lakomiec B, Chrapusta SJ (2001) Disseminating histologically benign multiple papilloma of the choroid plexus: case report. Folia Neuropathol 39:209–213

Jamjoom AA, Sharab MA, Jamjoom AB, Satti MB (2009) Rapid evolution of a choroid plexus papilloma in an infant. Br J Neurosurg 23:324–325

Jeibmann A, Wrede B, Peters O, Wolff JE, Paulus W, Hasselblatt M (2007) Malignant progression in choroid plexus papillomas. J Neurosurg 107:199–202

Jinhu Y, Jianping D, Jun M, Hui S, Yepeng F (2007) Metastasis of a histologically benign choroid plexus papilloma: case report and review of the literature. J Neuro-Oncol 83:47–52

Kahn EA, Luros JT (1952) Hydrocephalus from overproduction of cerebrospinal fluid, and experiences with other parillomas of the choroid plexus. J Neurosurg 9:59–67

Karim A, Fowler M, McLaren B, Cardenas R, Patwardhan R, Nanda A (2006) Concomitant choroid plexus papillomas involving the third and fourth ventricles: a case report and review of the literature. Clin Neurol Neurosurg 108:586–589

Kim TW, Jung TY, Jung S, Kim IY, Moon KS, Jeong EH (2012) Unusual radiologic findings and pathologic growth patterns on choroid plexus papillomas. J Korean Neurosurg Soc 51:272–275

Knierim DS (1990) Choroid plexus tumors in infants. Pediatr Neurosurg 16:276–280

Koh EJ, Wang KC, Phi JH, Lee JY, Choi JW, Park SH et al (2014) Clinical outcome of pediatric choroid plexus tumors: retrospective analysis from a single institute. Childs Nerv Syst 30:217–225

Koos B, Paulsson J, Jarvius M, Sanchez BC, Wrede B, Mertsch S et al (2009) Platelet-derived growth factor receptor expression and activation in choroid plexus tumors. Am J Pathol 175:1631–1637

Krieger MD, Panigrahy A, McComb JG, Nelson MD, Liu X, Gonzalez-Gomez I et al (2005) Differentiation of choroid plexus tumors by advanced magnetic resonance spectroscopy. Neurosurg Focus 18:E4

Krishnan S, Brown PD, Scheithauer BW, Ebersold MJ, Hammack JE, Buckner JC (2004) Choroid plexus papillomas: a single institutional experience. J Neuro-Oncol 68:49–55

Laarakker AS, Nakhla J, Kobets A, Abbott R (2017) Incidental choroid plexus papilloma in a child: a difficult decision. Surg Neurol Int 8:86

Lafay-Cousin L, Keene D, Carret AS, Fryer C, Brossard J, Crooks B et al (2011) Choroid plexus tumors in children less than 36 months: the Canadian pediatric brain tumor consortium (CPBTC) experience. Childs Nerv Syst 27:259–264

Lam S, Lin Y, Cherian J, Qadri U, Harris DA, Melkonian S et al (2013) Choroid plexus tumors in children: a population-based study. Pediatr Neurosurg 49:331–338

Lechanoine F, Zemmoura I, Velut S (2017) Treating cerebrospinal fluid rhinorrhea without dura repair: a case report of posterior fossa choroid plexus papilloma and review of the literature. World Neurosurg 108:990 e991–990 e999

Lena G, Genitori L, Molina J, Legatte JR, Choux M (1990) Choroid plexus tumours in children. Review of 24 cases. Acta Neurochir 106:68–72

Li L, Grausam KB, Wang J, Lun MP, Ohli J, Lidov HG et al (2016) Sonic Hedgehog promotes proliferation of Notch-dependent monociliated choroid plexus tumour cells. Nat Cell Biol 18:418–430

Longatti P, Basaldella L, Orvieto E, Dei Tos A, Martinuzzi A (2006) Aquaporin(s) expression in choroid plexus tumours. Pediatr Neurosurg 42:228–233

Losi-Guembarovski R, Kuasne H, Guembarovski AL, Rainho CA, Colus IM (2007) DNA methylation patterns of the CDH1, RARB, and SFN genes in choroid plexus tumors. Cancer Genet Cytogenet 179:140–145

Louis DN, Ohgaki H, Wiestler OD, Cavenee WK, Burger PC, Jouvet A et al (2007) The 2007 WHO classification of tumours of the central nervous system. Acta Neuropathol 114:97–109

Louis DN, Perry A, Reifenberger G, von Deimling A, Figarella-Branger D, Cavenee WK et al (2016) The 2016 World Health Organization classification of tumors of the central nervous system: a summary. Acta Neuropathol 131:803–820

Mallick S, Benson R, Melgandi W, Rath GK (2017) Effect of surgery, adjuvant therapy, and other prognostic factors on choroid plexus carcinoma: a systematic review and individual patient data analysis. Int J Radiat Oncol Biol Phys 99:1199–1206

Masson CB (1934) Complete removal of two tumors of the third ventricle with recovery. Arch Surg 28:527–537

Mazloom A, Wolff JE, Paulino AC (2010) The impact of radiotherapy fields in the treatment of patients with choroid plexus carcinoma. Int J Radiat Oncol Biol Phys 78:79–84

McGirr SJ, Ebersold MJ, Scheithauer BW, Quast LM, Shaw EG (1988) Choroid plexus papillomas: long-term follow-up results in a surgically treated series. J Neurosurg 69:843–849

Menon G, Nair SN, Baldawa SS, Rao RB, Krishnakumar KP, Gopalakrishnan CV (2010) Choroid plexus tumors: an institutional series of 25 patients. Neurol India 58:429–435

Merino DM, Shlien A, Villani A, Pienkowska M, Mack S, Ramaswamy V et al (2015) Molecular characterization of choroid plexus tumors reveals novel clinically relevant subgroups. Clin Cancer Res 21:184–192

Morshed RA, Lau D, Sun PP, Ostling LR (2017) Spinal drop metastasis from a benign fourth ventricular choroid plexus papilloma in a pediatric patient: case report. J Neurosurg Pediatr 20:471–479

Nagib MG, O'Fallon MT (2000) Lateral ventricle choroid plexus papilloma in childhood: management and complications. Surg Neurol 54:366–372

Niikawa S, Ito T, Murakawa T, Hirayama H, Ando T, Sakai N et al (1993) Recurrence of choroid plexus papilloma with malignant transformation--case report and lectin histochemistry study. Neurol Med Chir (Tokyo) 33:32–35

Nupponen NN, Paulsson J, Jeibmann A, Wrede B, Tanner M, Wolff JE et al (2008) Platelet-derived growth factor receptor expression and amplification in choroid plexus carcinomas. Mod Pathol 21:265–270

Ogiwara H, Dipatri AJ Jr, Alden TD, Bowman RM, Tomita T (2012) Choroid plexus tumors in pediatric patients. Br J Neurosurg 26:32–37

Ostrom QT, Gittleman H, Liao P, Rouse C, Chen Y, Dowling J et al (2014) CBTRUS statistical report: primary brain and central nervous system tumors diagnosed in the United States in 2007-2011. Neuro-Oncology 16(Suppl 4):iv1–i63

Packer RJ, Perilongo G, Johnson D, Sutton LN, Vezina G, Zimmerman RA et al (1992) Choroid plexus carcinoma of childhood. Cancer 69:580–585

Pandey S, Sharma V, Singh K, Ghosh A, Gupta PK (2016) Uncommon presentation of choroid plexus papilloma in an infant. J Pediatr Neurosci 11:61–63

Paul L, Madan M, Rammling M, Chigurupati S, Chan SL, Pattisapu JV (2011) Expression of aquaporin 1 and 4 in a congenital hydrocephalus rat model. Neurosurgery 68:462–473

Pencalet P, Sainte-Rose C, Lellouch-Tubiana A, Kalifa C, Brunelle F, Sgouros S et al (1998) Papillomas and carcinomas of the choroid plexus in children. J Neurosurg 88:521–528

Peyre M, Bah A, Kalamarides M (2012) Multifocal choroid plexus papillomas: case report. Acta Neurochir 154:295–299

Pianetti Filho G, Fonseca LF, da Silva MC (2002) Choroid plexus papilloma and Aicardi syndrome: case report. Arq Neuropsiquiatr 60:1008–1010

Picht T, Jansons J, van Baalen A, Harder A, Pietilae TA (2006) Infant with unusually large choroid plexus papilloma undergoing emergency surgery. Case report with special emphasis on the surgical strategy. Pediatr Neurosurg 42:116–121

Prendergast N, Goldstein JD, Beier AD (2018) Choroid plexus adenoma in a child: expanding the clinical and pathological spectrum. J Neurosurg Pediatr 21:428–433

Safaee M, Oh MC, Bloch O, Sun MZ, Kaur G, Auguste KI et al (2013a) Choroid plexus papillomas: advances in molecular biology and understanding of tumorigenesis. Neuro-Oncology 15:255–267

Safaee M, Clark AJ, Bloch O, Oh MC, Singh A, Auguste KI et al (2013b) Surgical outcomes in choroid plexus papillomas: an institutional experience. J Neuro-Oncol 113:117–125

Salunke P, Sahoo SK, Madhivanan K, Radotra BD (2014) A typical radiological presentation in a case of choroid plexus carcinoma. Surg Neurol Int 5:63

Sampath S, Nitin G, Yasha TC, Chandramouli BA, Devi BI, Kovoor JM (2008) Does choroid plexus tumour differ with age? Br J Neurosurg 22:373–388

Scala M, Morana G, Milanaccio C, Pavanello M, Nozza P, Garre ML (2017) Atypical choroid plexus papilloma: spontaneous resolution of diffuse leptomeningeal contrast enhancement after primary tumor removal in 2 pediatric cases. J Neurosurg Pediatr 20:284–288

Scholsem M, Scholtes F, Robe PA, Bianchi E, Kroonen J, Deprez M (2012) Multifocal choroid plexus papilloma: a case report. Clin Neuropathol 31:430–434

Serifoglu I, Oz II, Yazgan O, Caglar E, Guneyli S, Sunar Erdem CZ (2016) Spinal seeding of choroid plexus papilloma. Spine J 16:e421–e422

Shi YZ, Chen MZ, Huang W, Guo LL, Chen X, Kong D et al (2017a) Atypical choroid plexus papilloma: clinicopathological and neuroradiological features. Acta Radiol 58:983–990

Shi Y, Li X, Chen X, Xu Y, Bo G, Zhou H et al (2017b) Imaging findings of extraventricular choroid plexus papillomas: a study of 10 cases. Oncol Lett 13:1479–1485

Singh P, Khan A, Scott G, Jasper M, Singh E (2017) Lesson of the month 2: a choroid plexus papilloma manifesting as anorexia nervosa in an adult. Clin Med (Lond) 17:183–185

Spallone A, Pastore FS, Giuffre R, Guidetti B (1990) Choroid plexus papillomas in infancy and childhood. Childs Nerv Syst 6:71–74

St Clair SK, Humphreys RP, Pillay PK, Hoffman HJ, Blaser SI, Becker LE (1991) Current management of choroid plexus carcinoma in children. Pediatr Neurosurg 17:225–233

Sun MZ, Oh MC, Ivan ME, Kaur G, Safaee M, Kim JM et al (2014a) Current management of choroid plexus carcinomas. Neurosurg Rev 37:179–192;. discussion 192

Sun MZ, Ivan ME, Clark AJ, Oh MC, Delance AR, Oh T et al (2014b) Gross total resection improves overall survival in children with choroid plexus carcinoma. J Neuro-Oncol 116:179–185

Tabori U, Shlien A, Baskin B, Levitt S, Ray P, Alon N et al (2010) TP53 alterations determine clinical subgroups and survival of patients with choroid plexus tumors. J Clin Oncol 28:1995–2001

Tacconi L, Delfini R, Cantore G (1996) Choroid plexus papillomas: consideration of a surgical series of 33 cases. Acta Neurochir 138:802–810

Taggard DA, Menezes AH (2000) Three choroid plexus papillomas in a patient with Aicardi syndrome. A case report. Pediatr Neurosurg 33:219–223

Taylor MB, Jackson RW, Hughes DG, Wright NB (2001) Magnetic resonance imaging in the diagnosis and management of choroid plexus carcinoma in children. Pediatr Radiol 31:624–630

Thomas C, Ruland V, Kordes U, Hartung S, Capper D, Pietsch T et al (2015) Pediatric atypical choroid plexus papilloma reconsidered: increased mitotic activity is prognostic only in older children. Acta Neuropathol 129:925–927

Tong Y, Merino D, Nimmervoll B, Gupta K, Wang YD, Finkelstein D et al (2015) Cross-species genomics identifies TAF12, NFYC, and RAD54L as choroid plexus carcinoma oncogenes. Cancer Cell 27:712–727

Trifiletti RR, Incorpora G, Polizzi A, Cocuzza MD, Bolan EA, Parano E (1995) Aicardi syndrome with multiple tumors: a case report with literature review. Brain Dev 17:283–285

Unger MBE (1906) Zur kenntnis der primaren epithelgeschwulste der adergeflechte des dehirns. Arch Klin Chir 81:61–82

Wind JJ, Bell RS, Bank WO, Myseros JS (2010) Treatment of third ventricular choroid plexus papilloma in an infant with embolization alone. J Neurosurg Pediatr 6:579–582

Wolff JE, Sajedi M, Coppes MJ, Anderson RA, Egeler RM (1999) Radiation therapy and survival in choroid plexus carcinoma. Lancet 353:2126

Wolff JE, Sajedi M, Brant R, Coppes MJ, Egeler RM (2002) Choroid plexus tumours. Br J Cancer 87:1086–1091

Wrede B, Hasselblatt M, Peters O, Thall PF, Kutluk T, Moghrabi A et al (2009) Atypical choroid plexus papilloma: clinical experience in the CPT-SIOP-2000 study. J Neuro-Oncol 95:383–392

Zachary G, George J, Jaishri B, Peter B, Stephanie T (2014) Management of disseminated choroid plexus papilloma: a case study. Pediatr Blood Cancer 61:562–563

Chapter 11
Roles of the Choroid Plexus in CNS Infections

Christian Schwerk, Tobias Tenenbaum, and Horst Schroten

Abstract The choroid plexus (CP) is a highly vascularized endothelial-epithelial convolute located in the ventricular system of the brain. The CP is also the localization of the blood-cerebrospinal fluid barrier (BCSFB), which is formed by the CP epithelium. During infectious diseases of the central nervous system (CNS) the CP can play multiple roles, which will be covered by this review. To cause infection of the CNS pathogens need to cross into the brain, and the CP can present the site of entry across the BCSFB. For invading the CNS pathogens have evolved several mechanisms that can require the help of host immune cells. Infections with pathogens often also cause cell death in the CP, which can be responsible for impairment of barrier function. The CP reacts to the intruders by staging an inflammatory response involving the production of cytokines and chemokines and can function as an entry gate for host immune cells. Following entry of pathogens and host immune cells into the brain both players can cause severe damage leading to neurological sequelae. Additionally, impairment of important CP functions might be involved in further complications like hydrocephalus.

11.1 Introduction

Infections of the central nervous system (CNS) belong to the most serious diseases in animals and humans and present with a wide range of manifestations including meningitis, encephalitis and meningoencephalitis (Dando et al. 2014; Parikh et al. 2012). Although viral agents are the most common cause of CNS infection, bacterial infections have a higher potential to develop a serial illness with the possibility of permanent neurological sequelae (e.g. hearing loss, intellectual and cognitive

C. Schwerk (✉) · T. Tenenbaum · H. Schroten
Department of Pediatrics, Pediatric Infectious Diseases, Medical Faculty Mannheim, Heidelberg University, Mannheim, Germany
e-mail: Christian.schwerk@medma.uni-heidelberg.de; Tobias.tenenbaum@medma.uni-heidelberg.de; Horst.schroten@umm.de

© The American Physiological Society 2020
J. Praetorius et al. (eds.), *Role of the Choroid Plexus in Health and Disease*,
Physiology in Health and Disease, https://doi.org/10.1007/978-1-0716-0536-3_11

impairment) even after successful treatment (Grandgirard et al. 2013; Grimwood et al. 1995).

Generally, the CNS is rather well protected against challenge by pathogens from the environment. This protection is based on the fact that to enter the CNS pathogens need to cross certain barriers located between the blood and the brain. The best characterized of these barriers is the "classical" blood-brain barrier (BBB), which is located at the so-called microvascular unit. The microvascular unit in humans consists of the endothelial cells of the brain, which are connected to each other by tight junctions (TJs) and are supported by pericytes and the endfeet of astrocytes. Together, these cells present a considerable obstacle to invading pahogens (O'Brown et al. 2018; Daneman and Prat 2015).

Another one of these barriers is constituted by the blood-cerebrospinal fluid (CSF) barrier (BCSFB), which is seated at the choroid plexus (CP). The CP, located in the ventricular system of the brain, is best known as the organ responsible for producing and secreting about two thirds of the CSF (Liddelow 2015). An important requirement for this property is the extensive vascularization of the CP that, in concert with fenestrations in the endothelial cells of the CP, enables the production of CSF from the blood (Wolburg and Paulus 2010; Ghersi-Egea et al. 2018).

Similarly to the endothelium of the BBB, the epithelial cells of the CP are tightly connected to each other by TJ strands. This way the CP epithelium presents a "physical barrier" preventing a paracellular movement of high-molecular weight substances or leukocytes between the cells, including that of pathogens trying to enter the CNS. On the other hand, entry into the brain can also be achieved transcellularly directly through CP epithelial cells. This is hindered by a low pinocytotic activity and the expression of specifically located transporter systems by the CP epithelium, thereby providing a "biochemical barrier". Due to the combination of all these properties the CP epithelial cells can be considered as the morphological correlate of the BCSFB (Wolburg and Paulus 2010; Liddelow 2015).

During the progress of brain infection by pathogens the CP plays several roles, which will be covered in this chapter, some of which are rather passive, whereas others are more active. Importantly, the extensive vascularization of the CP provides a large surface not only for exchange between the blood and the CSF, but also for pathogens to enter the CNS. Although the properties of the CP epithelium comprising the BCSFB represent a significant "road block" for pathogens when trying to enter the brain, pathogens have developed an arsenal of strategies to overcome this barrier (Dando et al. 2014). In this regard, a role of the CP during CNS infection is to function as entry site for pathogens into the CNS.

Still, the functions of the CP go beyond the mere role of an entry gate. Maintenance of barrier function, or lack of, is of significant importance during the progress of CNS disease. At this stage, death of cells in the CP can play a major role, since it leads to weakening of the barrier function. Part of such an impairment of barrier function can be a direct response of the CP following recognition of foreign organisms. This function of the CP is mediated by host cell receptors that elicit cellular signaling pathways leading to the activation of inflammatory response genes. The host cell reply is a direct contribution of the CP to an inflammatory immune response that helps to defend against the invading pathogens.

Notably, the BCSFB can regulate and control the entry of immune cells into the CNS, thereby providing an immunological barrier separating the brain from the blood (Liddelow 2015; Wolburg and Paulus 2010; Shechter et al. 2013). As will be covered in this chapter this immune cell entry is also modulated by the CP during the course of a CNS disease. During this process the inflammatory response of CP cells plays an important role.

In addition to meningitis, encephalitis and meningoencephalitis, during an infection of the CNS further complications often occur, including neurological sequelae and hydrocephalus (Lucas et al. 2016). Besides being responsible for the presence of pathogens in the brain when serving as entry site, the CP might contribute in more ways to such additional complications, an aspect that will also be addressed.

11.2 The CP as Entry Gate: Mechanisms of Pathogens to Overcome the BCSFB

In a healthy state the BCSFB can protect the CP by blocking pathogen movement into the CNS. Under specific circumstances this barrier is either bypassed or broken down during the course of CNS infections. The CP is targeted by distinct groups of pathogens with most CNS infections caused by viruses, with an incidence of 20–30/ 100,000 per year (Michos et al. 2007; Rotbart 2000). CNS invading viruses can belong to many different families (Dahm et al. 2016), several of which have been shown to enter the CNS via the CP, including members of the *Retroviridae* and *Picornaviridae* (Falangola et al. 1995; Tabor-Godwin et al. 2010). Part of the family *Picornaviridae* is the genus Enterovirus, members of which often cause aseptic meningitis and encephalitis in neonates and young children (Rhoades et al. 2011).

Although bacteria cause CNS disease less frequently than viral pathogens, they provoke serious complications more commonly. The CP as entry gate has been considered for several bacterial species. These include Gram-negative bacteria as *Neisseria meningitidis* (*N. meningitidis*), *Haemophilus influenzae* and *Escherichia coli*, but also Gram-positive species including *Listeria monocytogenes* (*L. monocytogenes*) and different *Streptoccoccus* species (Lauer et al. 2018; Schwerk et al. 2015). Recently, infiltration of the CNS in rabbits by the Gram-positive bacterium *Bacilllus anthracis* was reported to occur mainly via the CP (Sittner et al. 2017).

Some evidence of entry into the brain across the BCSFB has also been reported for fungal pathogens and parasites, although CNS infection by the latter is rather uncommon. Parasites with involvement of the CP during brain infection include *Trypanosoma brucei*, *Angiostrongylus cantonensis* and *Toxoplasma gondii*. The most common fungus causing CNS infection would be *Cryptococcus neoformans* (*C. neoformans*) that affects immunocompromised patients, especially in case of an HIV infection. Commonly, *C. neoformans* is assumed to enter the CNS via the BBB, but in some instances choroid plexitis has been described during cryptococal brain infection [for recent reviews see (Schwerk et al. 2015; Lauer et al. 2018)].

When invading into the CNS by using the CP, the BCSFB needs to be overcome by pathogens, which can be accomplished by different strategies, which are not necessarily exclusive. The pathogens can cross the barrier between CP epithelial cells in a paracellular fashion or transcellularly directly through the cells. Furthermore, a so-called "Trojan horse" strategy can be applied, during which the pathogens enter the CNS inside of transmigrating host immune cells. In the following examples, pathogens that cross the BCSFB by applying these different strategies will be presented. A schematic overview of pathogens entering the CNS via the CP and their invasion strategies is presented in Fig. 11.1.

11.2.1 Paracellular Entry Between CP Epithelial Cells

The TJ strands connecting the CP epithelial cells tightly seal this cell layer preventing an uncontrolled flux between the cells. Therefore, to allow a paracellular migration of pathogens between the cells the TJs need to be opened. Whereas several examples for a paracellular progress of pathogens into the CNS have been described for the BBB (Dando et al. 2014), evidence for this mechanism at the BCSFB is less frequent. For African trypanosomes, localization at the CP was detected by electron microscopy (Wolburg et al. 2012). It has been shown that early during infection these parasites cross the CP epithelium. Expression of cytokines as tumor necrosis factor α (TNFα) could lead to opening of the junctions of the BCSFB enabling a paracellular entry of trypanosomes (Quan et al. 1999; Masocha and Kristensson 2012). Further supporting a possible paracellular mechanism is the observation that trypanosomes can interact with the TJ proteins claudin-1 and claudin-11, which are present at the CP epithelium (Mogk et al. 2014). Still, a transcellular progress through the cells is also possible, which has been described for several pathogens at the BCSFB.

11.2.2 Transcellular Entry Directly Through CP Epithelial Cells

To avoid the need to open the cell-contacts between the cells forming the epithelial cell layer pathogens can employ the strategy to proceed directly through the cells. A prerequisite for this mechanism is cellular entry by the pathogen, which can be achieved by different mechanisms. One of the best described examples is the facultative intracellular Gram-positive bacterium *L. monocytogenes*, the cause of the disease listeriosis that can lead to meningitis and meningoencephalitis in neonates, but also in immunocompromised and elderly persons (Vazquez-Boland et al. 2001; Radoshevich and Cossart 2018). In vivo studies have shown that *L. monocytogenes* targets the CP during infection. Interestingly, monocytes with

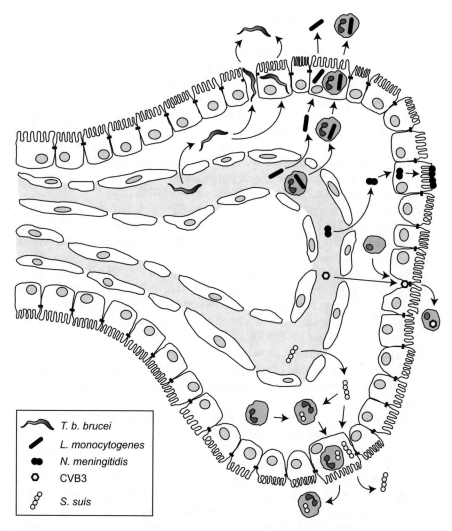

Fig. 11.1 A schematic depiction of examples for pathogens entering the CNS via the CP and their invasion strategies (for more details please refer to the text). The highly vascularized CP is characterized by blood vessels composed of a fenestrated endothelium and a layer of epithelial cells, which are tightly connected by TJs and constitute the BCSFB. It has been shown that African trypanosomes (*T. b. brucei*) pass the CP epithelium. Although evidence for a paracellular process exists, a transcellular crossing is also possible. The Gram-positive bacterium *L. monocytogenes* can enter the CNS inside of infected monocytes in a "Trojan horse" fashion, but in vitro *L. monocytogenes* can also directly invade CP epithelial cells. The latter mechanism has also been shown for the Gram-negative bacterium *N. meningitidis* that forms microcolonies at the apical side of CP epithelial cells following transcellular cossing of the CP epithelium. The Picornavirus CVB3 can bind to the TJ-associated CAR. A certain population of myeloid cells is highly susceptible to infection with CVB3 and serves as "Trojan horse" during traversal into the CNS. A "Trojan horse" mechanism has also been suggested for the Gram-positive bacterium *S. suis*, which was found inside of transmigrating PMN in an in vitro model. Alternatively, *S. suis* can directly invade CP epithelial cells

internal bacteria are often found (Berche 1995; Prats et al. 1992; Schluter et al. 1996).

L. monocytogenes has developed an intricate game plan to hijack host pathways for cell invasion, which requires the expression of a variety of virulence factors, including those of the internalin family. Two of the most important members of this protein family are Internalin (InlA) and Internalin B (InlB). Both InlA and InlB interact with receptors on the surface of host cells, which are E-cadherin (Ecad) for InlA and Met, the hepatocyte growth factor receptor, for InlB, respectively. Binding of InlA and InlB leads to activation of the receptor downstream signaling pathways, which enable the bacteria to hijack the endocytotic machinery of host cells, a process that ultimately enables *L. monocytogenes* to invade even into non-phagocytotic cells (Pizarro-Cerda et al. 2012; Radoshevich and Cossart 2018).

Invasion of *L. monocytogenes* into cells of the CP was investigated in an in vitro model of the BCSFB based on human epithelial CP papilloma (HIBCPP) cells, which enables analysis of infection by pathogens from both the basolateral "blood" and the apical "CSF" side (Dinner et al. 2016). Cell entry occurred in a polar fashion from the physiologically relevant "blood" side, which can be explained by the exclusive basolateral localization of Ecad and Met on HIBCPP cells (Grundler et al. 2013). The invasion process was Dynamin-dependent and relied on the activation of downstream mitogen activated protein kinase (MAPK) signaling, since specific inhibition of either Dynamin-mediated endocytosis or activation of MAPK signaling significantly decreased infection of HIBCPP cells (Dinner et al. 2017). *L. monocytogenes* can also invade into sheep CP epithelial cells, a process that requires the immunogenic surface protein C (IspC) of *L. monocytogenes* (Wang and Lin 2008).

The human-specific Gram-negative bacterium *N. meningitidis* is a colonizer of the upper respiratory tract. Although often present in the form of commensal strains, in susceptible individuals some strains of *N. meningitidis* can cause life threatening disease like sepsis and brain infections leading to meningitis (Stephens 2009). Different sites of entry into the CNS are under consideration for *N. meningitidis*, including postcapillary veins and venules located at subpial and subarachnoidal spaces (Join-Lambert et al. 2010; Christodoulides et al. 2002), but also the CP, since in human patients with meningococcal disease bacteria have been found in the lumen of blood vessels and associated CP epithelial cells (Guarner et al. 2004; Pron et al. 1997). Similar to *L. monocytogenes*, *N. meningitidis* invades HIBCPP cells, as model of the human BCSFB, in vitro in a polar fashion from the basolateral "blood" side, although the direct interactions between bacteria and host cells have not been elucidated yet. Immunofluorescence and electron microscopic analyses have shown the presence of transmigrated bacteria forming microcolonies on the apical cell surface in immediate neighborship to cells containing invaded bacteria, supporting the existence of a transcellular mechanism for crossing the CP epithelium (Schwerk et al. 2012).

The natural host of the Gram-positive bacterium *S. suis* are pigs, where the bacteria colonize the tonsils. As a zoonotic pathogen *S. suis* can cause CNS infection not only in pigs but also in humans (Staats et al. 1997). Analyses in infected pigs and mouse models have pointed to the CP as entry gate for *S. suis* into the brain (Sanford 1987; Williams and Blakemore 1990; Madsen et al. 2002; Dominguez-Punaro et al. 2007). In vitro, *S. suis* could invade into porcine and human models of the BCSFB consisting of CP epithelial cells, again preferentially from the basolateral side, indicating the existence of a transcellular mechanism for crossing the CP epithelium (Schwerk et al. 2012; Tenenbaum et al. 2009).

Entry into host cells for a transcellular migration has also been shown for viral pathogens. The Picornavirus Echovirus 30 (EV30) that causes aseptic meningitis and meningoencephalitis mainly in infants and young children invades directly into human CP epithelial cells in vitro (Schneider et al. 2012). In contrast to *L. monocytogenes*, *N. meningitidis* and *S. suis*, the virus EV30 can infect the HIBCPP cells from the basolateral as well as the apical cell side. Interestingly, following infection from the basolateral side, replicated viruses are also released from the apical "CSF" side. The infection capacity and replication rate is strain-dependent, since only specific EV30 outbreak strains were capable of inducing loss of CP barrier function, barrier morphology and causing viral dissemination (Dahm et al. 2018). Besides different Enterovirus strains, other viruses such as Mumps virus have been discussed as capable of affecting the BCSFB (Wolinsky et al. 1976). However, research in this field is scarce.

11.2.3   Entry in a "Trojan Horse" Fashion Inside of Host Immune Cells

Invasion of the CNS via a Trojan horse mechanism requires pathogens to be able to survive inside of infected immune cells. A good candidate for employing such a mechanism is *L. monocytogenes* that, as a facultative intracellular pathogen, is able to survive inside of infected monocytes. As mentioned previously in this chapter, during in vivo *studies L. monocytogenes* is often found inside monocytes in the CP of infected animals (Berche 1995; Prats et al. 1992; Schluter et al. 1996) and there is ample evidence for brain entry by the bacteria inside of monocytes (Drevets and Bronze 2008; Drevets et al. 2004; Join-Lambert et al. 2005). Further supporting the existence of a Trojan horse mechanism for *L. monocytogenes* is the observation that mice, which had been intravenously inoculated with bacteria, displayed brain infection despite gentamicin treatment, which causes elimination of blood borne bacteria not hiding inside of infected macrophages (Drevets et al. 2001). Another bacterial pathogen for which evidence for a "Trojan horse" mechanism has been found is *S. suis*. When in vitro infection assays of primary porcine choroid plexus epithelial cells (PCPEC) were performed in presence of porcine polymorphonuclear neutrophils (PMNs), transmigrating PMNs containing intracellular bacteria could be observed (Wewer et al. 2011).

Coxsackievirus B3 (CVB3), belonging to the family of *Picornaviridae*, is often responsible for aseptic meningitis and encephalitis in neonates (Rhoades et al. 2011).

Infection experiments in mice have suggested that CVB3 binds to the TJ-associated Coxsackie and Adenovirus receptor (CAR) of the CP epithelium. A nestin[+] population of myeloid cells recruited to the CP is highly susceptible to infection with CVB3 and traverses into the CNS thereby serving as a "Trojan horse" for CVB3 (Feuer et al. 2003; Tabor-Godwin et al. 2010). Invasion of the CNS inside of immune cells is also possible for HIV, since infected monocytes were found in the CP of patients (Petito et al. 1999; Falangola and Petito 1993).

11.3 Impairment of the BCSFB and Cell Death in the CP During CNS Infection

The tight connections between the CP epithelial cells are of major importance for maintaining the BCSFB at the CP. As mentioned before, during the course of a CNS infection these connections are not necessarily unchangeable. For example, during a paracellular progress of pathogens into the CNS, an opening of at least the TJs is required. Impairment of the BCSFB can have important consequences for the influx of undesired substances, including more pathogens, into the CNS.

One cellular event that can directly lead to loss of barrier function is death of cells in the CP. In vivo, lesions at the CP were detected in pigs that had been naturally challenged with *S. suis* or were experimentally infected (Sanford 1987; Williams and Blakemore 1990). Evidence that inflammatory events, such as infection with bacteria, can cause death of CP epithelial cells was found in vitro, where infection of PCPEC with *S. suis* from the apical cell side (equivalent to bacteria having entered the CNS) or treatment with the pro-inflammatory cytokine TNFα caused cell death by apoptotic and necrotic mechanisms (Tenenbaum et al. 2006; Schwerk et al. 2011). Along those lines, challenge of PCPEC led to the regulation of genes involved in programmed cell death and up-regulation of inflammatory response genes including TNFα (Schwerk et al. 2011). In the same model system *S. suis* and TNFα caused alterations of barrier function with concomitant induction of actin stress fibers and dislocation of TJ proteins (Tenenbaum et al. 2008; Zeni et al. 2007). Impairment of barrier function could also enable reseeding of bacteria from the CNS into the blood across the CP in a paracellular fashion (Tenenbaum et al. 2009).

An increased apoptosis of cells in the CP has also been observed during viral infections. Maedi-Visna Virus (MVV), a lentivirus causing systemic infection in sheep, activates extrinsic and intrinsic apoptosis pathways in sheep CP cells, thereby inducing cell death in the CP (Duval et al. 2002). Also, CVB3 initiates apoptotic cell death in the CP of infected mice (Tabor-Godwin et al. 2010; Puccini et al. 2014). As will be addressed later in this chapter, since the CP is essential in CSF production and immune regulation, those cell death processes might lead to additional complications during infectious disease of the CNS.

11.4 The CP Contributes to the Host Inflammatory Response Following Infection

During invasion of the CNS via the BCSFB the pathogens interact with cells of the CP, either directly or indirectly. Whereas indirect interactions can, for example, occur when a pathogen passes through the CP inside of an infected host immune cell in "Trojan horse" fashion, direct interactions usually involve binding of the causative agent to receptor molecules on the surface of host cells. Examples already mentioned in this chapter for those receptors are E-cad and Met, which serve as targets for *L. monocytogenes* to initiate entry in CP epithelial cells.

Other, highly important, receptors belong to the so-called pattern recognition receptor (PRR) family that recognize typical signatures of pathogens termed pathogen-associated molecular patterns (PAMPS) (Takeuchi and Akira 2010). Amongst the most important PRRs are the Toll-like receptors (TLRs), which can initiate downstream signaling after recognizing their respective ligands. Several different TLRs have been described dependent on their subcellular localization and ligand-specificity (Kawai and Akira 2010; Akira et al. 2006). Recognition of invading pathogens by host cell receptors in the CP can initiate downstream signaling that eventually elicits a direct contribution to the inflammatory immune response (Beutler 2009).

The cellular answer of the CP following recognition of foreign organisms can also help in staging a host cell response involving immune cells. It is well known that the CP plays an important role as gate for regulated entry of immune cells into the CNS (Liddelow 2015; Shechter et al. 2013; Wolburg and Paulus 2010). Noteworthy, the inflammatory response by the CP, possibly in combination with a pathogen-induced impairment of the BCSFB, is likely to impact on host immune cell transmigration across the CP. A schematic overview of contributions of the CP to the inflammatory response following infection is given in Fig. 11.2.

11.4.1 Direct Contribution to the Host Inflammatory Response

Interaction of pathogens with host cell receptors will lead to activation of downstream signal transduction pathways, which consist of cascades of consecutive protein modifications. These include, among others, the activity of specific MAPKs such as the extracellular signal regulated kinases 1 and 2 (Erk1/2) or the MAPK p38. Activity of these kinases will ultimately lead to the activation of transcription factors required to stage an inflammatory response by regulating the expression of inflammatory response genes (Krachler et al. 2011).

Some of the most important of the inflammatory response genes activated by transcription factors express a class of signaling molecules that, dependent on their origin and function, can be grouped into pro- or anti-inflammatory cytokines or,

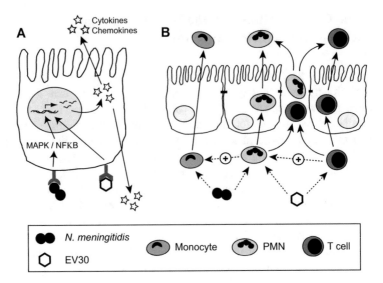

Fig. 11.2 A schematic overview showing possible contributions of the CP to the inflammatory response of the host following infection (for more details please refer to the text). (**a**) Surface receptors of CP epithelial cells recognizing pathogens, including bacteria (e.g. *N. meningitidis*) and virus (e.g. EV30), transmit signals to the nucleus leading to activation of inflammatory response genes as cytokines and chemokines. Signal transduction can involve distinct intracellular pathways including MAPK and NFκB signaling. Expressed cytokines and chemokines are secreted to the apical (CSF) and /or basolateral (blood) side of the CP epithelium. (**b**) The CP can serve as entry side for host immune cells (including monocytes, PMNs and T cells) into the brain during CNS infections. For PMNs and T cells both para- and transcellular pathways have been described. Transmigration rates are influenced by cytokines and chemokines, and host immune cells have been shown to affect each other. In an in vitro model transmigration of monocytes across the CP epithelium was enhanced by PMNs in the presence of *N. meningitidis*, and transmigration of PMNs was improved by the presence of T cells during challenge with EV30

when chemotactically active on immune cells, into chemokines (Zhang and An 2007). In vivo experiments with mice showed that following infection with *S. suis* the cytokines TNFα and interleukin 1β (IL1β) as well as the chemokine monocyte chemotactic protein 1 (MCP-1) can be detected in the CP on RNA level (Dominguez-Punaro et al. 2007). Up-regulation of several cytokines and chemokines, including TNFα, IL1β, IL6 and IL8, was confirmed by microarrays following challenge of PCPEC with *S. suis* from the apical "CSF side" in vitro. Similarly, employing an in vitro model of the BCSFB based on HIBCPP cells, basolateral infection with *N. meningitidis* was shown to activate expression and secretion of several cytokines and chemokines (Borkowski et al. 2014; Steinmann et al. 2013). Evaluation of gene expression and Gene Set Enrichment Analysis indicated that this pro-inflammatory response involved NFκB signaling and up-regulation of the IκBζ transcription factor. Activation of host cellular signal transduction seemed to be dependent on recognition by TLR2/TLR6 rather than TLR2/TLR1. TLR4 was not involved, but it has to be mentioned that HIBCPP cells express only low levels of TLR4 (Borkowski et al. 2014). Therefore, the lack of response to TLR4 in this model might not completely recapitulate the in vivo situation, since in rodents expression of TLR4 and response to inflammatory stimuli was detected in the CP (Chakravarty and Herkenham 2005; Rivest 2009).

The immune response of host cells to *L. monocytogenes* has been well researched in different cell types (Stavru et al. 2011). Concerning a contribution of the CP, during in vitro studies *L. monocytogenes* activated expression of several inflammatory response genes including IL6, IL8 and CXCL1-3 in HIBCPP cells. Interestingly, gene activation was differentially regulated by specific MAPK signaling pathways. Whereas maximal secretion of IL8 requires Erk1/2 signaling, inhibition of Erk1/2 kinases positively influenced expression of IL6 (Dinner et al. 2017).

Expression of inflammatory response genes in the CP has also been described after infection with viral pathogens. In this regard, infection with CVB3 caused expression of the chemokine C-C motif ligand 12 (CCL12) in the CP (Tabor-Godwin et al. 2010). In cell culture experiments, infection of human CP epithelial cells with EV30 also lead to the secretion of chemokines (CXCL1-3, CXCL10, CCL5, CCL20) and cytokines (IL6, IL7, macrophage colony-stimulating factor (MCSF)). Noteworthy, IL7 was selectively secreted in high amounts from the apical side towards the "CSF" compartment of the in vitro system (Schneider et al. 2012; Dahm et al. 2017). However, only a few publications have analyzed the CSF, which is produced by the CP epithelium, in EV30 meningitis patients. It could be demonstrated that IL-8, TNF-α and IL-1R2 levels were increased, but CXCL1-3, CXCL10, CCL5, CCL12 and CCL20 as well as IL6 and IL7 were not analyzed (Sulik et al. 2014). In a recent publication using a multiplex ELISA assay a distinct cytokine/chemokine expression profile in the CSF of Enterovirus and human Parechovirus (HPeV) infected children could be detected. Here, IFNγ, GRO/CXCL1, IL-8, MIP1a/CCL3, MIP1b/CCL4, MCP3/CCL7, IL12p40, EGF, FGF-2, Eotaxin/CCL11, GM-CSF, IL-5, IL-10, and MCP-1 were significantly higher in Enterovirus meningitis patients compared to both HPeV infected and controls (Fortuna et al. 2017).

11.4.2 CNS Entry of Host Immune Cells Across the CP During the Course of Infection

The CP is a well known route for immune cells to cross over into the CNS. In healthy individuals the BCSFB might allow immunosurveillance in the brain by enabling controlled entry of immune cells, in contrast to the BBB that is considered a barrier for leukocyte entry (Engelhardt et al. 2017; Shechter et al. 2013): When pathogens invade the brain, the host organism responds with recruitment of immune cells to the site of infection. Interestingly, the amounts of distinct immune cell populations and the chronological order of their appearance in the CSF vary dependent on the pathogen: bacterial CNS infections typically exhibit PMNs at early stages of disease, viral infections are, rather, associated with lymphocytes and monocytes, and parasites cause a presence of 10% or more eosinophils in the total leukocyte population (Graeff-Teixeira et al. 2009; Lucht et al. 1992; Meeker et al. 2012b).

Immune cells of the host have been detected in the CP during CNS infection by several pathogens in vivo (Lauer et al. 2018). These include the already mentioned examples of *L. monocytogenes* and CVB3, where in both cases a "Trojan horse" mechanism is probably taking place. Using the PCPEC and HIBCPP models of the porcine and human BCSFB, respectively, transmigration of immune cells was investigated under infectious conditions in vitro. In the PCPEC model, integrin CD11b/CD18-dependent transmigration of PMNs was observed, which was increased following infection with *S. suis* (Wewer et al. 2011). Comparable to invading pathogens, host immune cells can choose a paracellular or a transcellular route for crossing cellular barriers (Engelhardt and Wolburg 2004). Interestingly, transmigration of PMNs across PCPEC seemed to follow a paracellular route until the PMNs approached the TJs. At this point, "Funnel-like" structures originating from the apical membrane enable continuation of their journey by a transcellular mechanism (Wewer et al. 2011). Transmigration of PMNs was also shown across HIBCPP cells that had been infected with *S. suis* (de Buhr et al. 2017). In this situation transmigrated PMNs formed neutrophil extracellular traps (NETs), which are known to have antimicrobial function (Brinkmann et al. 2004; Fuchs et al. 2007).

Studies investigating immune cell transmigration across HIBCPP cells infected with *N. meningitidis* or EV30 showed elegantly via a new technique called focal ion bean scanning electron microcscopy (FIB-SEM) that PMNs and T-cells cross CP epithelial cells by both a paracellular and a transcellular mechanism. Noteworthy, in these models different types of immune cells can influence each other, e.g. transmigration of monocytes was enhanced by PMNs in HIBCPP cells challenged with *N. meningitidis*. The presence of naïve CD3+ T lymphocytes improved PMN crossing of HIBCPP cells infected with EV30 (Dahm et al. 2017, 2018; Steinmann et al. 2013). Transmigration of macrophages as well peripheral blood mononuclear cells was also supported by feline CP epithelial cells and explant cultures, and transmigration rates were greatly enhanced by addition of Feline Immunodeficiency Virus (Meeker et al. 2012a).

11.5   Additional Complications During CNS Infections: Role of the CP

After entering the CNS via the CP, pathogen-induced inflammation and the response of host immune cells can cause permanent damage leading to neurological sequelae including hearing or visual loss and intellectual impairment (Weber and Tuomanen 2007; Grandgirard et al. 2013; Grimwood et al. 1995). Still, the CP might be involved in causing additional complications in a more complex way than only providing a site of entry for pathogens and immune cells. As mentioned earlier in this chapter, apoptosis in the CP has been observed in in vitro and in vivo models after infection with certain pathogens as the viruses MVV and CVB3 (Duval et al. 2002; Puccini et al. 2014; Tabor-Godwin et al. 2010).

It has been suggested that this damage would have a significant impact on CP functions, e.g. its role during immune regulation and the production of CSF, which eventually could lead to hydrocephalus (Rhoades et al. 2011). Choroid plexitis or ventriculitis have been observed together with hydrocephalus during infections with pathogens known to invade the CP as *C. neoformans* and *L. monocytogenes* (Ben Shimol et al. 2012; Graciela Agar et al. 2009; Ito et al. 2008). *S. suis* has also been demonstrated to induce apoptotic and non-apoptotic mechanisms in choroid plexus epithelial cells and thereby could lead to CP dysfunction (Tenenbaum et al. 2006).

11.6 Conclusion

From the published literature it is obvious that the CP plays several important roles during infectious diseases of the CNS. The CP provides an entry site for pathogens into the brain and contributes to the inflammatory response by producing cytokines and chemokines and by regulating the transmigration of host immune cells. The presence of both pathogens and immune cells in the CNS can lead to severe damage that is responsible for permanent neurological sequelae. Additionally, impairment of BCSFB function can have serious consequences for the outcome of disease and the emergence of further consequences.

It should be noted that due to its roles during infections of the CNS the CP also serves as a target for therapeutic treatment options (Lehtinen et al. 2013; Dragunow 2013). Application of dexamethasone has been used as adjunctive therapy to reduce damage caused by the inflammatory response, but the response is not necessarily favorable (Brouwer et al. 2013; McIntyre et al. 1997). Furthermore, matrix metalloproteinase (MMP) inhibitors as well as inhibition of complement component 5 plus dexamethasone have shown positive results in murine models of meningitis (Kasanmoentalib et al. 2015; Liechti et al. 2015; Ricci et al. 2014).

Still, much more research is required to understand the roles of the CP during infections of the CNS in maximum detail. Acquiring this knowledge will certainly contribute in developing options for prevention of CNS infection and treatment during the course of disease. This includes strategies that directly target the CP, but also the generation of therapeutic substances that are able to traverse into the brain. The application of both in vitro and in vivo models of CNS infections should be of importance to accomplish these aims.

References

Akira S, Uematsu S, Takeuchi O (2006) Pathogen recognition and innate immunity. Cell 124 (4):783–801. https://doi.org/10.1016/j.cell.2006.02.015

Ben Shimol S, Einhorn M, Greenberg D (2012) Listeria meningitis and ventriculitis in an immunocompetent child: case report and literature review. Infection 40(2):207–211. https://doi.org/10.1007/s15010-011-0177-6

Berche P (1995) Bacteremia is required for invasion of the murine central nervous system by *Listeria monocytogenes*. Microb Pathog 18(5):323–336. https://doi.org/10.1006/mpat.1995.0029

Beutler B (2009) Microbe sensing, positive feedback loops, and the pathogenesis of inflammatory diseases. Immunol Rev 227(1):248–263. https://doi.org/10.1111/j.1600-065X.2008.00733.x

Borkowski J, Li L, Steinmann U, Quednau N, Stump-Guthier C, Weiss C, Findeisen P, Gretz N, Ishikawa H, Tenenbaum T, Schroten H, Schwerk C (2014) Neisseria meningitidis elicits a pro-inflammatory response involving I kappa B zeta in a human blood-cerebrospinal fluid barrier model. J Neuroinflammation 11:163. https://doi.org/10.1186/S12974-014-0163-X

Brinkmann V, Reichard U, Goosmann C, Fauler B, Uhlemann Y, Weiss DS, Weinrauch Y, Zychlinsky A (2004) Neutrophil extracellular traps kill bacteria. Science 303(5663):1532–1535. https://doi.org/10.1126/science.1092385

Brouwer MC, McIntyre P, Prasad K, van de Beek D (2013) Corticosteroids for acute bacterial meningitis. Cochrane Database Syst Rev 6:CD004405. https://doi.org/10.1002/14651858.CD004405.pub4

Chakravarty S, Herkenham M (2005) Toll-like receptor 4 on nonhematopoietic cells sustains CNS inflammation during endotoxemia, independent of systemic cytokines. J Neurosci 25(7):1788–1796. https://doi.org/10.1523/JNEUROSCI.4268-04.2005

Christodoulides M, Makepeace BL, Partridge KA, Kaur D, Fowler MI, Weller RO, Heckels JE (2002) Interaction of Neisseria meningitidis with human meningeal cells induces the secretion of a distinct group of chemotactic, proinflammatory, and growth-factor cytokines. Infect Immun 70(8):4035–4044

Dahm T, Rudolph H, Schwerk C, Schroten H, Tenenbaum T (2016) Neuroinvasion and inflammation in viral central nervous system infections. Mediat Inflamm 2016:1–16. https://doi.org/10.1155/2016/8562805

Dahm T, Frank F, Adams O, Lindner HA, Ishikawa H, Weiss C, Schwerk C, Schroten H, Tenenbaum T, Rudolph H (2017) Sequential transmigration of polymorphonuclear cells and naive CD3(+) T lymphocytes across the blood-cerebrospinal-fluid barrier in vitro following infection with Echovirus 30. Virus Res 232:54–62. https://doi.org/10.1016/j.virusres.2017.01.024

Dahm T, Adams O, Boettcher S, Diedrich S, Morozov V, Hansman G, Fallier-Becker P, Schadler S, Burkhardt CJ, Weiss C, Stump-Guthier C, Ishikawa H, Schroten H, Schwerk C, Tenenbaum T, Rudolph H (2018) Strain-dependent effects of clinical echovirus 30 outbreak isolates at the blood-CSF barrier. J Neuroinflammation 15(1):50. https://doi.org/10.1186/s12974-018-1061-4

Dando SJ, Mackay-Sim A, Norton R, Currie BJ, St John JA, Ekberg JA, Batzloff M, Ulett GC, Beacham IR (2014) Pathogens penetrating the central nervous system: infection pathways and the cellular and molecular mechanisms of invasion. Clin Microbiol Rev 27(4):691–726. https://doi.org/10.1128/CMR.00118-13

Daneman R, Prat A (2015) The blood-brain barrier. Cold Spring Harb Perspect Biol 7(1):a020412. https://doi.org/10.1101/cshperspect.a020412

de Buhr N, Reuner F, Neumann A, Stump-Guthier C, Tenenbaum T, Schroten H, Ishikawa H, Muller K, Beineke A, Hennig-Pauka I, Gutsmann T, Valentin-Weigand P, Baums CG, von Kockritz-Blickwede M (2017) Neutrophil extracellular trap formation in the Streptococcus suis-infected cerebrospinal fluid compartment. Cell Microbiol 19(2). https://doi.org/10.1111/cmi.12649

Dinner S, Borkowski J, Stump-Guthier C, Ishikawa H, Tenenbaum T, Schroten H, Schwerk C (2016) A choroid plexus epithelial cell-based model of the human blood-cerebrospinal fluid barrier to study bacterial infection from the basolateral side. J Vis Exp 111:54061. https://doi.org/10.3791/54061

Dinner S, Kaltschmidt J, Stump-Guthier C, Hetjens S, Ishikawa H, Tenenbaum T, Schroten H, Schwerk C (2017) Mitogen-activated protein kinases are required for effective infection of human choroid plexus epithelial cells by *Listeria monocytogenes*. Microbes Infect 19(1):18–33. https://doi.org/10.1016/j.micinf.2016.09.003

Dominguez-Punaro MC, Segura M, Plante MM, Lacouture S, Rivest S, Gottschalk M (2007) Streptococcus suis serotype 2, an important swine and human pathogen, induces strong systemic and cerebral inflammatory responses in a mouse model of infection. J Immunol 179 (3):1842–1854

Dragunow M (2013) Meningeal and choroid plexus cells--novel drug targets for CNS disorders. Brain Res 1501:32–55. https://doi.org/10.1016/j.brainres.2013.01.013

Drevets DA, Bronze MS (2008) *Listeria monocytogenes*: epidemiology, human disease, and mechanisms of brain invasion. FEMS Immunol Med Microbiol 53(2):151–165. https://doi. org/10.1111/j.1574-695X.2008.00404.x

Drevets DA, Jelinek TA, Freitag NE (2001) *Listeria monocytogenes*-infected phagocytes can initiate central nervous system infection in mice. Infect Immun 69(3):1344–1350. https://doi. org/10.1128/IAI.69.3.1344-1350.2001

Drevets DA, Dillon MJ, Schawang JS, Van Rooijen N, Ehrchen J, Sunderkotter C, Leenen PJ (2004) The Ly-6Chigh monocyte subpopulation transports *Listeria monocytogenes* into the brain during systemic infection of mice. J Immunol 172(7):4418–4424

Duval R, Bellet V, Delebassee S, Bosgiraud C (2002) Implication of caspases during maedi-visna virus-induced apoptosis. J Gen Virol 83(Pt 12):3153–3161. https://doi.org/10.1099/0022-1317-83-12-3153

Engelhardt B, Wolburg H (2004) Mini-review: transendothelial migration of leukocytes: through the front door or around the side of the house? Eur J Immunol 34(11):2955–2963. https://doi. org/10.1002/eji.200425327

Engelhardt B, Vajkoczy P, Weller RO (2017) The movers and shapers in immune privilege of the CNS. Nature Immunol 18(2):123–131. https://doi.org/10.1038/ni.3666

Falangola MF, Petito CK (1993) Choroid plexus infection in cerebral toxoplasmosis in AIDS patients. Neurology 43(10):2035–2040

Falangola MF, Hanly A, Galvao-Castro B, Petito CK (1995) HIV infection of human choroid plexus: a possible mechanism of viral entry into the CNS. J Neuropathol Exp Neurol 54 (4):497–503

Feuer R, Mena I, Pagarigan RR, Harkins S, Hassett DE, Whitton JL (2003) Coxsackievirus B3 and the neonatal CNS: the roles of stem cells, developing neurons, and apoptosis in infection, viral dissemination, and disease. Am J Pathol 163(4):1379–1393. https://doi.org/10.1016/S0002-9440(10)63496-7

Fortuna D, Cardenas AM, Graf EH, Harshyne LA, Hooper DC, Prosniak M, Shields J, Curtis MT (2017) Human parechovirus and enterovirus initiate distinct CNS innate immune responses: pathogenic and diagnostic implications. J Clin Virol 86:39–45. https://doi.org/10.1016/j.jcv. 2016.11.007

Fuchs TA, Abed U, Goosmann C, Hurwitz R, Schulze I, Wahn V, Weinrauch Y, Brinkmann V, Zychlinsky A (2007) Novel cell death program leads to neutrophil extracellular traps. J Cell Biol 176(2):231–241. https://doi.org/10.1083/jcb.200606027

Ghersi-Egea JF, Strazielle N, Catala M, Silva-Vargas V, Doetsch F, Engelhardt B (2018) Molecular anatomy and functions of the choroidal blood-cerebrospinal fluid barrier in health and disease. Acta Neuropathol 135(3):337–361. https://doi.org/10.1007/s00401-018-1807-1

Graciela Agar CH, Orozco Rosalba V, Macias Ivan C, Agnes F, Juan Luis GA, Jose Luis SH (2009) Cryptococcal choroid plexitis an uncommon fungal disease. Case report and review. Can J Neurol Sci 36(1):117–122

Graeff-Teixeira C, da Silva AC, Yoshimura K (2009) Update on eosinophilic meningoencephalitis and its clinical relevance. Clin Microbiol Rev 22(2):322–348. https://doi.org/10.1128/CMR. 00044-08

Grandgirard D, Gaumann R, Coulibaly B, Dangy JP, Sie A, Junghanss T, Schudel H, Pluschke G, Leib SL (2013) The causative pathogen determines the inflammatory profile in cerebrospinal fluid and outcome in patients with bacterial meningitis. Mediat Inflamm 2013:1–12. https://doi. org/10.1155/2013/312476

Grimwood K, Anderson VA, Bond L, Catroppa C, Hore RL, Keir EH, Nolan T, Roberton DM (1995) Adverse outcomes of bacterial-meningitis in school-age survivors. Pediatrics 95 (5):646–656

Grundler T, Quednau N, Stump C, Orian-Rousseau V, Ishikawa H, Wolburg H, Schroten H, Tenenbaum T, Schwerk C (2013) The surface proteins InlA and InlB are interdependently required for polar basolateral invasion by *Listeria monocytogenes* in a human model of the blood-cerebrospinal fluid barrier. Microb Infect 15(4):291–301. https://doi.org/10.1016/j.micinf.2012.12.005

Guarner J, Greer PW, Whitney A, Shieh WJ, Fischer M, White EH, Carlone GM, Stephens DS, Popovic T, Zaki SR (2004) Pathogenesis and diagnosis of human meningococcal disease using immunohistochemical and PCR assays. Am J Clin Pathol 122(5):754–764. https://doi.org/10.1309/A7M2-FN2T-YE6A-8UFX

Ito H, Kobayashi S, Iino M, Kamei T, Takanashi Y (2008) *Listeria monocytogenes* meningoencephalitis presenting with hydrocephalus and ventriculitis. Intern Med 47(4):323–324

Join-Lambert OF, Ezine S, Le Monnier A, Jaubert F, Okabe M, Berche P, Kayal S (2005) *Listeria monocytogenes*-infected bone marrow myeloid cells promote bacterial invasion of the central nervous system. Cell Microbiol 7(2):167–180. https://doi.org/10.1111/j.1462-5822.2004.00444.x

Join-Lambert O, Morand PC, Carbonnelle E, Coureuil M, Bille E, Bourdoulous S, Nassif X (2010) Mechanisms of meningeal invasion by a bacterial extracellular pathogen, the example of Neisseria meningitidis. Prog Neurobiol 91(2):130–139. https://doi.org/10.1016/j.pneurobio.2009.12.004

Kasanmoentalib ES, Valls Seron M, Morgan BP, Brouwer MC, van de Beek D (2015) Adjuvant treatment with dexamethasone plus anti-C5 antibodies improves outcome of experimental pneumococcal meningitis: a randomized controlled trial. J Neuroinflammation 12:149. https://doi.org/10.1186/s12974-015-0372-y

Kawai T, Akira S (2010) The role of pattern-recognition receptors in innate immunity: update on toll-like receptors. Nat Immunol 11(5):373–384. https://doi.org/10.1038/ni.1863

Krachler AM, Woolery AR, Orth K (2011) Manipulation of kinase signaling by bacterial pathogens. J Cell Biol 195(7):1083–1092. https://doi.org/10.1083/jcb.201107132

Lauer AN, Tenenbaum T, Schroten H, Schwerk C (2018) The diverse cellular responses of the choroid plexus during infection of the central nervous system. Am J Physiol Cell Physiol 314 (2):C152–C165. https://doi.org/10.1152/ajpcell.00137.2017

Lehtinen MK, Bjornsson CS, Dymecki SM, Gilbertson RJ, Holtzman DM, Monuki ES (2013) The choroid plexus and cerebrospinal fluid: emerging roles in development, disease, and therapy. J Neurosci 33(45):17553–17559. https://doi.org/10.1523/JNEUROSCI.3258-13.2013

Liddelow SA (2015) Development of the choroid plexus and blood-CSF barrier. Front Neurosci 9:32. https://doi.org/10.3389/fnins.2015.00032

Liechti FD, Bachtold F, Grandgirard D, Leppert D, Leib SL (2015) The matrix metalloproteinase inhibitor RS-130830 attenuates brain injury in experimental pneumococcal meningitis. J Neuroinflammation 12:43. https://doi.org/10.1186/s12974-015-0257-0

Lucas MJ, Brouwer MC, van de Beek D (2016) Neurological sequelae of bacterial meningitis. J Infect 73(1):18–27. https://doi.org/10.1016/j.jinf.2016.04.009

Lucht F, Cordier G, Pozzetto B, Fresard A, Revillard JP (1992) Evidence for T-cell involvement during the acute phase of echovirus meningitis. J Med Virol 38(2):92–96

Madsen LW, Svensmark B, Elvestad K, Aalbaek B, Jensen HE (2002) *Streptococcus suis* serotype 2 infection in pigs: new diagnostic and pathogenetic aspects. J Comp Pathol 126(1):57–65. https://doi.org/10.1053/jcpa.2001.0522

Masocha W, Kristensson K (2012) Passage of parasites across the blood-brain barrier. Virulence 3 (2):202–212. https://doi.org/10.4161/viru.19178

McIntyre PB, Berkey CS, King SM, Schaad UB, Kilpi T, Kanra GY, Perez CM (1997) Dexamethasone as adjunctive therapy in bacterial meningitis. A meta-analysis of randomized clinical trials since 1988. JAMA 278(11):925–931

Meeker RB, Bragg DC, Poulton W, Hudson L (2012a) Transmigration of macrophages across the choroid plexus epithelium in response to the feline immunodeficiency virus. Cell Tiss Res 347 (2):443–455. https://doi.org/10.1007/s00441-011-1301-8

Meeker RB, Williams K, Killebrew DA, Hudson LC (2012b) Cell trafficking through the choroid plexus. Cell Adhes Migr 6(5):390–396. https://doi.org/10.4161/cam.21054

Michos AG, Syriopoulou VP, Hadjichristodoulou C, Daikos GL, Lagona E, Douridas P, Mostrou G, Theodoridou M (2007) Aseptic meningitis in children: analysis of 506 cases. PLoS One 2(7):e674. https://doi.org/10.1371/journal.pone.0000674

Mogk S, Meiwes A, Shtopel S, Schraermeyer U, Lazarus M, Kubata B, Wolburg H, Duszenko M (2014) Cyclical appearance of African trypanosomes in the cerebrospinal fluid: new insights in how trypanosomes enter the CNS. PLoS One 9(3):e91372. https://doi.org/10.1371/journal.pone.0091372

O'Brown NM, Pfau SJ, Gu C (2018) Bridging barriers: a comparative look at the blood-brain barrier across organisms. Genes Dev 32(7–8):466–478. https://doi.org/10.1101/gad.309823.117

Parikh V, Tucci V, Galwankar S (2012) Infections of the nervous system. Int J Crit Illn Inj Sci 2 (2):82–97. https://doi.org/10.4103/2229-5151.97273

Petito CK, Chen H, Mastri AR, Torres-Munoz J, Roberts B, Wood C (1999) HIV infection of choroid plexus in AIDS and asymptomatic HIV-infected patients suggests that the choroid plexus may be a reservoir of productive infection. J Neurovirol 5(6):670–677

Pizarro-Cerda J, Kuhbacher A, Cossart P (2012) Entry of Listeria monocytogenes in mammalian epithelial cells: an updated view. Cold Spring Harb Perspect Med 2(11):a010009. https://doi.org/10.1101/cshperspect.a010009

Prats N, Briones V, Blanco MM, Altimira J, Ramos JA, Dominguez L, Marco A (1992) Choroiditis and meningitis in experimental murine infection with Listeria monocytogenes. Eur J Clin Microbiol Infect Dis 11(8):744–747

Pron B, Taha MK, Rambaud C, Fournet JC, Pattey N, Monnet JP, Musilek M, Beretti JL, Nassif X (1997) Interaction of Neisseria meningitidis with the components of the blood-brain barrier correlates with an increased expression of PilC. J Infect Dis 176(5):1285–1292

Puccini JM, Ruller CM, Robinson SM, Knopp KA, Buchmeier MJ, Doran KS, Feuer R (2014) Distinct neural stem cell tropism, early immune activation, and choroid plexus pathology following coxsackievirus infection in the neonatal central nervous system. Lab Investig 94 (2):161–181. https://doi.org/10.1038/labinvest.2013.138

Quan N, Mhlanga JD, Whiteside MB, McCoy AN, Kristensson K, Herkenham M (1999) Chronic overexpression of proinflammatory cytokines and histopathology in the brains of rats infected with Trypanosoma brucei. J Comp Neurol 414(1):114–130

Radoshevich L, Cossart P (2018) Listeria monocytogenes: towards a complete picture of its physiology and pathogenesis. Nat Rev Microbiol 16(1):32–46. https://doi.org/10.1038/nrmicro.2017.126

Rhoades RE, Tabor-Godwin JM, Tsueng G, Feuer R (2011) Enterovirus infections of the central nervous system. Virology 411(2):288–305. https://doi.org/10.1016/j.virol.2010.12.014

Ricci S, Grandgirard D, Wenzel M, Braccini T, Salvatore P, Oggioni MR, Leib SL, Koedel U (2014) Inhibition of matrix metalloproteinases attenuates brain damage in experimental meningococcal meningitis. BMC Infect Dis 14:726. https://doi.org/10.1186/s12879-014-0726-6

Rivest S (2009) Regulation of innate immune responses in the brain. Nat Rev Immunol 9 (6):429–439. https://doi.org/10.1038/nri2565

Rotbart HA (2000) Viral meningitis. Semin Neurol 20(3):277–292. https://doi.org/10.1055/s-2000-9427

Sanford SE (1987) Gross and histopathological findings in unusual lesions caused by Streptococcus suis in pigs. II. Central nervous system lesions. Can J Vet Res 51(4):486–489

Schluter D, Chahoud S, Lassmann H, Schumann A, Hof H, Deckert-Schluter M (1996) Intracerebral targets and immunomodulation of murine Listeria monocytogenes meningoencephalitis. J Neuropathol Exp Neurol 55(1):14–24

Schneider H, Weber CE, Schoeller J, Steinmann U, Borkowski J, Ishikawa H, Findeisen P, Adams O, Doerries R, Schwerk C, Schroten H, Tenenbaum T (2012) Chemotaxis of T-cells after infection of human choroid plexus papilloma cells with Echovirus 30 in an in vitro model of the blood-cerebrospinal fluid barrier. Virus Res 170(1–2):66–74. https://doi.org/10.1016/j. virusres.2012.08.019

Schwerk C, Adam R, Borkowski J, Schneider H, Klenk M, Zink S, Quednau N, Schmidt N, Stump C, Sagar A, Spellerberg B, Tenenbaum T, Koczan D, Klein-Hitpass L, Schroten H (2011) In vitro transcriptome analysis of porcine choroid plexus epithelial cells in response to *Streptococcus suis*: release of pro-inflammatory cytokines and chemokines. Microbes Infect 13 (11):953–962. https://doi.org/10.1016/j.micinf.2011.05.012

Schwerk C, Papandreou T, Schuhmann D, Nickol L, Borkowski J, Steinmann U, Quednau N, Stump C, Weiss C, Berger J, Wolburg H, Claus H, Vogel U, Ishikawa H, Tenenbaum T, Schroten H (2012) Polar invasion and translocation of Neisseria meningitidis and *Streptococcus suis* in a novel human model of the blood-cerebrospinal fluid barrier. PLoS One 7(1):e30069. https://doi.org/10.1371/journal.pone.0030069

Schwerk C, Tenenbaum T, Kim KS, Schroten H (2015) The choroid plexus-a multi-role player during infectious diseases of the CNS. Front Cell Neurosci 9:80. https://doi.org/10.3389/fncel. 2015.00080

Shechter R, London A, Schwartz M (2013) Orchestrated leukocyte recruitment to immune-privileged sites: absolute barriers versus educational gates. Nat Rev Immunol 13(3):206–218. https://doi.org/10.1038/nri3391

Sittner A, Bar-David E, Glinert I, Ben-Shmuel A, Weiss S, Schlomovitz J, Kobiler D, Levy H (2017) Pathology of wild-type and toxin-independent *Bacillus anthracis* meningitis in rabbits. PLoS One 12(10):e0186613. https://doi.org/10.1371/journal.pone.0186613

Staats JJ, Feder I, Okwumabua O, Chengappa MM (1997) *Streptococcus suis*: past and present. Vet Res Commun 21(6):381–407

Stavru F, Archambaud C, Cossart P (2011) Cell biology and immunology of *Listeria monocytogenes* infections: novel insights. Immunol Rev 240(1):160–184. https://doi.org/10. 1111/j.1600-065X.2010.00993.x

Steinmann U, Borkowski J, Wolburg H, Schroppel B, Findeisen P, Weiss C, Ishikawa H, Schwerk C, Schroten H, Tenenbaum T (2013) Transmigration of polymorphnuclear neutrophils and monocytes through the human blood-cerebrospinal fluid barrier after bacterial infection in vitro. J Neuroinflammation 10:31. https://doi.org/10.1186/1742-2094-10-31

Stephens DS (2009) Biology and pathogenesis of the evolutionarily successful, obligate human bacterium Neisseria meningitidis. Vaccine 27(Suppl 2):B71–B77. https://doi.org/10.1016/j. vaccine.2009.04.070

Sulik A, Kroten A, Wojtkowska M, Oldak E (2014) Increased levels of cytokines in cerebrospinal fluid of children with aseptic meningitis caused by mumps virus and echovirus 30. Scand J Immunol 79(1):68–72. https://doi.org/10.1111/sji.12131

Tabor-Godwin JM, Ruller CM, Bagalso N, An N, Pagarigan RR, Harkins S, Gilbert PE, Kiosses WB, Gude NA, Cornell CT, Doran KS, Sussman MA, Whitton JL, Feuer R (2010) A novel population of myeloid cells responding to coxsackievirus infection assists in the dissemination of virus within the neonatal CNS. J Neurosci 30(25):8676–8691. https://doi.org/10.1523/ JNEUROSCI.1860-10.2010

Takeuchi O, Akira S (2010) Pattern recognition receptors and inflammation. Cell 140(6):805–820. https://doi.org/10.1016/j.cell.2010.01.022

Tenenbaum T, Essmann F, Adam R, Seibt A, Janicke RU, Novotny GE, Galla HJ, Schroten H (2006) Cell death, caspase activation, and HMGB1 release of porcine choroid plexus epithelial cells during *Streptococcus suis* infection in vitro. Brain Res 1100(1):1–12. https://doi.org/10. 1016/j.brainres.2006.05.041

Tenenbaum T, Matalon D, Adam R, Seibt A, Wewer C, Schwerk C, Galla HJ, Schroten H (2008) Dexamethasone prevents alteration of tight junction-associated proteins and barrier function in porcine choroid plexus epithelial cells after infection with *Streptococcus suis* in vitro. Brain Res 1229:1–17. https://doi.org/10.1016/j.brainres.2008.06.118

Tenenbaum T, Papandreou T, Gellrich D, Friedrichs U, Seibt A, Adam R, Wewer C, Galla HJ, Schwerk C, Schroten H (2009) Polar bacterial invasion and translocation of Streptococcus suis across the blood-cerebrospinal fluid barrier in vitro. Cell Microbiol 11(2):323–336. https://doi.org/10.1111/j.1462-5822.2008.01255.x

Vazquez-Boland JA, Kuhn M, Berche P, Chakraborty T, Dominguez-Bernal G, Goebel W, Gonzalez-Zorn B, Wehland J, Kreft J (2001) Listeria pathogenesis and molecular virulence determinants. Clin Microbiol Rev 14(3):584–640. https://doi.org/10.1128/CMR.14.3.584-640.2001

Wang L, Lin M (2008) A novel cell wall-anchored peptidoglycan hydrolase (autolysin), IspC, essential for *Listeria monocytogenes* virulence: genetic and proteomic analysis. Microbiology 154(Pt 7):1900–1913. https://doi.org/10.1099/mic.0.2007/015172-0

Weber JR, Tuomanen EI (2007) Cellular damage in bacterial meningitis: an interplay of bacterial and host driven toxicity. J Neuroimmunol 184(1–2):45–52. https://doi.org/10.1016/j.jneuroim.2006.11.016

Wewer C, Seibt A, Wolburg H, Greune L, Schmidt MA, Berger J, Galla HJ, Quitsch U, Schwerk C, Schroten H, Tenenbaum T (2011) Transcellular migration of neutrophil granulocytes through the blood-cerebrospinal fluid barrier after infection with *Streptococcus suis*. J Neuroinflammation 8:51. https://doi.org/10.1186/1742-2094-8-51

Williams AE, Blakemore WF (1990) Pathogenesis of meningitis caused by *Streptococcus suis* type 2. J Infect Dis 162(2):474–481

Wolburg H, Paulus W (2010) Choroid plexus: biology and pathology. Acta Neuropathol 119 (1):75–88. https://doi.org/10.1007/s00401-009-0627-8

Wolburg H, Mogk S, Acker S, Frey C, Meinert M, Schonfeld C, Lazarus M, Urade Y, Kubata BK, Duszenko M (2012) Late stage infection in sleeping sickness. PLoS One 7(3):e34304. https://doi.org/10.1371/journal.pone.0034304

Wolinsky JS, Klassen T, Baringer JR (1976) Persistence of neuroadapted mumps virus in brains of newborn hamsters after intraperitoneal inoculation. J Infect Dis 133(3):260–267

Zeni P, Doepker E, Schulze-Topphoff U, Huewel S, Tenenbaum T, Galla HJ (2007) MMPs contribute to TNF-alpha-induced alteration of the blood-cerebrospinal fluid barrier in vitro. Am J Physiol Cell Physiol 293(3):C855–C864. https://doi.org/10.1152/ajpcell.00470.2006

Zhang JM, An J (2007) Cytokines, inflammation, and pain. Int Anesthesiol Clin 45(2):27–37. https://doi.org/10.1097/AIA.0b013e318034194e

Chapter 12
Hydrocephalus

Marianne Juhler

Abstract Hydrocephalus is an extremely frequent condition either as a disease entity on its own or as a complication or sequelae to other brain diseases. The tradition for many decades to divide hydrocephalus into "obstructive" and "communicating" is being re-evaluated, as developments in imaging technology are greatly improving visualization of obstruction sites and as it is now increasingly perceived that turnover and absorbtion of CSF occur by several probably parallel physiologies.

New technology for measuring ICP allows mobile measurements during full mobilization and under normal everyday conditions and provides the possibility to approach ICP measurement in normal or "pseudo-normals" (patients with other diseases but without any clinical suspicion of an ICP or CSF circulation disorder). This is changing our perception of reference values for ICP, which are highly dependent on body posture and probably considerably lower than previously assumed.

Treatment is still based on two surgical principles; CSF diversion via valve-regulated drains from the ventricles into an extracranial absorbtion site (most frequently the peritoneum) or endoscopically created communication between the ventricular system and the subarachnoid space; most frequently into the basal cisterns. Although hydrocephalus is not one disease, but a group of conditions spanning age groups from the neonate to the elderly with several underlying pathologies and a diversity of clinical presentations, these two surgical procedures remain the only treatment options. This restriction of treatment options to just two despite the obvious diversity of hydrocephalus is probably one of the major explanations for significant shortcomings of treatment resulting in high surgical revision rates. The need for new therapy concepts based on improved and more detailed physiological understanding of normal brain water physiology and different types of hydrocephalic pathophysiology is highly needed.

M. Juhler (✉)
Neurocentret, Rigshospitalet, University Clinic of Neurosurgery, Copenhagen, Denmark
e-mail: Marianne.juhler@regionh.dk

© The American Physiological Society 2020
J. Praetorius et al. (eds.), *Role of the Choroid Plexus in Health and Disease,*
Physiology in Health and Disease, https://doi.org/10.1007/978-1-0716-0536-3_12

12.1   Definition and Classification

Hydrocephalus is defined as an abnormally increased volume of cerebrospinal fluid (CSF) within the cerebral ventricular system. The definition does not relate to any underlying cause or any other specifics, and hydrocephalus is thus not one disease but a group of very different conditions with the common trait that they lead to a dilated ventricular system (Rekate 2009; Rigamonti et al. 2014). Consequently, hydrocephalus may be classified according to e.g. age-groups, clinical symptoms or underlying pathology, and a particular case of hydrocephalus may be defined by more than one—and usually several—criteria. The overlapping criteria imply that generally accepted hydrocephalus terminology used on daily basis may be ambiguous. As classification is a reflection of our perception of etiology and pathogenesis of the condition, classification determines our choice of both diagnostic procedures and treatment (Fig. 12.1). Equally important, classification has implications for analysis and comparison of research data. These are the reasons why this chapter emphasizes the importance of clinical classification.

Even in daily clinical work, hydrocephalus is classified by several criteria

- Age (infantile—juvenile—adult)
- Clinical presentation
- Level of ICP (high or "normal pressure")
- Anatomy—communicating vs. non-communicating
- Congenital vs. acquired
- Underlying cause/etiology

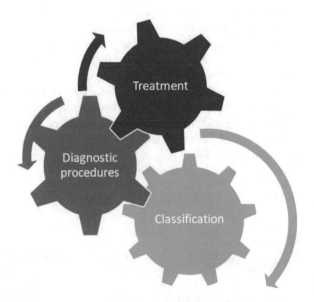

Fig. 12.1 Classification reflects the way we perceive a disease entity. Classification is therefore a strong determinant of our clinical actions for diagnostic procedures, interpretation of clinical data and choice of treatment

Classification based on age usually divides hydrocephalic conditions into infantile hydrocephalus, juvenile hydrocephalus, adult hydrocephalus and hydrocephalus in the elderly. Apart from the obvious chronology relating to a general concept of changing physiology across age groups, this terminology also builds strongly on different clinical presentations being typical for each age group.

In infants and very young children, the cranium can expand because of the still flexible calvarial bones and unclosed cranial sutures. Increasing head circumference is thus the hallmark of infantile hydrocephalus, and many infants have only subtle other symptoms (Kahle et al. 2016). Other frequent symptoms are irritability and crying probably because of headache and discomfort, downward gaze because of compression of the upper brainstem, and a spectrum of delayed psychomotor development from very severe handicap to no/only insignificant symptoms (Kulkarni et al. 2018; Paulsen et al. 2015; Schmidt et al. 2018). Perinatal brain haemorrhage of the germinal matrix, infection and congenital malformations are the most frequent etiologies underlying infantile hydrocephalus. Pathoanatomically and physiologically, obstruction of the CSF pathways causing increase of the intracranial pressure (ICP) is very often evident; however some cases are still classified as "communicating" if the applied imaging fails to clearly demonstrate a cause and site of obstruction (Rekate 2009).

Once the cranial sutures are closed, and the head is no longer malleable, ICP increase can no longer be attenuated by cranial expansion. Juvenile and adult hydrocephalus thus have similar clinical characteristics when the onset is *acute or sub-acute* increase in ICP. The main symptoms are headache, nausea, vomiting, dizziness and foggy vision (obscurations) and varying degrees of decreased consciousness. The symptoms are caused by elevated ICP and are worsened by lying down because ICP increases in horizontal positions (Holmlund et al. 2018; Petersen et al. 2016; Qvarlander et al. 2013). If untreated, ICP can become very high with coma and in the end lethal exitus. Development of clinical symptoms can also span a longer period of time with gradual progression. In such sub-chronic and chronic conditions, impaired intellectual and mental functions become the main symptoms. In children, this can give rise to stagnating mental development. Squint or other oculo-motor symptoms are also frequent. In adults, insidious onset of hydrocephalus leads to a clinical picture resembling hydrocephalus in the elderly/"normal pressure hydrocephalus". In both cases, some degree of headache and dizziness often accompany these symptoms. In children, the underlying cause is often a tumor obstructing CSF pathways (Paulsen et al. 2017; Chap. 10). In adults, a variety of causes lead to so-called "secondary NPH"; i.e. following subarachnoid haemorrhage, posttraumatic or related to neoplastic disease. Based on imaging studies, some cases of hydrocephalus in both children and adults seem to be caused by a hitherto undiagnosed congenital or early acquired anomaly; typically aquaductal stenosis or cystic malformation related to the fourth ventricle (Barami et al. 2018; Daou et al. 2016; Ibáñez-Botella et al. 2017; Osborn and Preece 2006; Rodis et al. 2016).

Hydrocephalus in the elderly is typically described by the clinical picture of "normal pressure hydrocephalus" (NPH) characterized by a very slow onset with a triad symptomatology consisting of characteristic gait impairment (gait apraxia with balance disturbance), urinary urge or urge incontinence and "sub-cortical" dementia (Rigamonti et al. 2014). Intracranial pressure in the vast majority of cases is much lower than in the above mentioned age groups, and hydrocephalus in the elderly is thus almost synonymous with "normal pressure hydrocephalus" (NPH). When the underlying cause in unknown as in most cases, this type of hydrocephalus is often referred as idiopathic NPH (iNPH) (Daou et al. 2016; Bergsneider et al. 2005). The differential diagnosis within a spectrum of degenerative brain diseases causing similar symptoms and increased ventricular size secondary to brain atrophy is a particular challenge. The two most common are Alzheimer's disease and microvascular brain disease. The diagnosis and clinical management is made even more difficult by co-existence of iNPH and degenerative brain pathology in the same patient referred to as NPH co-morbidities (Bech et al. 1999; Klinge et al. 2005).

Using a classification based on age groups is challenged when the relationship between the patient's current age and the hydrocephalus initiating event are not congruent—sometimes even separated by many years. In an infant with symptoms occurring soon after birth, a classification of congenital hydrocephalus is intuitively logical. However, in a mature adult, hydrocephalus is not naturally classified as a congenital or pediatric type, even if there is a history of the condition from early childhood. It becomes even more challenging, when diagnosis is made in adulthood with imaging or clinical characteristics suggesting a congenital or very early acquired disorder; e.g. in previously healthy young adult with a very large head, chronically dilated ventricles and a cystic posterior fossa anomaly on imaging studies.

Classification based on clinical symptoms is intertwined with age-group related symptoms, whether symptoms are caused by high or less elevated ICP, or whether symptoms are acute or chronic. In very general terms, hydrocephalus in children is often caused by an identifiable obstruction of CSF pathways with high ICP and an acute or sub-acute onset. Obstructive high pressure hydrocephalus is thus the most frequent condition in children, but not synonymous with pediatric hydrocephalus.

In equally general terms, hydrocephalus in the elderly leads to slow development of symptoms over many months or even years and in many cases, there is no visible obstruction with current imaging technology. iNPH is often used synonymously with slow development of typical triad symptoms and presumed communication between CSF compartments. However, the same gradually developing "typical NPH" symptoms are also seen in adults and elderly patients, when a visible obstruction is seen on imaging or when obstruction—though not demonstrable by current diagnostics—is very likely following e.g. a haemorrhagic or traumatic event with chronic disruptive effects on CSF pathways following the initial insult (Daou et al. 2016; Eide 2018; Ibáñez-Botella et al. 2017).

Classification based on etiologies is often made to distinguish between haemorrhagic or infectious insults, posttraumatic hydrocephalus, hydrocephalus related to neoplastic disease, or congenital malformations of the CSF pathways. For clearly obstructive etiologies—e.g. a space occupying process obstructing the ventricular system—the explanation for the resulting hydrocephalus seems straight-forward, as passage of CSF through the ventricular system is blocked. However, when there is no visible obstruction, the pathophysiological mechanism leading to hydrocephalus is much less clear. Although these conditions are often termed "communicating hydrocephalus", there must be an impediment for CSF exchange, and in current state of the art clinical management and research, the distinction between obstructive and communicating hydrocephalus is being severely challenged. Probably the two main reasons for this are to be sought firstly in the enormous development of imaging technologies demonstrating much more anatomic detail, so that e.g. obstructing membranes (Blitz et al. 2017; Fushimi et al. 2006) and directional CSF flow (Capel et al. 2018; Garnotel et al. 2018; Lokossou et al. 2018) can be visualized; and secondly in the increasing knowledge about CSF pathway physiology. Since the 1960s the traditional physiological explanation for communicating hydrocephalus has been a blockage of CSF absorbtion into the saggital sinus via the arachnoid villi (Broadbelt and Stoodley 2016). However, it is becoming increasingly clear, that there are other routes of CSF exchange, absorbtion and regulation, and that CSF clearence through the villi is one among probably several parallel systems (Amiry-Moghaddam et al. 2010; Greitz 2006, 2007; Mestre et al. 2018a, b; Ohene et al. 2018; Oi and Di Rocco 2006; Rasmussen et al. 2018, Vindedal et al. 2016).

If there is a hierarchy of CSF clearance systems or if some are in reserve to be activated when a primary system is destroyed is completely unknown. Thus there seem to be several ways that a given disease process can result in "communicating" hydrocephalus. In addition, it is unknown if different types of insults disturb CSF regulation physiologies in similar or different ways. For instance, a haemorrhagic insult to the CSF pathways (e.g. SAH) vs. an infectious insult (e.g. meningitis) can both result in acute hydrocephalus during the acute disease and chronic hydrocephalus as a long term outcome. Both can cause physical obstruction, inflammatory reactions, scarring and adhesions in CSF the subarachnoid space, interference with absorbtion mechanisms, disturbance of molecular pathways, etc.—but they may do this to different degrees, with different main attack points and in different combinations.

In current practice, the overlapping diagnostic criteria present a real problem. The same patient may meet several classification criteria, and two different patients may share several classification criteria. This holds the risk of unclear definitions. Figures 12.2 and 12.3 provides two case illustrations of this problem.

12.2 Imaging

Imaging of the brain is a mainstay in diagnosis. The imaging modality is preferably MRI both because it provides high imaging quality with very good anatomical detail and because it does not expose the patient to ionizing radiation. MRI requires the subject to remain still during a 30–60 min scan time, as movement artifacts may make the images useless by distortion. Thus the patient has to be reasonably cooperative. If the patient is able to remain still only for a short time, the imaging

Fig. 12.2 One patient can meet several classification criteria. *N*ormal *P*ressure *H*ydrocephalus (NPH). This hydrocephalic condition is defined by typical clinical triad, large cerebral ventricles on imaging and ICP levels that are normal or only slightly elevated. However, this condition also meets the following classification criteria: (a) Adult hydrocephalus, (b) Acquired, (c) Idiopathic or secondary, (d) Communicating or obstructive. In addition, the same clinical presentation can be seen in aquaductal stenosis (so-called "arrested hydrocephalus"), where the initiating event was in childhood with no or few initial symptoms—presenting decades later with the clinical picture of NPH

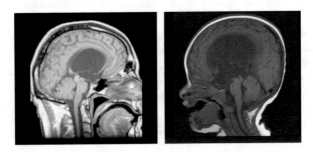

Fig. 12.3 Different patients may share several classification criteria. Left: MRI of 35 years old woman. Normal intellectual development, but always "a bit clumsy" with a large head. Increasing symptoms over several months with headache, dizziness, reduced intellectual sharpness and slightly impaired balance. Right: MRI of 2 month old baby boy. Increasing head circumference, crying spells, interrupted sleep, regurgitation of his food, episodes of "sun-set" eyes. Both have: (a) Obstructive hydrocephalus, (b) Aquaductal stenosis, (c) Congenital hydrocephalus

protocol can however be abbreviated to contain only the most necessary imaging for clinical decision-making, which is particularly useful in follow-up and complication diagnosis in children. Alternatively, a full MRI can be carried out under general anaesthesia.

CT is still widely used. With the most recent spiral scan protocol; this type of imaging is much faster than MRI lasting less than a minute for a full imaging sequence. These qualities make this imaging type easy to use in emergencies and in un-cooperative patients. In many countries, MRI is still less accessible than CT. However, the anatomical detail is not optimal for visualization of small and delicate obstructions and may result in an erroneous diagnosis of communicating hydrocephalus. In addition, CT uses ionizing radiation/X-ray for imaging, and as hydrocephalus in most cases is a permanent condition with usually several imaging studies over time, many patients will accumulate a significant exposure to ionizing radiation. This is a particular draw-back in children both because of the vulnerability of the developing brain and because of the accumulated exposure over an entire lifetime.

Imaging should in some cases include the spinal column; particularly in congenital hydrocephalic conditions, where malformations and additional pathology in the medulla and spinal CSF spaces may contribute to the hydrocephalus. Spinal dysraphism, tethered cord and syringomyelia are examples of this (Cinalli et al. 2019).

The neurosurgeon or neurologist first of all uses the images to determine the size of the ventricular system. Evan's index is the ratio between the maximum width of the frontal horns and maximal internal diameter of skull on the same axial image. The upper limit for the normal index is 0.3. The index can be assessed on both on CT and MRI. The ratio varies with age and as ventricular enlargement also occurs "*ex vacuo*" in brain atrophy, an Evans index >0.3 is not necessarily indicative of hydrocephalus. Secondly, the type and if possible the etiology for hydrocephalus is taken into consideration. (Agarwal et al. 2016; Langner et al. 2017; Raybaud 2016; Yamada and Kelly 2016). Table 12.1 and Fig. 12.4 list the most important radiological features.

Ultrasound is readily used to visualize the ventricular system in small children as long as the fontanel is still open. It is particularly useful in this age group, as it requires no cooperation from the patient, is completely non-invasive and can be performed repeatedly bed-side or in out-patient clinics. In older children and adults, ultrasound can be used to measure the optic nerve sheath diameter (ONSD), for which there is increasing evidence as an intra-individual indicator for ICP useful for follow up of individual patients (Robba et al. 2018; WUPN Workshop 2018). As the temporal bone is thin in most people, it is also possible to obtain an ultrasound window here aiming at the third ventricle which can be used to assess its width (WUPN Workshop 2018). These methods are promising as non-invasive imaging indicators of ICP but still not fully developed or in general use in the neurosurgical community.

Dynamic imaging of CSF flow with MR is possible, but difficult to interpret and is only routinely used in few institutions. This methodology is under continuing development (Bunck et al. 2012; Heidari Pahlavian et al. 2016; Yildiz et al. 2017).

Brain venograms are attracting increasing attention as a useful tool in diagnosis of conditions with elevated ICP. This is particularly relevant in the clinical syndrome of Idiopathic Intracranial Hypertension (IIH, pseudotumor cerebri) and in "external

Table 12.1 Assessment of clinical neuro-imaging in hydrocephalus

Size of ventricular system	Sites of obstruction	Space-occupying obstructive lesions	Indicators of elevated ICP	Differential diagnosis
Evans ratio • By far the most commonly used measure for ventricular size	• Foramen of Monroi • Aquaduct • Fourth ventricle • Other sites in ventricles • Posterior fossa outlets • Extra ventricular/cisternal obstructions	• Tumor • Cystic malformations • Intraventricular haematoma	• Transependymal edema • Compressed sulci • Bulging corpus callosum • Bulging of third ventricle boundaries • Narrow callosal angle ($<90°$)	*Signs of brain atrophy/degenerative brain disease* • White matter pathology • Wide sulci • Narrow gyri • Wide callosal angle

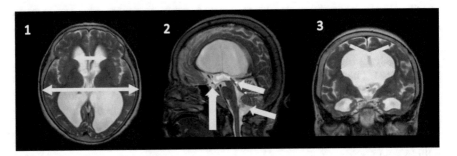

Fig. 12.4 Radiological assessment of hydrocephalus. T2 weighted MRI: the case is a young adult with radiologically severe hydrocephalus. On the axial image (1), the measurements for calculation Evans index are shown. On the sagittal image (2), the thick arrow shows downward bulging of the anterior floor of the third ventricle and the upper small arrow shows indicates occlusion of the aquaduct. The lower arrow indicates where to look for obstruction of fourth ventricle outlets. The coronal image (3) shows the callosal angle and the width of the ventricles at the level of Foramen Monroi

hydrocephalus" (subarachnoid CSF accumulation) in children (Bateman and Napier 2011; Rekate 2009; Sainz et al. 2019), where venous hypertension seems a logical pathogenesis. Increased venous pressure occurs if there is a stenosis of one or more venous sinus with a resulting pressure gradient and may be corrected by stenting the stenotic venous sinus (Aguilar-Pérez and Henkes 2015; Zhou et al. 2018). In other cases there is a uniform elevation of venous pressure without stenosis or pressure gradient, and in such cases stenting is not logical, and a systemic pathology should be looked for.

12.3   Measurement of ICP and CSF Circulation in Clinical Use

ICP measurement is another mainstay in clinical management of hydrocephalus and is performed both for primary diagnostic purposes and to guide ICP regulating treatment. Currently, only invasive methods are available, as non-invasive methods are—at best—semi-quantitative (Khan et al. 2017; Narayan et al. 2018; Xu et al. 2016). In principle, ICP is measurable at any point along the neuro-axis and in any CNS compartment. In practice, spinal measurements are only in the CSF compartment, whereas intracranial measurements can be performed in the extra-dural, subdural, paremchymal and intraventricular compartments. There are small measurement differences between the measurements sites, but in general, they all correspond well (Zacchetti et al. 2015).

The ICP signal pulsates with the cardiac cycle, fluctuates with the respiratory pattern and in addition has a slower "inherent oscillation" in characteristic macro-patterns. The ICP signal is thus complex, and even basic analysis is multidimensional with assessment of mean ICP, pulse wave amplitude and macro-pattern morphology (Figs. 12.5 and 12.6).

Normal mean ICP is quoted to be 8–15 mmHg in many textbooks and clinical reference programs (Mollan et al. 2018; Brain Trauma Foundation Guidelines 2016). However, measurements in normal persons without any confirmed or suspected ICP or CSF anomaly are extremely scarce because of the invasive nature of ICP measurement. Development of ICP measurement technology provides the possibility to measure in fully mobilized persons outside the hospital environment. Measurements with this technology in patients without suspected ICP/CSF pathology ("pseudo-normal cohort") have shown values that are considerably lower that the generally quoted range and which are strongly dependent on body position. According to these new data, ICP in humans is probably between 0 and +5 mmHg in the horizontal position and between −5 and 0 mmHg in the vertical position (Andresen et al. 2016; Pedersen et al. 2018). If further established as reference values, these findings will change reference guidelines for interventions with ICP regulation; e.g. valve settings in hydrocephalus shunts and ICP management in ICU settings.

Intracranial measurement is confined to neurosurgery departments, as it requires insertion of the transducer through a cranial burrhole. In all other settings including clinical neurology, ICP is assessed as the CSF opening pressure via a lumbar puncture. Lumbar CSF pressure is thus used in neurological guidelines (Mollan et al. 2018). When using this procedure, the investigator must be aware that the curled-up position necessary for the lumbar puncture artificially increases the measured fluid pressure (Andresen et al. 2016). The investigator must also be aware that the lumbar puncture opening pressure provides only data for a maximum duration of approximately 30 min and is a day-time procedure. These are important time related limitations, both because ICP is higher during night time and in some patients is only pathological during sleep and also because ICP changes dynamically e.g. with body

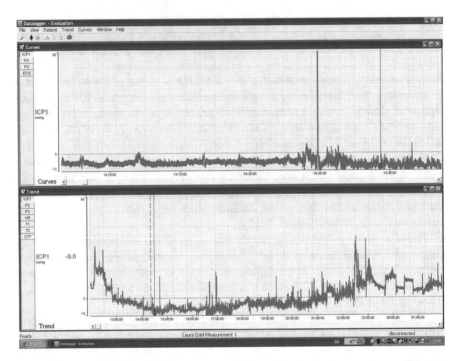

Fig. 12.5 Normal ICP signal. Measurement with transducer in brain parenchyma. The even amplitude band with a uniform pattern and mean ICP around—5 mmHg (daytime, upright position) characterizes the normal ICP signal

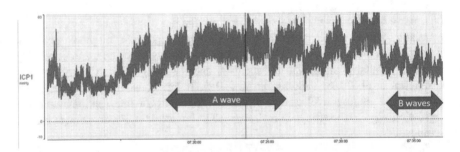

Fig. 12.6 Abnormal ICP signal. Measurement with transducer in brain parenchyma. Elevated mean ICP, high amplitude and non-uniform pattern characterizes this very abnormal ICP signal. Abnormal macro-patterns are divided into A- and B waves originally described by Lundberg (1960). This terminology is still the basis of clinical and research ICP signal analysis

positions and Valsalva physiology during day-time. ICP assessment via lumbar puncture is thus a very limited window to the true ICP.

Measurement of CSF absorbtion can also be performed as a clinically diagnostic procedure. Mostly, it is used in assessment of patients with NPH symptomatology because ICP is not elevated in this type of hydrocephalus; distinction between ventricular dilatation because of abnormal CSF dynamics and brain atrophy is rarely possible; and because the clinical symptoms resembling degenerative brain diseases makes differential diagnosis difficult. The rationale for this type of additional investigation is that reduced absorption of CSF is the underlying cause. The basic principle is to infuse a small amount of mock CSF with simultaneous recording of the resulting pressure changes. If absorbtion is normal, the resulting pressure increase is small; whereas impaired absorbtion results in a higher pressure increase for the same volume of infusate. The relative increase is calculated as the resistance to CSF outflow; R-out or as the CSF conductance (Czosnyka et al. 2012; Eklund et al. 2007; Jacobsson et al. 2018; Kasprowicz et al. 2016).

12.4   Treatment

Before the mid-1950s surgical treatment of hydrocephalus was described in sporadic case reports or very small series with a high mortality and uncertain benefit in the survivors. The American neurosurgeon Walter Dandy pioneered both understanding of hydrocephalus pathophysiology and hydrocephalus surgery with clinical observations, experimental studies and several mainly endoscopic surgeries between 1915 and 1946 (Blitz et al. 2018). Hydrocephalus surgery was far from generally accepted, only performed in very select institutions and then abandoned because of high morbidity and mortality as well as technical shortcomings of endoscopic equipment.

Hydrocephalus surgery became a standard neurosurgical procedure when the principle behind a CSF shunt with one-way passage and valve-bearing drainage from the cerebral ventricles into an extracranial absorbtion site was invented in 1955 by the engineer John Holter and the neurosurgeon Eugene Spitz, as the former had a son with neonatal hydrocephalus. This treatment principle (Fig. 12.7) quickly spread worldwide because it is both simple and immediately effective, and it is nowadays one of the most frequently performed neurosurgical procedures. This surgical procedure has been life- and brain-saving for hundreds of thousands of patients. For the first two decades, the standard surgical technique was placement of the distal catheter via the facial vein into the jugular vein. From the 1970s it became standard to place the distal catheter in the peritoneum.

Despite its success, there are implant-related shortcomings and complications which persist over the decades despite developments in valve technology and surgical methodology. The implant durability is limited to a median implant survival of a few years (3–6 years depending mainly on population age and previous shunt history) because of short term complications dominated by implant infections, tube/valve blockage and disconnections and long term complications with overdrainage, blockage by tissue ingrowth into the ventricular catheter, slowly accumulating debris in the valve house or disconnections. Consequently, the average shunt patient will face a future with several shunt revisions (Kofoed Månsson et al. 2017). Currently, the knowledge and technology to select different draining principles for different ages, different patient types and different hydrocephalus etiologies does not exist.

Fig. 12.7 Simplified drawing of a hydrocephalus shunt. The ventricular drain is inserted via a burrhole in the skull. The most frequently used position is the frontal horn. The valve mechanism (asterisk) is placed subcutaneously in the tubing on the skull. The drainage tube is the further passed distally in the subcutis and inserted into the peritoneal cavity. The mechanical function of hydrocephalus valves is based on the same principles as industrial valves, but on a much smaller scale and with functions parameters suitable for pressure in the human CNS. The most widely used principle is a differential pressure valve with either a fixed or adjustable opening pressure. Additional components (anti-siphons and gravitational devices) can be added to overcome the effect of excessive drainage in the upright position (over-drainage)

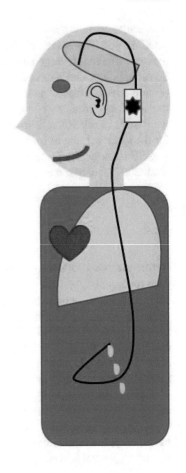

Perhaps this shortcoming in surgical understanding of the disease is an important part of the complication profile.

In the 1990s endoscopic surgery for hydrocephalus was re-introduced, which was made possible by improvements in endoscopic technology. The principle is to fenestrate the ventricular system from the inside thus creating an egress for the CSF into the subarachnoid space, where it can circulate out and be absorbed. Accordingly this type of operation is only feasible if the fenestration bypasses an obstruction in the CSF pathways, i.e. the success rate is highly dependent on selection of patients with obstructive hydrocephalus types. The usual access for this type of surgery is via the frontal horn via the foramen Monroi into the third ventricle, which is fenestrated through its anterior floor. Therefore, the procedure is generally known by its abbreviation ETV (Endoscopic Third Ventriculostomy). Durability statistics for ETVs are emerging only very recently. They indicate that closure of the outlet or other inability to manage the hydrocephalus occurs within 3–6 months probably as a combination of immediate and delayed primary failures and that ETVs functioning beyond this seems to be durable.

Hydrocephalus Treatment and the Choroid Plexus As described above, current treatment options for hydrocephalus are surgical and aimed at bypassing a blockage in the CSF flow pathways and/or leading the CSF to a site with better absorbtion. Surgical strategy is thus directed almost exclusively at improving CSF clearance. Surgical modification of CSF secretion is sometimes attempted by endoscopic cauterization of the choroid plexus (Warf 2013), but the procedure is probably only indicated in infantile hydrocephalus and not widely used because of uncertainties about its long-term efficiency and durability (Karimy et al. 2016; Muir et al. 2016; Weil et al. 2016).

Pharmacological treatment may in theory be aimed both at reducing CSF production and improving its clearance. The only principles in clinical use are diuretic substances. Carbonicanhydrase inhibitor (Diamox®), which was previously used as a diuretic, is known to reduce choroid plexus secretion of CSF, but is may have additional unexplored effects on CSF clearance and absorption. Unfortunately, Diamox®, has little and uncertain clinical effect on hydrocephalus seemingly regardless of type and etiology, but has effect and is used clinically to treat IIH and glaucoma. The use of loop-diuretics is very questionable, and clinical series and trials with other pharmacological principles like osmotic agents, anti-inflammatory agents and fibrinolysis have so far unfortunately not shown clinically feasible effects (Del Bigio and Di Curzio 2016).

12.5 Conclusion

Current management of hydrocephalus is purely surgical creating a CSF diversion either intracranially with endoscopic fenestration between the ventricular and cisternal compartments or with implantation of a valve-regulated drainage tube (shunt surgery) to an extracranial absorbtion site. The endoscopic fenestration is very efficient in cases where a blockage can be by-passed in this way; however, this limitation applies to only a segment of much less than 50% of all hydrocephalus cases. In contrast, the drainage solution is not dependent on etiology or blockage site, and thus can be used in all types of hydrocephalus with immediate relief. This solution, however, carries other complications and limitations with the consequence that overall the median durability of shunt surgery is just 4–6 years.

Currently, the only pharmacological treatment is the old diuretic carbo-anhydrase inhibitor (Diamox®). This is used with reasonably good results in the hydrocephalus-related condition IIH, but is not very effective in for hydrocephalus treatment.

To improve treatment of hydrocephalus, we face the following challenges and need clinical, experimental and translational research dealing with

- Better perception and visualization of obstruction sites
- Better definition of clinical predictive factors
- Biomarkers and genetic markers for parallel CSF/water transport pathways— which may be individual

- Development of adjuvant treatment
  - could be molecular water or electrolyte transport modifiers
  - the aim could be stand-alone pharmacological treatment
  - or it could be to act as an adjuvant to improve ETV success rate

We also need shunts mimicking normal physiology, but before we can do that, we need

- Documented reference values for normal ICP
  - Including individual and age-related variations
- Knowledge about ICP and CSF regulation in humans
  - Including individual and age-related variations
- Improved understanding how shunts affect CSF and ICP physiology

References

Agarwal A, Bathla G, Kanekar S (2016) Imaging of communicating hydrocephalus. Semin Ultrasound CT MR 37(2):100–108. https://doi.org/10.1053/j.sult.2016.02.007

Aguilar-Pérez M, Henkes H (2015) Treatment of idiopathic intracranial hypertension by endovascular improvement of venous drainage of the brain. Ophthalmologe 112(10):821–827. https://doi.org/10.1007/s00347-015-0136-1. In German

Amiry-Moghaddam M, Hoddevik EH, Ottersen OP (2010) Aquaporins: multifarious roles in brain. Neuroscience 168(4):859–861. https://doi.org/10.1016/j.neuroscience.2010.04.071

Andresen M, Hadi A, Juhler M (2016) Evaluation of intracranial pressure in different body postures and disease entities. Acta Neurochir Suppl 122:45–47. https://doi.org/10.1007/978-3-319-22533-3_9.

Barami K, Chakrabarti I, Silverthorn J et al (2018) Diagnosis, classification, and management of fourth ventriculomegaly in adults: report of 9 cases and literature review. World Neurosurg 116:e709–e722. https://doi.org/10.1016/j.wneu.2018.05.073. Epub 2018 May 17

Bateman GA, Napier BD (2011) External hydrocephalus in infants: six cases with MR venogram and flow quantification correlation. Childs Nerv Syst 27(12):2087–2096. https://doi.org/10.1007/s00381-011-1549-z

Bech RA, Waldemar G, Gjerris F et al (1999) Shunting effects in patients with idiopathic normal pressure hydrocephalus; correlation with cerebral and leptomeningeal biopsy findings. Acta Neurochir (Wien) 141(6):633–639

Bergsneider M, Black PM, Klinge P et al (2005) Surgical management of idiopathic normal-pressure hydrocephalus. Neurosurgery 57(3 Suppl):S29–S39

Blitz AM, Aygun N, Herzka DA et al (2017) High resolution three-dimensional MR imaging of the skull base: compartments, boundaries, and critical structures. Radiol Clin N Am 55(1):17–30. https://doi.org/10.1016/j.rcl.2016.08.011

Blitz AM, Ahmed AK, Rigamonti D (2018) Founder of modern hydrocephalus diagnosis and therapy: Walter Dandy at the Johns Hopkins Hospital. J Neurosurg 1:1–6. https://doi.org/10.3171/2018.4.JNS172316. [Epub ahead of print]

Brain Trauma Foundation (2016) Guidelines for the management of severe traumatic brain injury. https://braintrauma.org/uploads/03/12/Guidelines_for_Management_of_Severe_TBI_4th_Edition.pdf

Broadbelt A, Stoodley M (2016) An anatomical and physiological basis for CSF pathway disorders. In: Mallucci C, Sgouros S (eds) Cerebrospinal fluid disorders. CRC, Boca Raton, FL. ISBN 978-0824728335. https://www.taylorfrancis.com/books/e/9780429114342

Bunck AC, Kroeger JR, Juettner A et al (2012) Magnetic resonance 4D flow analysis of cerebrospinal fluid dynamics in Chiari I malformation with and without syringomyelia. Eur Radiol 22 (9):1860–1870. https://doi.org/10.1007/s00330-012-2457-7

Capel C, Baroncini M, Gondry-Jouet C et al (2018) Cerebrospinal fluid and cerebral blood flows in idiopathic intracranial hypertension. Acta Neurochir Suppl 126:237–241. https://doi.org/10.1007/978-3-319-65798-1_48.

Cinalli G, Memet Ozek M, Sainte-Rose C (eds) (2019) Pediatric hydrocephalus. Springer, Cham. ISBN 978-3-319-27248

Czosnyka M, Czosnyka Z, Agarwal-Harding KJ, Pickard JD (2012) Modeling of CSF dynamics: legacy of Professor Anthony Marmarou. Acta Neurochir Suppl 113:9–14. https://doi.org/10.1007/978-3-7091-0923-6_2. PMID: 22116414

Daou B, Klinge P, Tjoumakaris S et al (2016) Revisiting secondary normal pressure hydrocephalus: does it exist? A review. Neurosurg Focus 41(3):E6. https://doi.org/10.3171/2016.6.FOCUS16189

Del Bigio MR, Di Curzio DL (2016) Nonsurgical therapy for hydrocephalus: a comprehensive and critical review. Fluids Barriers CNS 13:3. https://doi.org/10.1186/s12987-016-0025-2

Eide PK (2018) The pathophysiology of chronic noncommunicating hydrocephalus: lessons from continuous intracranial pressure monitoring and ventricular infusion testing. J Neurosurg 129 (1):220–233. https://doi.org/10.3171/2017.1.JNS162813

Eklund A, Smielewski P, Chambers I et al (2007) Assessment of cerebrospinal fluid outflow resistance. Med Biol Eng Comput 45(8):719–735

Fushimi Y, Miki Y, Takahashi JA et al (2006) MR imaging of Liliequist's membrane. Radiat Med 24(2):85–90

Garnotel S, Salmon S, Balédent O (2018) Numerical cerebrospinal system modeling in fluid-structure interaction. Acta Neurochir Suppl 126:255–259. https://doi.org/10.1007/978-3-319-65798-1_51. PubMed PMID: 29492571

Greitz D (2006) Reprint of: radiological assessment of hydrocephalus: new theories and implications for therapy. Neuroradiol J 19(4):475–495

Greitz D (2007) Paradigm shift in hydrocephalus research in legacy of Dandy's pioneering work: rationale for third ventriculostomy in communicating hydrocephalus. Childs Nerv Syst 23 (5):487–489. Epub 2007 Mar 17

Heidari Pahlavian S, Bunck AC, Thyagaraj S (2016) Accuracy of 4D flow measurement of cerebrospinal fluid dynamics in the cervical spine: an in vitro verification against numerical simulation. Ann Biomed Eng 44(11):3202–3214

Holmlund P, Eklund A, Koskinen LD et al (2018) Venous collapse regulates intracranial pressure in upright body positions. Am J Physiol Regul Integr Comp Physiol 314(3):R377–R385. https://doi.org/10.1152/ajpregu.00291.2017

Ibáñez-Botella G, González-García L, Carrasco-Brenes A et al (2017) LOVA: the role of endoscopic third ventriculostomy and a new proposal for diagnostic criteria. Neurosurg Rev 40 (4):605–611. https://doi.org/10.1007/s10143-017-0813-4

Jacobsson J, Qvarlander S, Eklund A et al (2018) Comparison of the CSF dynamics between patients with idiopathic normal pressure hydrocephalus and healthy volunteers. J Neurosurg 1:1–6. https://doi.org/10.3171/2018.5.JNS173170

Kahle KT, Kulkarni AV, Limbrick DD et al (2016) Hydrocephalus in children. Lancet 387:788–799

Karimy JK, Duran D, Hu JK et al (2016) Cerebrospinal fluid hypersecretion in pediatric hydrocephalus. Neurosurg Focus 41(5):E10

Kasprowicz M, Lalou DA, Czosnyka M et al (2016) Intracranial pressure, its components and cerebrospinal fluid pressure-volume compensation. Acta Neurol Scand 134(3):168–180. https://doi.org/10.1111/ane.12541. Epub 2015 Dec 15. Review. PMID: 26666840

Khan MN, Shallwani H, Khan MU et al (2017) Noninvasive monitoring intracranial pressure – a review of available modalities. Surg Neurol Int 8:51. https://doi.org/10.4103/sni.sni_403_16. eCollection 2017

Klinge P, Marmarou A, Bergsneider M et al (2005) Outcome of shunting in idiopathic normal-pressure hydrocephalus and the value of outcome assessment in shunted patients. Neurosurgery 57(3 Suppl):S40–S52

Kofoed Månsson P, Johansson S, Ziebell M et al (2017) Forty years of shunt surgery at Rigshospitalet, Denmark: a retrospective study comparing past and present rates and causes of revision and infection. BMJ Open 7(1):e013389. https://doi.org/10.1136/bmjopen-2016-013389

Kulkarni AV, Sgouros S, Leitner Y et al (2018) International Infant Hydrocephalus Study Investigators. International Infant Hydrocephalus Study (IIHS): 5-year health outcome results of a prospective, multicenter comparison of endoscopic third ventriculostomy (ETV) and shunt for infant hydrocephalus. Childs Nerv Syst 9:2391–2397. https://doi.org/10.1007/s00381-018-3896-5.

Langner S, Fleck S, Baldauf J et al (2017) Diagnosis and differential diagnosis of hydrocephalus in adults. Rofo 189(8):728–739. https://doi.org/10.1055/s-0043-108550

Lokossou A, Balédent O, Garnotel S et al (2018) ICP monitoring and phase-contrast MRI to investigate intracranial compliance. Acta Neurochir Suppl 126:247–253. https://doi.org/10.1007/978-3-319-65798-1_50. PubMed PMID: 29492570

Lundberg N (1960) Continuous recording and control of ventricular fluid pressure in neurosurgical practice. Acta Psychiatr Neurol Scand Suppl 36(149):1–193

Mestre H, Hablitz LM, Xavier AL et al (2018a) Aquaporin-4-dependent glymphatic solute transport in the rodent brain. Elife 7:e40070. https://doi.org/10.7554/eLife.40070

Mestre H, Tithof J, Du T (2018b) Flow of cerebrospinal fluid is driven by arterial pulsations and is reduced in hypertension. Nat Commun 9(1):4878. https://doi.org/10.1038/s41467-018-07318-3.

Mollan SP, Davies B, Silver NC et al (2018) Idiopathic intracranial hypertension: consensus guidelines on management. J Neurol Neurosurg Psychiatry 89(10):1088–1100. https://doi.org/10.1136/jnnp-2017-317440. Epub 2018 Jun 14

Muir RT, Wang S, Warf BC (2016) Global surgery for pediatric hydrocephalus in the developing world: a review of the history, challenges, and future directions. Neurosurg Focus 41(5):E11

Narayan V, Mohammed N, Savardekar AR et al (2018) Noninvasive intracranial pressure monitoring for severe traumatic brain injury in children: a concise update on current methods. World Neurosurg 114:293–300. https://doi.org/10.1016/j.wneu.2018.02.159

Ohene Y, Harrison IF, Nahavandi P et al (2018) Non-invasive MRI of brain clearance pathways using multiple echo time arterial spin labelling: an aquaporin-4 study. Neuroimage 188:515–523. https://doi.org/10.1016/j.neuroimage.2018.12.026

Oi S, Di Rocco C (2006) Proposal of "evolution theory in cerebrospinal fluid dynamics" and minor pathway hydrocephalus in developing immature brain. Childs Nerv Syst 22(7):662–669. Epub 2006 May 10. Review. PubMed PMID: 16685545

Osborn AG, Preece MT (2006) Intracranial cysts: radiologic-pathologic correlation and imaging approach. Radiology 239(3):650–664

Paulsen AH, Lundar T, Lindegaard KF (2015) Pediatric hydrocephalus: 40-year outcomes in 128 hydrocephalic patients treated with shunts during childhood. Assessment of surgical outcome, work participation, and health-related quality of life. J Neurosurg Pediatr 16(6):633–641. https://doi.org/10.3171/2015.5.PEDS14532

Paulsen AH, Due-Tønnessen BJ, Lundar T et al (2017) Cerebrospinal fluid (CSF) shunting and ventriculocisternostomy (ETV) in 400 pediatric patients. Shifts in understanding, diagnostics, case-mix, and surgical management during half a century. Childs Nerv Syst 33(2):259–268. https://doi.org/10.1007/s00381-016-3281-1

Pedersen SH, Lilja-Cyron A, Andresen M et al (2018) The relationship between intracranial pressure and age-chasing age-related reference values. World Neurosurg 110:e119–e123. https://doi.org/10.1016/j.wneu.2017.10.086

Petersen LG, Petersen JC, Andresen M et al (2016) Postural influence on intracranial and cerebral perfusion pressure in ambulatory neurosurgical patients. Am J Physiol Regul Integr Comp Physiol 310(1):R100–R104. https://doi.org/10.1152/ajpregu.00302.2015

Qvarlander S, Sundström N, Malm J et al (2013) Postural effects on intracranial pressure: modeling and clinical evaluation. J Appl Physiol 115(10):1474–1480. https://doi.org/10.1152/japplphysiol.00711.2013

Rasmussen MK, Mestre H, Nedergaard M (2018) The glymphatic pathway in neurological disorders. Lancet Neurol 17(11):1016–1024. https://doi.org/10.1016/S1474-4422(18)30318-1

Raybaud C (2016) MR assessment of pediatric hydrocephalus: a road map. Childs Nerv Syst 32 (1):19–41. https://doi.org/10.1007/s00381-015-2888-y

Rekate HL (2009) A contemporary definition and classification of hydrocephalus. Semin Pediatr Neurol 16:9–15

Rigamonti D, Juhler M, Wikkelsø C (2014) The differential diagnosis of normal pressure hydrocephalus. In: Rigamonti D (ed) Adult hydrocephalus. Cambridge University Press, Cambridge,. on-line ISBN 9781139382816. https://doi.org/10.1017/CBO9781139382816

Robba C, Santori G, Czosnyka M et al (2018) Optic nerve sheath diameter measured sonographically as non-invasive estimator of intracranial pressure: a systematic review and meta-analysis. Intensive Care Med 44(8):1284–1294. https://doi.org/10.1007/s00134-018-5305-7

Rodis I, Mahr CV, Fehrenbach MK et al (2016) Hydrocephalus in aqueductal stenosis--a retrospective outcome analysis and proposal of subtype classification. Childs Nerv Syst 32 (4):617–627. https://doi.org/10.1007/s00381-016-3029-y

Sainz LV, Zipfel J, Kerscher SR, Weichselbaum A et al (2019) Cerebro-venous hypertension: a frequent cause of so-called "external hydrocephalus" in infants. Childs Nerv Syst 35 (2):251–256. https://doi.org/10.1007/s00381-018-4007-3

Schmidt LB, Corn G, Wohlfahrt J et al (2018) School performance in children with infantile hydrocephalus: a nationwide cohort study. Clin Epidemiol 22(10):1721–1731. https://doi.org/10.2147/CLEP.S178757

Vindedal GF, Thoren AE, Jensen V et al (2016) Removal of aquaporin-4 from glial and ependymal membranes causes brain water accumulation. Mol Cell Neurosci 77:47–52. https://doi.org/10.1016/j.mcn.2016.10.004

Warf BC (2013) The impact of combined endoscopic third ventriculostomy and choroid plexus cauterization on the management of pediatric hydrocephalus in developing countries. World Neurosurg 79(2 Suppl):S23.e13–S23.e15. https://doi.org/10.1016/j.wneu.2011.02.012. Epub 2011 Nov 7. Review. PMID: 22120411

Weil AG, Westwick H, Wang S et al (2016) Efficacy and safety of endoscopic third ventriculostomy and choroid plexus cauterization for infantile hydrocephalus: a systematic review and meta-analysis. Childs Nerv Syst 32(11):2119–2131

WUPN Workshop (2018). https://www.erasmus.gr/microsites/1147/scientific-program

Xu W, Gerety P, Aleman T et al (2016) Noninvasive methods of detecting increased intracranial pressure. Childs Nerv Syst 32(8):1371–1386. https://doi.org/10.1007/s00381-016-3143-x

Yamada S, Kelly E (2016) Cerebrospinal fluid dynamics and the pathophysiology of hydrocephalus: new concepts. Semin Ultrasound CT MR 37(2):84–91. https://doi.org/10.1053/j.sult.2016.01.001

Yildiz S, Thyagaraj S, Jin N (2017) Quantifying the influence of respiration and cardiac pulsations on cerebrospinal fluid dynamics using real-time phase-contrast MRI. J Magn Reson Imaging 46 (2):431–439. https://doi.org/10.1002/jmri.25591

Zacchetti L, Magnoni S, Di Corte F et al (2015) Accuracy of intracranial pressure monitoring: systematic review and meta-analysis. Crit Care 19:420. https://doi.org/10.1186/s13054-015-1137-9

Zhou D, Meng R, Zhang X (2018) Intracranial hypertension induced by internal jugular vein stenosis can be resolved by stenting. Eur J Neurol 25(2):365–e13. https://doi.org/10.1111/ene.13512

Printed in the United States
by Baker & Taylor Publisher Services